Forests:
Nature, People, Power

Forests:
Nature, People, Power

Edited by

Martin Doornbos,
Ashwani Saith
and
Ben White

ISBN: 0-631-22188-3

Blackwell Publishers Ltd
108 Cowley Road
Oxford OX4 1JF, UK

Blackwell Publishers Inc
350 Main Street
Malden, Massachusetts 02148, USA

British Library Cataloguing in Publication Data has been applied for

Library of Congress Cataloging in Publication Data has been applied for

This text was originally published as special issue 31.1 of *Development and Change*, published by Blackwell Publishers on behalf of the Institute of Social Studies, The Hague.

Typeset by Downdell, Oxford
Printed and bound Great Britain
By MPG Books, Bodmin, Cornwall

This book is printed on acid-free paper

Contents

1 Forest Lives and Struggles: An Introduction

Martin Doornbos, Ashwani Saith and Ben White

It is hardly necessary to justify a volume focusing on the forests of Africa, Asia and Latin America. Forests, on the ground and in social theory, are now highly contested spaces, the arenas of struggles and conflicts, where both trees and forest dwellers usually find themselves on the losing side. The high visibility of these unresolved conflicts has spurred theoretical development, policy interventions and institutional change. But it is in the nature of such induced phenomena that they follow rather than precede the events which catalyse them. As such, the possibility remains that the insights and capabilities they create may lag too far behind the pace of destruction and change.

Most of the thirteen chapters selected for this volume are based on recent field research, rich in detail and nuanced in interpretation. Many of them call into question received wisdoms, and in doing this discover unexpected turns in forest paths, life cycles or landscape trajectories. They also highlight in various ways the complex articulations of local processes and global forces in tropical forest struggles. Taken together, they show how social science research on forests (as on natural resource management issues generally) has come of age, moving far beyond the crude 'tragedy of the commons' and 'prisoner's dilemma' approaches of the 1970s and early 1980s.

As editors we feel it is useful to highlight four dimensions emerging from these contributions. The first concerns the remarkably wide array of ongoing forest struggles, conflicts and movements, spanning virtually every part of the South, involving a spectrum of stakeholders with diverse interests with respect to forests. While such struggles go back a long way, often echoing earlier struggles in the North, their range and intensity has increased in recent decades. Second is the rise of wider national, regional and global concern over the powerful and now palpable negative externalities emanating from the destruction of forests; this has induced responses which could potentially temper these destructive processes, and there is now a genuine concern over long-term sustainability. The horizon of the long term has been frequently and dramatically foreshortened by such high visibility and broad spectrum crisis events as the forest fires of Southeast Asia. The media industry has propelled these distant events into the immediate perception of otherwise insulated constituencies in the North which are vital in long-term problem solving.

Third, parallel to the struggles over forest lives and livelihoods, and over these externalities, are intense debates over the nature and future, use and abuse, of nature. Forests form one of the arenas for these theoretical interactions, contestations and advances. Various strands of theoretical debates and discourse intersect within forest space: issues concerning the emergent rights-based approach to development; questions of environmental accounting, accountability and sustainability; gender aspects; the discourse over indigenous knowledge; the nature and consequences of resource-based economic growth; emerging governance and resource management issues at regional and global levels; new theoretical approaches addressing a crucial range of issues concerning institutional change whether spurred by spontaneous events and processes or initiated by state interventions. Finally, there is the cutting issue of resolution of trade-offs and conflicts, of 'solutions' to the problems of forests and those who live in and depend upon them. Does the new focus, the recent highlighting of the place of nature in development, throw up answers to the questions it raises?

Theory and Process

For most received economic theory, nature constitutes a pre-theory concept. In their original state, water, air, timber, fish, land and such like are 'free' or 'spontaneous' gifts of nature, incorporating no prior human processing or expenditure of human labour. They are openly available to all without social or economic restriction, and as they are in plentiful supply, there is no market for them. As such, in economic accounting, they become costless, and are beyond the domain of economic theorizing. There are passing references to such an initial, primitive, natural phase of human and economic development in the works of most early writers, some of whom acknowledge the very speculative and tentative basis for the characterization of societies of this phase. This simple exclusion of nature from economics is suddenly overtaken by a later phase where it is as simply postulated that these free goods have now become unfree, and have taken an economic incarnation as products. This transformation is not itself the subject of explanation in economic theory. Once treated as a set of economic products, nature can be dealt with in much the same manner as other products, subject to some qualifications. Quite apart from the problems arising from market failure, this treatment of nature effectively shuts theory off from the recognition, and therefore the dynamic analysis of some crucial processes of resource exploitation. The key to entering this fertile, though unsettled, theoretical space lies in posing the question unasked by mainstream economic theorizing and the social sciences generally, viz., that enquiring into the conditions underlying the demise of nature as a free good and its reincarnation as an economic one. Implicitly, it juxtaposes on the initial state of plenty, a new scenario involving the emergence of scarcity, markets and prices. But it is precisely in the interim, the unrecognized space between these two postulated states or phases of nature that many vital concerns lie buried.

Nature has thus long remained silenced and invisible in social science theory, yet fully engaged and exploited in socio-economic reality. The stresses and pressures arising from the latter sphere, however, have frequently induced and catalysed theoretical insights, innovations and advances. Valuable as these extensions have been, they have lagged far behind the situations that have created the space for their accommodation in theory. They have also encountered resistance, if only because most mainstream social science approaches — in their innate drive to be holistic and consistent — have left little room for the role of 'new', i.e., untheorized phenomena. Often, accommodating nature could amount to abandoning several other received tenets, and disturbing the comprehensive, internally consistent power of a theoretical system. Further, while various dimensions of the relationship between nature and economy have been separately theorized, these partial analyses do not add up to an integrated statement on the nature and role of nature.

While this is evident in all disciplines, it is perhaps most clear in the domain of economics. For instance, the relationship between nature, technology and economy entered classical economic theory in the form of a breakthrough — the so-called law of diminishing marginal returns to the intensification of inputs on a given amount of land. Marxian writing also elicited one commonality between nature and labour, namely, that both were exploited, exhausted and spat out by the capitalist process. That the fertility of the soil was depleted by intensive cultivation was also acknowledged; but the implications of this for the sustainability of long-run growth were recognized only via intermediate depressionary effects through their impact on the distribution of income. The powerful insight embedded in this relationship remained unexcavated. Indeed, its significance was ignored and also misread, inviting the critique that the growth process could be, and indeed was, protected from the braking effect of this law by the shift away from agricultural inputs to other natural resources, such as forest exploitation and mining, which were not subject to this 'law'. Industrialism subsequently took the spotlight further away from this issue as planning theory adopted a fixed coefficient production function approach, exemplified in the input–output matrix. There was also a rising faith in the power of the unbound Prometheus to pre-empt, or to override, any emergent shortages of natural resources through processes of substitution and induced technical change. Both as paradigm and as process, industrialization came to dominate nature and to exercise power over it. The relationship between the trinity of nature, technology and economy remains an unsettled one, even though the technological fixities, certainties and optimism of the past have been effectively eroded.

A second illustration is provided by the late discovery of the existence and far-reaching implications of externalities, literally being effects which the internally-closed theoretical system of market economics had to define away if it was to maintain its central propositions. With respect to deforestation, these could take the form of flooding, loss of topsoil, gullying and the loss of cultivable land further down the watershed due to deforestation in the higher reaches; or through the loss of wild life systems supported by the forests. But externalities could also arise at a global level, as through the loss of biodiversity, through global warming, or through the costs of forest fires. Typically, these external effects began to knock on the doors of conventional theory only in the late stages of industrialization.

Third, the crisis of the commons, forests included, lay at the root of new institutional theorizing attempting partly to explain the process of natural losses, and partly to propel the idea that a solution required the prior full assignation of private property rights in previously 'free' or communally-held natural resources. In a different vein, by shutting out the space between the free and the priced states of nature, social struggles in the evolution of property rights in nature were sidelined. Fourth, at a broader level, the entire basis of the modern growth paradigm has been protected by the unwillingness to recognize the role of the environment. It is interesting that while the first systematic recognition of the need to measure national income while maintaining capital intact was made early this century, the insight was translated and absorbed into national income accounting systems exclusively with respect to the depreciation of capital in the form of machines and constructed infrastructure; nature and the spontaneous exploitation of natural resources do not enter the scene. Even now, the deeply misleading notion of gross domestic product has virtually universal currency, while environmentally-adjusted national accounting systems remain the specialized hobby-horse of a small minority. This persists despite clear and repeated demonstrations that the application of the principle of maintaining capital intact to include natural resources would undermine the power of the modern paradigm of economic growth measured in money-metric terms. Fifth, these various insights combine in different ways to generate the powerful, though embryonic, concept of sustainable development, and its relationship with the environment.

Other disciplines within the social sciences cannot boast of greater advances, but neither do they have to admit to so many dead ends and missed opportunities. It is worth noting that the breadth and intensity of short-term crises and perceived longer-term sustainability have set challenges in the fields of social science theory and development policy which have inexorably brought the different disciplines closer in generating satisfactory explanations. Nature, and the sustainability of local and global development processes, provide prime examples of stimuli inspiring such theoretical reconstructions.

Life Cycles and Contestations

The two chapters with which we begin introduce many of the issues mentioned above. They show that there is clearly no standard life cycle of stages through which all forests pass. Official and mainstream versions of forest and landscape history may present serious misreading of forest histories, as illustrated in the West African cases of Leach and Fairhead where 'ancient' forests now appear themselves to have been created from savanna in a cyclical rather than linear process. Although notions of 'standard' unilinear evolutionary forest trajectories must be avoided, the empirical study of forest life cycles in their numerous variations can be extremely enlightening, as shown by many of the chapters in this volume. Certain common moments and variants in forest life cycles deserve mention.

One such moment is the intervention of states to establish a system of property rights over forest land, often first declaring 'virgin' forests occupied by indigenous/traditional users as 'empty' or 'waste' lands (Nygren). Another is the attempt to establish systems of scientific forest management, involving models of technocratically controlled silviculture. These themselves constitute a development discourse, involving unusually long-term planning perspectives (usually longer than the human life-span) and an obsession with maximization and stabilization of woody growth as the sole purpose of the forest's existence to serve the needs of urban consumers, industry or export (Sivaramakrishnan, Klooster).

State regulations and policies on forest management (discussed in many of the chapters, but see particularly McCarthy on Indonesia) often exhibit huge divergences between the control that states claim to exercise and the confused reality, in which legally 'state-owned' forest lands have for generations been places where large numbers of people have lived, obtained livelihoods and allocated use-rights among themselves, whether in collusion or contestation with local officialdom and forest rangers (Sato). Governmental command and control structures often tend to lapse into chaotic local level arrangements involving *de facto* sub-privatizations and alienation in favour of specific interest groups. People living in and near forests may indeed experience seemingly haphazard oscillations in state policies. At times people who do not live in forests are encouraged to move there and cut them down (for industrial uses, for export, for ranching, for peasant farming); at other times it is not the trees which must move for the people, but the people who must move to make room for the trees, cajoled or forced to move away. Such 'forest cleansing' exercises are primarily for the benefit of distant rather than local populations, as forests are seen as vehicles for wildlife conservation, for protection of watersheds, or even for absorption of carbon generated by industrial emissions in far-off continents, in what is now called 'carbon offset forestry'.

Examples of what may happen when trees rather than local livelihoods become the objects of 'conservation' are provided by Sato's account of

Karen people living in the 'ambiguous lands' in buffer zones between areas designated for forest conservation and farming, and Cohen's depiction of the upland Akha populations of the Lao People's Democratic Republic who have been shifted to settlements on lower slopes and incorporated into lowland commercial Tai agriculture as an impoverished, opium-addicted wage labour force.

In broader historical perspective, therefore, the notion of forest life cycles may help to better situate longer-term policy shifts and reversals: forest policy thus undergoes its own trajectories and life cycles. Whether currently new policy departures will prove effective in extending the life-span of the world's remaining forests, and in countering the formidable interests in exploiting them for shorter-term gain, remains to be seen.

It is important to recognize the different kinds of struggles being waged, from different actor positions and perspectives, and within rapidly changing contexts. One point emerging from the contributions to this volume is the great complexities of social and institutional relations involved in forest use and forest policy making at different times and places.

The process of transformation of the forest from free good to fully commoditized good sets up stresses and tensions which induce regulatory regimes for controlling access to and rights over forest space and resources. These may be run for and by estate holders wishing to assert their claims on the forest as their property, overriding or rejecting the claims of forest dwellers and fringe communities. But such regimes may also be state-bureaucratic, where the state claims superior rights over all others. The simultaneity and overlapping of such layers of rights, each perhaps self-proclaimed, asserted, legitimized and protected, forms the basis of an interactive matrix of stakeholders each vying and contesting the claims of the other while attempting to derive from the forest its particular use and exchange values. The spectrum of stakeholders is further widened through the inclusion of other specific constituencies which, while not having direct interests of access or ownership, are nevertheless engaged through suffering some of the negative externalities of the general processes of deforestation. In such complex contexts, the vesting of formal or *de jure*, superior rights in distant state power may be largely ineffective if it is unable to daily monitor the routes into and out of the forest of the various petty claimants to the forest; or if the bureaucratic machinery set up for regulation and monitoring forest use is itself corrupted by forest contractors seeking large-scale open access.

There is thus no single process or unified scheme or path of commoditization, bur rather a multiplicity of forest lives, trajectories and struggles. Forest dwellers seek direct use values and express a desire to protect, but also to exploit their habitat in line with their livelihood strategies; fringe communities fight to retain access to the forest for land clearance and cultivation, for fuel and firewood, for various natural products and for hunting; forest contractors focus on extracting maximum profits in the shortest possible time by converting nature's cumulative stock into a flow of saleable products; regulating bureaucracies play all sides so as to maximize their purchase on this flow of resources, acting as forest gatekeepers, whether on the edge of the forest itself or as government bureaucrats awarding permissions for or simply condoning large-scale violations; and the State, in some abstract form, pronounces on the entire process in many, often conflicting voices, and formulates legalities and procedures governing the forest as a whole, this in the name of present and future generations of 'society', using the legitimizing idiom and vocabulary of modern science, management theory, and social obligation. Even this is an unrealistic simplification. Forest dwellers do not come in standardized homogenized forms; rather they comprise complex hierarchically structured groups and communities which may simultaneously display aspects of group solidarity, unity and shared objectives with respect to perceived threats from the rest of the world, but which nevertheless manifest internal stresses and frictions arising from inequalities in power, status and economic strength. The same applies to fringe communities. The observable paradox is that complexity militates against the emergence of any unified 'solution', while the mosaic of specific struggles and solutions pressed for by various stakeholders at different levels, ranging from the local to the global, often leaves the mainstream juggernaut of commoditization rolling on unabated. Institutional failure thus ensures that the negative externalities of market failure will be, or are actually being, realized.

As Ekoko's chapter from Cameroon illustrates, the set of actors may extend to include supranational agencies such as the World Bank through its efforts to promote redefinition of forest ownership and management relations in individual countries. If one were to explore relations between 'state' and 'community' with respect to control of forest resources, it would be important to identify a whole range of other actors pushing that relationship in different directions in different cases.

Just as forest life cycles and lines of contestation differ from region to region and from case to case, analytical perspectives also tend to differ, often in complex dialogue with 'reality'. The key issue may, for example, be presented as one of state dominance restricting or privileging access to various social groups as has been argued in several studies here, in particular the two chapters on India (Sivaramakrishnan, Sundar). In contrast, others would claim that the central issue is one of the effects of networks of local patronage and power undermining 'well intentioned' state forest policies, which are in principle sound and available to protect weaker groups. In yet other contexts, such as Indonesia, the problem is posed differently, as one of a 'weak strong' state, which has never built up (or inherited) a strong and autonomous forest department as in India, having more or less sold out to large commercial interests exploiting national forest resources at the cost of livelihoods of weaker sections and the country's environmental stability. What strategies then suggest themselves: strengthening community control of forest resources and participation in forest management, or strengthening the role of the state?

The next four chapters provide critical analysis, from various viewpoints, of new and emerging strategies of 'decentralized' and 'community-based' forest management in different parts of the world, in which the objectives of natural resource conservation on the one hand, and local livelihoods and welfare on the other are no longer seen as incompatible but rather as integrated and mutually reinforcing. Various questions are raised about both the equity and effectiveness of such new forms. Brown and Rosendo's account of interactions between rubber tappers and environmental organizations in Rondônia, Brazil, presents a case in which the alliance between environmental NGOs and grassroots organizations of local forest users achieved some political success (through legal recognition of rights of access to forest resources, and the establishment of 'extractive' forest reserves), but has failed to achieve economic objectives as rubber tappers continue to lead impoverished lives.

'Joint' forest management, as Sundar reminds us in the case of India, is neither always as new, nor always as 'joint' as one might expect; community involvement in forest management may remain at the level of rhetoric as far as participation in the setting of basic agendas is concerned. 'Community-based' forestry embraces a wide range of set-ups, in which government is often reluctant to relinquish centralized control despite a rhetoric of decentralization and participation. In the Philippines, as Gauld argues, the new policy discourse of community-based forest management is still shaped mainly by efforts to maintain centralized control over forest resources and by a political economy still orientated towards commercial timber production on principles of 'scientific management'.

While 'community' becomes more important in policy discourse on forest management, Gauld reminds us, there is no coherent alternative theoretical model of community behaviour to replace the individualistic assumptions of 'tragedy of the commons' and 'prisoner's dilemma' approaches. As a result, 'community' is often understood at least implicitly in reductionist terms: as socially, politically and economically homogeneous, masking a whole range of internal imbalances of interests and power (including gender interests). These issues tend to manifest themselves where state 'withdrawal' or re-definition of government roles in relation to resource management leave new 'empty' spaces that may be occupied by other actors with interests in forest exploitation. Local élites, as in Klooster's account of tree theft in a Mexican community forestry co-management scheme, may invoke the 'weapons of the not-so-weak' to reproduce at local level the power and privilege of forestry élites, engaging in large-scale internal corruption. Community management as strategy thus has definite limitations, beginning with fundamental questions about who defines and constitutes 'community', closely followed by other queries about mechanisms and imbalances of 'community power' or empowerment, all of which take us back to the issue of the local exercise of power and patronage networks. Difficult questions then arise: if local divisions (whether of gender, class, ethnicity or whatever) insert themselves negatively in 'community-based' and 'social' forestry approaches, does this not call for a reconsideration of the role of the state and perhaps for reversal of currently popular trends to decentralization? Taking the studies in this volume together we can perhaps conclude that the verdict is not yet in on the potential of community-based management as a mainstay of forest resource management for the new millennium.

The later (and in many cases final) stages of forest life cycles have many variants, but again some common patterns may be discerned. Eventually, late if not too late in life, forests may be released from the vice of accumulation, their product no longer necessary to feed the industrial machine. In some cases, death is the release; the loss of interest in the forest is coterminous with the existence of the forest itself, and the forest is exploited to extinction. What stops, or controls, this process before the point of extinction? The answer may lie in economic factors — it ceases to be economically viable or profitable to continue with the exploitation of forest resources. It may also reflect a grudging response extracted from the powers that be — whether public or private — by protest and opposition movements involving forest communities or other protagonists. In principle, it could also come through the state, from a dawning realization of the folly of persisting on a path fraught with societal injustice and risks associated with a blindfolded lurch into a nature-destroying growth process.

In other cases, exploitation leads to crisis and disasters, as in the recent Indonesian case. Harwell's analysis of responses to the calamitous forest fires of Kalimantan highlights the common view of states and development agencies that 'natural' disasters can be overcome or prevented only by more development, including the use of high technology; at the same time, high technology itself can be a double-edged sword, as in the case of GIS technology which may also be appropriated to undermine hegemonic states' control of information. Meanwhile, other forests are destined for a quieter, more graceful old age. At present, this latter path is mainly reserved for forests in the wealthier nations of the North, whose demands for timber no longer place pressures on local forests but are met largely by imports. The forest therefore becomes available for other uses, as in Knight's study of the emergence of 'forest tourism' in Japan where domestic forests become important sites for the recreation of Japan's urban middle classes.

It is improbable that the trajectories and life cycles of forests observable in the case of the early developers of the North will manifest themselves in roughly similar fashion amongst the late developing economies of the South. While early industrialization indeed led to rapid deforestation in parts of the developed economies, the conversion of free forests into costly timber invited a shift to importation from the forests of poor countries. In general, this switch occurred well short of a stage of acute deforestation. The shift to imports has acted to intensify the demands placed on the domestic forests of the developing economies arising from their own patterns of urbanization, infrastructural development and industrialization. In addition, in many countries, the inability of these developmental processes to generate adequate alternative sources of employment and entitlement means that the forest is retained as a safety valve in the form of a land frontier. The processes of deforestation in the South have therefore been of a different order of intensity, and continue to run on unabated, with few of the demand-reduction factors and forces — as evident in the early developing economies — yet in sight. Export-orientated, market-led growth based on resource and labour intensive activities, as promoted recently by international development agencies, has added further fuel to this fire.

Thus, paradoxically, Japan's monocultural plantation forests become the sites of the urban recreant's search for 'nature', although they themselves are of course completely artificial creations designed originally for the maximization of timber production. The juxtaposition of the Japanese and Indonesian cases to close the volume highlights again the interconnectedness of things where forests are concerned; as Japan's forest recreants return from the quiet of the forest to the comfort of their urban homes, the Indonesian forests from which a good part of their houses and furnishings originated go up in smoke, in one of the largest conflagrations the world has ever seen.

In sum, we think this collection points to a fascinating and dynamic, if sobering, field for understanding both changing directions in state–society relations and changing trends in managing vital natural resources. The field of analysis emerging relates to the area of environmental politics or political ecology rather than to the study of 'forests' *per se*; the 'lessons' it potentially contains may be as relevant to the analysis of political-institutional frameworks for, and contests over, other kinds of resources as to specifically forest contexts.

2 Development Discourses and Peasant–Forest Relations: Natural Resource Utilization as Social Process

Anja Nygren

INTRODUCTION

Environmental hazards in the Third World have attracted increasing attention among scientists, conservationists and development experts in the past two decades. Dozens of books and articles have been published on deforestation in Amazonia, environmental degradation in sub-Saharan Africa, and loss of biodiversity and genetic base in tropical central and south America. This has clear connections to contemporary world politics where environmental issues are becoming one of the main arenas of debate, and where tropical forests are gaining the symbolic value of a 'heritage of the world community'. Research focusing on deforestation became important only when the phenomenon was politically defined as a global problem. The same holds true with the current discourse on biodiversity, global climatic change, and sustainable development (Adams, 1990; Roe, 1993, 1995; Sachs, 1993; Taylor and Buttle, 1992).

There is a large literature analysing the relationships of global forces and local processes in tropical deforestation and environmental degradation (Blaikie and Brookfield, 1987; Millikan, 1992; Redclift and Goodman, 1991; Schmink, 1995; Utting, 1993). An increasing number of scholars have addressed the multifaceted impact of the larger social structures and political-economic processes on local strategies of resource use (Adams, 1990; Neumann, 1997; Nugent, 1993; Peet and Watts, 1996; Ribot, 1995). They also stress that in order to understand deforestation as a social process, it is necessary to recognize the unequal relations of power and how they relate to resource access and control. As remarked by Blaikie and Brookfield (1987), environmental problems in the Third World are less a problem of poor management, over-population, or ignorance, as of social action, political-economic constraints and global inequalities.

This requires increasing attention to the broader issues involving development and power, as well as to the complex relationship between social structure and the cultural construction of nature (Gandy, 1996). We cannot analyse people's resource management strategies as something determined solely by the local culture; rather we need to explore them in relationship to historically shaped relations of production and power (Agarwal, 1992; Moore, 1993). Natural resource utilization is a social process in which different interest groups, with diverse and often conflicting intentions, confront each other at local, regional, national and global levels (Schmink and Wood, 1992). The social relations of resource utilization are historically and politically constructed, and the concepts of sustainability and development change over time and between different social and cultural actors.

A spate of recent studies have looked at the changes caused by state, market and development interventions in the natural resource management strategies, property rights and landscape composition of forest-dependent local peoples. Ghimire (1994), Neumann (1997), Ribot (1995) and Vivian (1994), among others, examine how African states and global environmental organizations have established governing mechanisms to strengthen their control over natural resources and over the people traditionally dependent on them. Peluso offers a sensitive analysis of the conflicts between forest-dependent people, foresters and the state over the use of forests in Java (Peluso, 1992a) and Indonesia generally (Peluso, 1992b, 1996), while Fairhead and Leach (1994), Fortmann (1995), Rocheleau et al. (1995) and Roe (1995) analyse a series of environmental narratives that embody a history of Western images and ideologies about the local processes of deforestation and environmental crises in Africa. Most of these studies, however, focus their attention on an analysis of the natural resource management strategies of traditional forest dwellers, with a long history of forest utilization; and of how their conceptions of sustainable use of forests differ from the views of Western developers. As such, these studies offer valuable insights into how 'outsider' developers and state officials have tried to exercise their authority over the peoples of the forests, and over their natural resources.[1] Less attention has been paid to the natural resource use strategies and environmental conceptions of the migrant settlers colonizing tropical forests. As noted by Nugent (1993: 20–34), the peasant colonists of Amazonia have frequently been represented in the literature as maladaptive land invaders and haphazard forest destroyers, who pose an immediate threat to the few remaining indigenous groups. Through categorical cultural representations, the forest-clearing activities of non-Indian colonists have been attributed to their primordial 'land hunger' or to their cultural 'forest phobia' (Joly, 1989; Sandner 1981/2), with little reference to contextual factors — including colonization policies, land tenure regimes, and market forces — which have combined to reinforce

1. This does not mean that in the studies mentioned local people are portrayed as environmentally-benign 'forest peoples', with idealistic tones. The forest utilization practices of the studied populations are carefully linked to the broader social structure and environmental history.

a land use pattern of forest conversion in most land colonization areas (Browder, 1995).[2] It is just these multifaceted demands of wider agrarian policies and global development discourses on the livelihood strategies of peasant colonists which may offer valuable angles to understanding the natural resource utilization strategies and forest-relations of the peasant forest-fellers, with their bargaining positions in the larger society and in a far-reaching global economy.

This chapter presents a case study of peasants as forest-fellers in the context of Central America. It analyses the history of a Costa Rican rural community, Alto Tuis (a pseudonym), as a history of change seen through the process of forest utilization from the epoch of pioneer colonization to the present day. Its aim is to unravel the multiple causes of deforestation and the plurality of cultural constructions of forests by the local peasants at different stages in the history of Alto Tuis. The simple starting point is that deforestation involves much more than the physical act of felling trees. It is a process of change in the people's land tenure and land-use systems, in their social stratification and power relations, and in their environmental perceptions and cultural constructions — a process of change that has to be examined from a diachronic perspective. The case study focuses particularly on those structural changes in people's livelihood strategies that led to increasing deforestation in the area, and on the historically-changing relationships between local conceptions of forests and global discourses on natural resource utilization and development.

Political Ecology: From Local to Global

In order to see the history of deforestation within a particular social and cultural context, there is a need to move beyond the old opposition between 'virtuous peasants' and 'vicious states' to a perspective in which both the peasantry and the state are seen as internally differentiated (Bernstein, 1990: 69; Mitchell, 1990). A political ecology approach, combined with cultural interpretations, offers a fruitful alternative.[3] This study emphasizes that environmental changes are inextricably linked to social and political processes and that social relations of production are central to an

2. For inspiring case studies on colonist peasants in Amazonia, see Bedoya Garland (1994), Cleary (1993), Nugent (1993, 1997), Schmink and Wood (1992) and Townsend (1995).
3. Bryant (1992) identifies the contextual sources of environmental change, conflicts over access and the political ramifications of environmental change as the three critical areas of political ecology. However, as noted by Peet and Watts (1996: 3–13) and Peluso (1992b: 51), political ecology can hardly be considered a theory, *per se*. It is rather a wide-ranging research field, pluralist in its orientations and lacking an explicit theoretical coherence. A political ecology inspired perspective has been succesfully utilized in analyses of the historical circumstances leading to local patterns of resource use and control (Blaikie and Brookfield, 1987; Hecht and Cockburn, 1989; Painter and Durham, 1994; Peluso,

understanding of deforestation. In order to grasp the larger impact of struggles over forest and power, local natural resource management practices are analysed in relation to world-wide processes by progressing from a local to a global perspective. The initial interest is in the local resource users and their social relationships; the analysis then broadens to the corresponding economic and political events on the regional, national and global scales that affect local systems of production and systems of signification (Agarwal, 1992; Schmink and Wood, 1992). An examination of the historical dimensions of resource conflicts is essential to an understanding of contemporary struggles, in order to reveal the changing social relations of production and cultural constructions related to local landscape (Fairhead and Leach, 1994; Peluso, 1992a, 1992b; Rocheleau et al., 1995).

According to Watts (1989: 4), the history of deforestation can be seen as a process of change in which 'productive resources, property rights, and authority are struggled over'. In this struggle, local people alter their production strategies, as well as their perceptions of the environment, within a social context which is structured, but not determined. According to the changes in their natural and political ambience, they try to create strategies of survival and resistance in order to improve their control over the utilization of natural resources. However, a one-sided actor-orientated paradigm, according to which no matter how degraded people might be, they preserve a certain potential for creativity and space for manoeuvre (Torres, 1992; Valestrand, 1991; Verschoor, 1994), is not sufficient. By stressing people's capacity to invent and create, actor-orientated researchers tend to remove agents from structures and to replace determinism with voluntarism. At the same time, they easily forget that the central causes of environmental degradation and rural deprivation are to be found in land tenure relations, market dependencies, organization of economies, and social relations of production (Bebbington, 1993; Millikan, 1992).

PEASANT–FOREST RELATIONS IN COSTA RICA

The case study which follows is based on anthropological research carried out during 1990–2. The primary information is based on ethnographic field material consisting of 150 hours of tape-recorded interviews, together with dozens of informal meetings, daily conversations and participant observations. This material was supplemented by existing archive and statistical material, ministerial documents, law texts, and historical documents, as well

1992b; Schmink and Wood, 1992), plurality of perceptions of environment-development problems (Agarwal, 1992; Moore, 1993; Zimmerer, 1993), local resistance to state control and global environmental management (Ghimire, 1994; Neumann, 1997; Peluso, 1992a), trajectories of social movements (Escobar and Alvarez, 1992; Gadgil and Guha, 1994), and criticisms of development (Escobar, 1995; Hass, 1993; Rocheleau et al., 1995; Watts, 1989).

as by the interpretation of aerial photographs of the region that were taken in 1965, 1978 and 1988. Of course, there were methodological problems in reconstructing the local life-histories and cultural conceptions of forests from information gathered principally through ethnographic interviews. In this respect, it is important to note that retrospective human memory is highly selective, and in the case of historical narrations the past tends to be negotiated from the framework of the present. This is especially true of oral history, which bears many metaphoric messages and acts as a channel through which we express our cultural constructions of history and of the social world around us (Chambon, 1995; Fairhead and Leach, 1994; Townsend, 1995). The in-depth interviews proved to be the best way to obtain diachronic information about life in Alto Tuis, given the scarcity of historical documentation. The acquisition of diverse environmental narratives and constructions of the past in the interviews with the local people offered valuable information about their historical consciousness and social memory, and how the transformations in social relations of production, in national agrarian policies, and in global political economy were culturally perceived.

Context and Positions

Costa Rica, the third smallest country in the isthmus of Central America, comprises 51,000 km^2 of territory and 3.3 million inhabitants. It is often portrayed as a global leader in environmental protection and biodiversity conservation (Calvo, 1990; Holl et al., 1995: 1549). More than 25 per cent of the national territory has been set aside as protected areas — one of the highest proportions in the world. During the 1992 United Nations Conference on Environment and Development (UNCED), Costa Rica received an international award for its conservation policies, and the country was praised as a model for sustainable development world-wide (Boza, 1993).

Paradoxically, for most of the second half of the twentieth century, Costa Rica experienced one of the highest rates of deforestation in the world, a trend which has been reversed only in recent years. In 1940, approximately 75 per cent of the country's total land area was under forest cover; by 1995 the comparable figure was 25 per cent. Discounting the forest area inside national parks and wildlife areas, where no forest management is permitted, less than 5 per cent of Costa Rica's land area is now under productive primary forests (Segura et al., 1997).

The community of Alto Tuis, situated in the mountains of Turrialba in eastern Costa Rica, has a highly skewed land tenure and a complex social structure. Most of the population are small-scale peasants, with a few big cattle raisers and absentee land speculators; the rest are landless labourers. As small-scale cash-croppers of coffee and sugar cane, the peasants of Alto

Tuis have not been pure 'subsistence' producers for a long time; their production is closely linked to global markets and policies (Nygren, 1995a).

The following analysis looks at the history of deforestation in Alto Tuis. It begins with the epoch of land colonization and pioneers' slash-and-burn agriculture for basic food cropping, progresses to the next generation's timber sales and forest clearing for coffee and sugar cane, and thereafter to the conversion of forest to pasture. It then arrives at today's clandestine deforestation, discourse on forest conservation, and emerging resource conflicts. For each time period, the peasants' activities in the forest and their cultural conceptions of local environment are linked with contemporary agrarian policy, social structure, and ideological discourse on rural development. In this way, the study illustrates how relations between local people and the forests are embedded in global discourses on environment and development, and how the local exploitation of forest resources and subsequent resource scarcity relate to wider issues of forest, power and development. At the same time, it illustrates how concepts of the 'proper' utilization of natural resources change over time and between different social actors. Local constructions of deforestation are, therefore, not seen as mere reflections of cultural experience and history, but as the constitution of past and present power relations.

Inward Colonization and Agricultural Frontiers

The land colonization of Alto Tuis arose from the political perception, common in many Latin American societies, that forest areas occupied by indigenous populations constitute an empty space and a resource frontier to be exploited for the national wealth. Until the twentieth century small groups of Cabécar Indians lived in the area. Their livelihood strategies were based on the multiple use of natural resources, such as hunter-gathering, small-scale swidden agriculture, and mixed home gardening (Morrison and León, 1951).

At the end of nineteenth century, a railway was built from central Costa Rica to the Atlantic Coast. Facilitated by the improved networks, many big coffee and sugar cane haciendas were established in the Valley of Turrialba, around Alto Tuis. Most of them were owned by English and German proprietors, encouraged by the Costa Rican government in the hope that this would advance international trade. The haciendas required a large labour force; because of the small number of the indigenous population, landless Mestizo people from central Costa Rica were persuaded to move to this eastern periphery with the promise of employment on the haciendas (Hall, 1976: 33–41).

Stimulated by this governmental colonization policy, some of the labourers on the estates were inspired to try their luck as pioneers in the surrounding mountains. According to agrarian legislation of the time, virgin forests were state-owned wastelands that colonists could appropriate by clearing the land for cultivation (Salas Víquez, 1985). The first struggle over land in Alto Tuis was thus between the Cabécar Indians and pioneering peasants. In this struggle, the hunter-gathering Cabécar were defined as 'primitive' forest-dwellers with no rights of possession in the land. The colonization meant not only a struggle over the physical occupation of space, but also a cultural struggle over knowledge and power regarding the 'rational' use of natural resources. The communal land use system of the Cabécar was seen as inferior to the conception of land as a private commodity, and their hunter-gatherer livelihood strategies as inferior to forest felling for agriculture (Nygren, 1995b: 55–7).

The inward migration of the peasants was anything but spontaneous.[4] The movement was closely linked to the agrarian policy of that time, and to the profound transformation in the Costa Rican agrarian structure. First, the government promoted the formation of coffee estates in the Valley of Turrialba by liberalizing the privatization of the national forest lands, giving free land to the haciendas in return for their deforestation activities and the construction of the necessary infrastructure for their estate agriculture. Thereafter, the state began to favour the creation of peasant settlements on the surrounding hinterlands as a safety valve for landless people and as a reserve of untouched natural resources to be exploited (Salas Víquez, 1985). The socially constructed role of peasants was, at this time, to act as pioneers converting virgin forests for the agricultural development of the country.

The peasant settlements also served as a way of ensuring a seasonal supply of labour for the haciendas, especially for the peak time of the coffee harvest. The costs of the social reproduction of labour were minimal in the case of these seasonal wage workers, who also had their own small-scale agriculture. A similar land colonization strategy was applied throughout Costa Rica (Roseberry, 1991; Salas Víquez, 1985), but was especially prevalent in Turrialba, where most of the coffee and sugar cane haciendas were concentrated and where inward colonization was most conspicuously linked to the latifundia–minifundia complex.

The land colonization of Alto Tuis was not brought about solely by poor peasants, however. Although the first colonists were principally landless poor searching for a piece of land for cultivation, they were joined by land speculators and absentee landlords. The social structure became complex and the process of land claiming chaotic. Pioneers sold their lands to

---

4. Augelli (1987), Sandner (1981/2) and Vargas Ulatea (1986) argue that the remoteness and the relative distance of the colonized areas from the moderating state institutions made the Costa Rican land colonization spontaneous. However, these studies pay little attention to the links between the land-use changes in a particular region and the agrarian policies of the country. When analysing the land colonization of Alto Tuis within a wider context of Costa Rican agrarian change, the whole question of spontaneous versus organized colonization seems deficient.

second-wave colonists and moved to a new frontier as part of a two-step migration. Many big landowners appropriated the land occupied by pioneering peasants, simply informing them that the land had been taken over. Given that costs for registering land and applying for the land title were prohibitively high, there was little that the poor peasants could do (Morrison and León, 1951).

The close connection of the colonization of Alto Tuis with the haciendas of the Valley can also be seen in the fact that the first colonists did not go very far away. Although they sought freedom from tenancy on the haciendas, they did not want to be relegated to a distant jungle with no connections, as they had from the outset begun to produce maize and beans and needed access to markets. This was encouraged through governmental grants and credits in order to meet the growing demand for basic crops in the national markets.

Many newcomers settled in close proximity to the pioneers, giving rise to a form of community. This was no static village settlement, for there were colonists of different origins, and a constant flow of people coming and going. Absentee landlords bought the parcels of smallholders, combining them into large landholdings, whilst at the same time, the land was fragmented via the customary land tenure arrangements. The whole system was dynamic, with intense competition for land, multiple disputes over productive resources, and rapid replication of the same sort of highly skewed land tenure as had been characteristic in the colonists' homeland.

It is thus not viable to analyse the history of Alto Tuis from a static perspective, assuming homogeneity among the pioneer colonists. In his study on Pejibaye, southern Costa Rica, Sewastynowicz (1986) argues that the first colonists developed an egalitarian society with no differentiation in social status. From this basis, Sewastynowicz assumes that on a pioneer frontier even the least promising members of society had the opportunity for upward social mobility, if they only had the right psychological motivation and good judgement. In reality, however, there were great differences in the colonists' status right from the outset, and the state played an important role in regulating the opportunities of different social actors. Sewastynowicz thus seeks the principal motive for inward migration in each colonist's individual desire for upward social mobility, and pays little attention to colonization as a social process.[5] At least in Alto Tuis, it was mostly the early land speculators who profited from the land claims made in this uninhabited jungle. The poor

5. Jones (1989) sees the same kind of individualistic, get-rich-quick motive behind the agrarian colonization in Costa Rica. He argues that the Costa Rican land colonization was driven by something much like the American 'frontier spirit', with the peasant colonists drawn to the uninhabited forests by the promise of prosperity (ibid.: 66). Such a view neglects the links between the expansion of agrarian capitalism and the agricultural colonization of the country, and pays little attention to the role of the peasants in the larger society.

colonists failed to radically improve their situation because they lacked the capital required for intensive cultivation or for extensive cattle raising, the strategy used by most of the absentee landlords.

Pioneer Deforestation: Struggle over Access

The deforestation of Alto Tuis began during the pioneer colonization: both peasants and absentee landlords were active deforesters. This was related to the contemporary agrarian policy, in which the legal basis for securing ownership of the national wasteland was to 'make improvements' on the land, and in which the value of the farm was calculated according to the jungle area cleared. The land tenure legislation, credit systems and public discourse on rural development were all based on the notion that forest land has no value until it is cleared for agriculture (Utting, 1993: 34–40).

In most cases, it was the peasants who carried out the labour-intensive task of forest felling; the big absentee landlords asked the local peasants to clear their forest as contract work or through land tenancy. This opportunity for contract work, together with seasonal wage work on the haciendas, had important impacts on the social organization of labour in Alto Tuis. Many of the non-monetary labour arrangements common in other pioneer regions never developed in Alto Tuis, which was so closely involved with the haciendas. Working festivals were arranged only when communal buildings were being built, and *peonada,* a reciprocal labour exchange typical among the peasants of lowland Panama and Colombia (Gudeman and Rivera, 1990; Heckadon-Moreno, 1984), was usually practised only among close kinsmen. Between non-relatives, wage labour was the most common working relationship. This situation was characteristic of many similar peasant settlements in Latin America, where the close proximity of coffee haciendas provided a demand for seasonal wage labour (Duncan and Rutledge, 1977; LeGrand, 1984).

The intensity of the pioneer deforestation in Alto Tuis cannot be justified by the need to clear new land to satisfy the food requirements of the expanding local population. In the first place, food agriculture in the area was never developed beyond the needs of the local population, and secondly, a considerable part of the Alto Tuis forest was cleared to accommodate speculative cattle raising. As most of the colonists never acquired title for their claims, they preferred to clear as much as possible in order to secure land ownership through deforestation. Those pioneers who had sufficient forest in reserve considered it better to fell a new plot of virgin forest for slash-and-burn agriculture than to return to the once-cultivated fallow. Although the felling of a new piece of forest was more costly, it was a means of appropriating the land, given the prevailing agrarian legislation and land occupation rights — a clear case of land tenure arrangements impacting on the rate of deforestation.

The pioneers of Alto Tuis identified themselves first and foremost as colonists, who 'tamed the wilderness of the jungle'. Removing the forest was a dangerous and labour-intensive job, and most of the pioneers remembered their lives during colonization as being hard and melancholic. While present-day environmental extensionists (*extensionistas*) idealize the life of traditional peasants as 'living in equilibrium with nature',[6] in the pioneers' world-view the forest was a jungle, intact and wild, that remained outside human control and outside the active circle of women. Women took care of fuelwood gathering, but on the secondary scrublands; it was not considered appropriate for them to go alone or to work in the dense forest. Forest felling was an activity only for men, and even for them, much mental and technical capacity was needed to accomplish such an arduous task.

The forest represented a source of unpredictable rains, thunder and storms in the pioneers' world-view, and it was also a place of malevolent supernatural beings. Alto Tuis' oral history includes many collective memories about the apparition of supernatural spirits to forest travellers, as well as charms to protect workers in the forest. This connects to the peoples' life-history as colonists who opened the virgin forests for agricultural capitalism: as they were closely incorporated into the market economy on the haciendas' hinterland, the pioneers of Alto Tuis perceived in the dense forest the constant risk of losing control over economic and symbolic resources. Deforestation meant access to participation in the great competition for physical and cultural authority over the remaining un-appropriated natural resources.

Logging and Cash Cropping

A remarkable point in the agrarian transformation of Alto Tuis was the construction of the penetration road in the late 1950s, a typical mark of the politicians' desire to incorporate the remote peripheries into the market economy (Jones, 1989). The construction of the road led to intensive forest exploitation in Alto Tuis. Both the peasants and the agricultural entrepreneurs were active loggers. However, for many poor peasants logging was a part of their survival strategy, while the bulk of the profit from timber sales went to some of the richest peasants and absentee landowners.

To this generation of active loggers, the forest was a non-renewable natural resource, whose high-quality timber could be logged for markets and the land brought under cultivation. This view was conditioned by the

---

6. A more thorough analysis of the environmental extensionists' conceptions of Costa Rican peasants and their environmental relations is presented in Nygren (1998). For portraits presented by development experts on the local environmental histories and crises in various parts of Africa, see the studies by Fairhead and Leach (1994, 1995), Neumann (1997), Roe (1991) and Rocheleau et al. (1995).

contemporary political environment, in which the peasants were encouraged to deforest the land in order to increase the country's agricultural wealth. The arguments of present-day foresters that the traditional peasants had an intrinsic antipathy toward the forests, or that they lacked a 'forest culture', are simply not adequate in this context (Nygren, 1998). In a situation where clearing the forest for agriculture was defined as the only reasonable land use strategy in the tropics, local people had little room to create sustainable forms of forestry.

The construction of the road also led to intensive cash cropping of coffee and sugar cane in Alto Tuis. This brought increasing articulation to national and international marketing systems, as well as increasing control over local production by state institutions. According to the green revolution ideology of that time, the people were urged to change from traditional polycultures to monocultural plantations and from green manure to agrochemicals. The majority of the Alto Tuis peasants became producers of the primary commodities of coffee and sugar for export. Simultaneously, their food cropping activities were marginalized and the participation of women in agricultural decision making diminished.

The question of peasant marginalization cannot, however, be categorized simply as food cropping versus export cropping: a number of factors were involved, such as the quantity of land, capital and labour required by each crop, and the forms of its commercialization. Due to its labour-intensity, coffee production was one of the few activities in which the Alto Tuis peasants had any chance of competing with the large-scale producers[7] — although this meant that they had to refrain from calculating a real value for their own work. Coffee production for fluctuating world markets was a risky operation for the small peasants. However, in the context of an aggressive agro-export policy, where most credits and grants were shifted to the capital-intensive coffee and sugar cane production while the basic grain producers were left to depend on their own resources, it was often the only opportunity.

The beginning of coffee cultivation did not have any noteworthy impact on deforestation in Alto Tuis, as the area planted for coffee by each household was not very large. The problem was rather that the people remained in a vulnerable position in regard to price control. Their freedom over the production process diminished when the agricultural agents began to set ever-changing requirements with regard to the quantity and quality of the coffee. The same was true of the transition from domestic to commercial sugar cane production, but in contrast to coffee, commercial sugar cane growing was possible only for a minority of the Alto Tuis households.

---

7. A considerable part of the coffee cultivation in Costa Rica was in the hands of peasants right from the outset, while the agricultural entrepreneurs controlled the processing and exporting stages (Roseberry, 1991). For a discussion on food cropping versus export cropping in Central America, see Brockett (1988), Jansen (1994), Maxwell and Fernando (1989), Utting (1993) and Williams (1986).

The orientation toward cash cropping complicated the agrarian transformation in Alto Tuis. The expansion of export agriculture did not lead to a direct transformation of Alto Tuis peasants into wage labourers, but a small coffee-growing peasantry did emerge. Most of them were not independent peasants but rather producers who supplemented their livelihood with seasonal wage work on the haciendas, and with diverse economic activities in the formal and informal sectors (Nygren, 1995b: 90–8). The complexity of these local livelihood strategies has often been underestimated in classical frontier theories, which treat the peasantry as a broad analytical category in the transformation from the 'pre-capitalist' to the 'capitalist' mode of production (Foweraker, 1981, 1982; Galli, 1981; Roseberry, 1978). Recent ethnographic analyses of tropical frontiers reveal instead the multidimensional nature of local livelihoods and the considerable importance of the informal economy, such as small-scale forest extractivism, mixed gardening, timbering, and ambulatory trading, in addition to capitalist cash cropping and wage working.[8] In Alto Tuis, as in many similar communities, the people were constantly moving between monetized and non-monetized, as well as between formal and informal spheres of the local economy.

Extensive Cattle Raising

The 1970s saw increasing conversion of forests to pasture lands all over Costa Rica. Extensive cattle raising programmes, supported by international aid agencies, were designed to increase the country's beef export to the hamburger markets of the United States (Brockett, 1988: 48–9). The cattle expansion effected profound changes in the landscape and social structure of Alto Tuis. The traditional practice by which large landowners lent a parcel of their land to tenants ceased, and many minifundists lost their opportunities for seasonal wage work, as the cattle ranchers needed little labour.

Most of those in Alto Tuis who began with commercial cattle husbandry could not afford a herd big enough for profitable beef cattle raising. They turned to dairy-cattle, despite difficulties in marketing the milk and low meat prices. The internationally sponsored cattle raising policy favoured large landholders, while many peasants lost control of their land and of the production process (Nygren, 1995a). Ecologically, large-scale cattle raising offered little prospect of sustainability in regions like Alto Tuis, where most of the pastures are situated on steep slopes and suffer from serious soil erosion.

---

8. For detailed analyses of the multiplicity of production systems and livelihood strategies in tropical peasant communities, see Cleary (1993), Leach (1994), Nugent (1993, 1997), Peluso (1996) and Schmink and Wood (1992). For an interesting study of the re-creation of a Peruvian peasant economy within the framework of international capitalist expansion and national agricultural policies, see Painter (1991).

According to information provided by aerial photographs, the extent of pasture land in Alto Tuis increased 197 per cent during the period 1965–78, signifying that 54 per cent of the total farm area was under pasture. Throughout the country, two-thirds of agricultural land was devoted to cattle raising at an economic return far below that obtained through the cultivation of export or food crops, not to speak of the ecological costs. Inequalities in land tenure and resource ownership only increased. The overall social problem is illustrated by the fact that in the late 1970s, 64 per cent of the agro-export and beef cattle farms in Costa Rica were larger than 200 ha. Together, they occupied 90 per cent of the total area in production and received 88 per cent of the agricultural credits (Guess, 1977: 12–14).[9]

From Deforestation to Forest Conservation? Diverging Policies and Perspectives

In the 1980s, Costa Rican agrarian policies underwent a U-turn, with strict restrictions placed on forest clearing. Deforestation was constructed as a global problem and tropical forests as 'the lungs of the world', whose protection is essential for the survival of the planet (Nygren, 1998). The practice of deforestation was condemned, affecting especially the poor peasants who owned the most marginal lands with limited agricultural potential. For many big landowners, the new legislation meant only greater bureaucracy: if the land was suitable for agriculture, there were no restrictions to its deforestation. Moreover, many big landowners carried out their forest clearing clandestinely, as did smallholders. In practice, the new legislation had little impact on forest clearing in Costa Rica: the rate of deforestation continued at an average of 50,000 ha per year throughout the 1980s. In Alto Tuis, only 25 per cent of the total land area remained under forest cover in 1988, signifying that almost all of the primary forests had disappeared and the secondary forests survived only in the steepest areas (Nygren, 1995b: 111–18).

Where the new policy was put into effect, forest issues were translated into rhetorical speeches in favour of national parks, with considerable support from international aid agencies and conservation foundations. Forestry was made synonymous with forest protection; it passed rapidly through a stage of formal acceptance via political decision making and public involvement, but had little to do with sustainable alternatives to deforestation, such as empowering rural people to local forestry actions (Carrière, 1991). The close connection between social inequality, rural poverty and deforestation

9. Edelman (1994) warns of exaggerating the 'hamburger connection' as the only cause of deforestation in Costa Rica. Although probably a major force behind forest clearing, the process of deforestation is too complex to be explained solely as 'cattle eating the forest'.

was ignored by the politicians, who stressed that the key to reducing deforestation lay in increased state control based on careful vigilance and strict sanctions for clandestine deforesters.

Today, Costa Rican peasants are being urged to engage in community forestry, agroforestry, and traditional polycultures. The forest law is very strict with regard to deforestation, and there are governmental incentives for reforestation. The international development agencies are placing increasing emphasis on forest protection and environmental conservation in their aid distribution for Costa Rica, while the Costa Rican government is negotiating a reduction of a part of its foreign debt by means of conservation (Segura et al., 1997: 114–15). This turn-round illustrates how the conceptions of reasonable use of natural resources are historically and politically constructed.

According to Costa Rican state officials, the main problem underlying the country's deforestation is the peasants' lack of cultural awareness of the value of conservation (Nygren, 1998). Yet the social identity of these migrant peasants has always been linked to forest clearing for agriculture; their world-view corresponds to their life-history as colonists, who gained the ownership of the land by felling its forests. In Costa Rican political discourse, to be a peasant (*campesino*) means more than just engaging in a particular mode of production. The concept is also endowed with stereotyped cultural characteristics — the *campesino* as uneducated forest destroyer and unruly client of development. Institutional power establishes itself not only in the structures of distribution and accumulation, but also in the cultural constructions of social representation and social order.

The historical differentiation in the cultural construction of reasonable use of natural resources can also be noted at the local level when comparing the environmental perceptions of different generations and genders in Alto Tuis. In contrast to the pioneer colonists, the young producers of Alto Tuis today know little about the forest. They devote themselves to small-scale coffee production or cattle raising, combined with agricultural wage work, but their life is hard as they cling to agriculture on tired lands. This generation is the most critical of the living conditions and environmental situation in Alto Tuis. They lament the fact that their grandfathers as pioneers felled quantities of trees and left them to rot, while their fathers as loggers sold all the fine timber at prices that were too low. Their own future looks unpredictable, with the increasing concentration of land in the hands of a few absentee landowners.

The relationship of these young producers to the forest is distant. It is no longer a source for gathering, because remaining forests are scarce, or secondary growth, for few native fruit trees and medicinal herbs remain. Most of the young people in Alto Tuis consider the forest a prohibitive place, whose utilization is restricted by the forest law. For the youth and children, the virgin forest is a curious relic from their fathers' and grandfathers' time. As children of an era of aggressive environmental education, they are the most interested in reforestation in Alto Tuis. However, they do not perceive this to mean large-scale tree planting, as proposed by most of the reforestation programmes; for them, reforestation means the planting of fruit trees on the field borders or as shade trees in pastures.

In terms of the gendered differences in forest-relations, it is worth noting that although the women are responsible for natural healing in Alto Tuis, it has always been the task of men to go in search of natural remedies in the dense jungle. Today, the women of Alto Tuis feel that serious illnesses are difficult to cure because of the scarcity of powerful remedies such as wild herbs and vines, and they lament that they have to search for fuelwood at ever greater distances. In the times of pioneer colonization, strategic gender differences with respect to critical productive resources were reflected in the fact that forest felling was considered an exclusively male activity; today, it is decision making in the coffee and sugar cane production which is the exclusive domain of men, because it is the men who usually own the cash crop cultivations. Women's relations to coffee production are confined to picking, a task which is stereotyped by many men as an ideal activity for women because of their 'patient character and innate handiness'.

This plurality in social relations of production and cultural constructions of environment has attracted scant attention in classical deforestation literature (Bunker, 1985; Foweraker, 1981; 1982). As Cleary (1993: 331–8) notes, in deforestation studies which assume a macro-structural approach, ethnographic diversity in the environmental relations of the people is easily forgotten. The peasantry is constructed as a monolithic category in a regional branch of a national sub-system of a global system, without taking into account the multiple cultural constructions of nature among the local actors, especially those differentiations revolving around the productive inequalities mediated by class, gender, ethnicity, and age.[10]

FOREST, POWER AND DEVELOPMENT

When analysing the environmental history of Alto Tuis from a diachronic perspective, the Alto Tuis landscape proves to be saturated with historical

10. See, for instance, the studies by Agarwal (1992), Guha (1990) and Leach (1994), on gender, domestic politics, and struggles over the environment at the household and community levels, as well as the studies by Moore (1993), Peluso (1992b, 1996) and Zimmerer (1993) on the diversity of perceptions of nature and environmental degradation at the local level. Rural communities tend to be politically fractured and socially differentiated in complex ways. Homogenic labels, often used to categorize rural people as 'indigenous', 'traditionalist', or 'subsistence peasants', easily mask the existing class, age, gender, and ethnic differences within the communities; on this, see Long (1996), Neumann (1997) and Pigg (1992).

struggles over the territory and its productive resources, as well as over the symbolic meanings of diverse production systems. In these struggles, local peasants find themselves confronted with capricious policies concerning the 'proper' utilization of forest and land. A major conflict in this respect arose in 1990, when a multinational coffee corporation bought various farms in Alto Tuis and began to clear the remaining forests on these farms for coffee plantations. The people of Alto Tuis were strongly opposed to this action and began a struggle to stop it. They claimed that the area felled by the estate consisted of dozens of hectares of forest, and that the estate had acquired the permission to fell the forest by corrupt means.

The point of the protest was not so much to preserve the remaining forest, but to show that the control and the benefits of forest resources belong to the local community. It was part of a growing social concern about the ability of outsiders to appropriate ever-dwindling local resources and to define themselves as arbiters and agents of development. The protest action was a demonstration of resistance against the estate's hegemonic conception of progress, and a visible attempt by the local people to claim their right to reconstruct the meanings of development. The protestors pursued a strategy of non-violence in their struggle: they criticized the obvious inequality of citizens before the law, claiming that the forest law only penalizes poor peasants, while large landowners and timber companies are allowed to fell quantities of forest. A local conservation group was established, and the estate's forest clearing activities were publicized through the media, with great concern being expressed over its impact on the region's environmental situation.

The community was not completely united in its opposition to the forest clearing, however. The leaders were two peasant union activists; they had the tacit support of the majority of the small and medium-sized coffee growers, who felt that the estate's expansion was a threat to their own coffee growing. On the other hand, the absentee cattle ranchers and land speculators associated the estate's expansion with modernization and an increase in land prices that could also profit them, while the landless labourers who worked at the estate thought it better to withdraw from the quarrel for fear of losing their jobs.

Officials in the Costa Rican Forest Service interpreted the whole protest as an expression of envy, claiming that the local people did not understand the benefits that the estate would bring to their community through employment opportunities. They argued vehemently that the cleared area was only a couple of hectares, and emphasized the amount of money which the estate had to pay for the cutting licence, as well as for timber taxes. The question of the legality of the action was left open. Papers in the archives of the Forest Service indicate that 29 ha was cleared. Since the area was not entirely suitable for agriculture, forest officials required in their authorization that if the area was cleared for coffee, soil erosion should be controlled with terraces. By the end of 1992, at least, the estate had not fulfilled this requirement, although the entire felled area was already under coffee cultivation.

Against the background of cases such as this, where big entrepreneurs and transnational companies are authorized to clear forest for extensive agriculture, the public demand in Costa Rica for environmental education to counter the cultural idiosyncrasy of the peasants toward natural resources sounds curious, to say the least. In fact, the current greening of the development discourse has concealed more than it has elucidated regarding knowledge of environmental questions. The defence of the environment has often been separated from social rights, and environmental questions have become a pretext for political intervention in rural communities. The establishment of protected areas, ecotourism businesses, and agreements for biodiversity trade are marketed as being free of controversy simply because they are 'green' (Carrière, 1995). On the other hand, the boom in local participation has shifted the discourse on rural development towards an understanding of local culture, with little interest in the social circumstances that construct the people's land use practices and knowledge systems (Nygren, 1998). The people of Alto Tuis know that their history is not made at the local level alone, but that any change in the current situation requires dialogue and struggle at different levels, from local to global.

In the ever-changing discourses of what resources are to be utilized, how and by whom, developers frequently define themselves as exponents of the right knowledge, while peasants are labelled as 'target groups' in need of education. First, the people of Alto Tuis were urged to clear the forest for development; today they are advised to plant trees in the name of development. Where once they were encouraged to log their forests completely, today they are 'educated' to value the trees as renewable resources that require management and yield cash. A similar *volte face* is to be found in agriculture. During the green revolution, the Alto Tuis peasants were urged to change their polycultures to efficient monocultures, but now polycultures are praised as highly sustainable production systems. Their traditional land conservation practices of intercropping and agroforestry were labelled as primitive compared with the use of agrochemicals; now the same agroforestry systems are hailed as one of the most ecologically sound production systems. In this struggle between different production systems and world-views, local livelihood strategies are continuously labelled as outmoded and obstructive to development.

The people of Alto Tuis have responded to this by challenging the developers' expertise as haphazard advice which is likely to change dramatically according to the vicissitudes of development policies. They contest the progressive character of development projects by pointing out that developers' promises are never fulfilled. In the current boom of 'natural products' and 'local environmental wisdom' the people of Alto Tuis feel themselves used. Where only a decade ago, physicians condemned their use of wild plants as medicines, there is now a stream of ethnopharmacologists wanting to be guided to the remote mountains in search of natural remedies. A local peasant, Don Rodrigo, was scornful of the whole circus, relating how a Cabécar Indian had just begun to sell wild vines and roots to a homeopath in the nearby town of Turrialba — the more rich in mould and the more bitter in taste, he laughed, the more money one could get for the plant. Such stories illustrate the ignorance of the developers who have no notion of the wider social and political context in which the 'utility' of natural resources and local environmental knowledge is continuously reconstructed.

Any attempt to improve the peasants' marginalized production conditions will fail unless there are other changes — in the land tenure system, the knowledge–power stratification, and the social violence against local action. The developers' humanistic assurance of 'working on behalf of the rural poor' offers no easy alternative, because it does not necessarily safeguard the rights and dignity of all citizens, and because the ethics of this approach, which is sharply separated from politics, is highly voluntaristic (Bernstein, 1991: 163–4; Ribot, 1995). There is a need for radical social and political change within which an alternative perception of social concern, cultural representation, and local action is possible. At its best, the struggle for more sustainable development means an increasing plurality of social actors and social movements, revising the one-sided view of development in favour of strategies that permit a new orientation to questions of knowledge and power through a more diverse conceptualization of the social reality.

CONCLUSION

This study has examined the history of deforestation in the community of Alto Tuis as a process of environmental and social change, involving continually shifting concepts of the reasonable use of natural resources, and the historically constructed relationships of development and power. The clearing of Alto Tuis' forests was promoted through agrarian legislation, land tenure arrangements, market policies and ideological discourse of colonization as modernization. In the case of the cash cropping peasants on the margins of the haciendas, no clear distinction between local and global could be made. Since Alto Tuis was incorporated into global systems from the outset, its deforestation could only be explained within those larger social structures that altered the local livelihood strategies over time.

Struggles over the utilization of natural resources in Alto Tuis consisted of a complex set of social actors and the dynamic articulation of different economies and cultures. These included hunter-gatherer Cabécar Indians, heterogeneous groups of pioneering peasants engaging in the market economy and supplementing their own agricultural production with seasonal wage work, capitalist land speculators, absentee cattle raisers, and timber dealers exploiting local productive resources, as well as different development agents pursuing their multiple development goals. The agrarian transformation was a dynamic process, in which traditional production systems, land tenure arrangements, and environmental perceptions articulated with new ones.

The relationship of local resource management practices to global processes varies according to the history of colonization in the communities in question, their incorporation into the global economy, and the capitalist mode of production. In the case of Alto Tuis, no clear-cut distinctions between conservationists and forest exploiters could be drawn, but as noted by Touraine (1981: 8) 'all historical societies have transformed their relations with the environment, this is the very definition of their historicity'. The local-level analysis of the social relations of production and the symbolic meanings of livelihood shows the complex interaction between global forces and local resource management practices in the history of Alto Tuis, while the ethnographic analysis of the local life-histories and cultural constructions of forests marks the inevitable encroachment of global development discourses upon the everyday life of the peasants.

As a whole, the agrarian transformation of Alto Tuis brought increasing state control over the production processes and livelihood strategies of the local peasants. Through cash cropping of coffee and sugar cane they became vulnerable to fluctuating prices on world markets and to ever-changing demands with regard to product quality. Through cattle expansion many of them lost ultimate control over production processes and became totally landless rural poor. The current model of agrarian development offers little hope for a better future. Despite the rhetoric of sustainable development, neoliberal non-traditional agriculture may lead to increasing rural deprivation unless attention is paid to the inegalitarian structures of resource access and power at local, national, and global levels (Barham et al., 1992; Utting, 1994). Currently emerging social movements are attempting to alter the course of hegemonic developmentalism through alternative strategies of social action and ways of making politics. In this struggle critical inquiry also has an important task. It must go beyond the relativizing of narratives to challenge the authoritative social discourse and find new ways to analyse the complex relationships between local problems and global forces.

REFERENCES

Adams, W. M. (1990) *Green Development: Environment and Sustainability in the Third World.* London: Routledge.

Agarwal, B (1992) 'The Gender and Environment Debate: Lessons from India', *Feminist Studies* 18: 119–59.

Augelli, J. P. (1987) 'Costa Rica's Frontier Legacy', *The Geographic Review* 77(1): 1–16.

Barham, B., M. Clark, E. Katz and R. Schuurman (1992) 'Nontraditional Agricultural Exports in Latin America', *Latin American Research Review* 27(2): 43–82.

Bebbington, A. (1993) 'Modernization from Below: An Alternative Indigenous Development?', *Economic Geography* 69(3): 274–92.

Bedoya Garland, E. (1994) 'The Social and Economic Causes of Deforestation in the Peruvian Amazon Basin: Natives and Colonists', in M. Painter and W. H. Durham (eds) *The Social Causes of Environmental Destruction in Latin America*, pp. 217–48. Ann Arbor, MI: The University of Michigan Press.

Bernstein, H. (1990) 'Taking the Part of Peasants?', in H. Bernstein, B. Crow, M. Mackintosh and C. Martin (eds) *The Food Question*, pp. 69–79. London: Earthscan.

Bernstein, R. J. (1991) *The New Constellation: The Ethical-Political Horizons of Modernity/ Postmodernity*. Cambridge: Polity Press.

Blaikie, P. and H. Brookfield (1987) *Land Degradation and Society*. London: Methuen.

Boza, M. (1993) 'Conservation in Action: Past, Present and Future of the National Park System of Costa Rica', *Conservation Biology* 7(2): 239–47.

Brockett, C. (1988) *Land, Power and Poverty: Agrarian Transformation and Political Conflict in Central America*. London: Unwin Hyman.

Browder, J. (1995) 'Redemptive Communities: Indigenous Knowledge, Colonist Farming Systems, and Conservation of Tropical Forests', *Agriculture and Human Values* XII(1): 17–30.

Bryant, R. L. (1992) 'Political Ecology: An Emerging Research Agenda in Third World Studies', *Political Geography Quarterly* 11(1): 2–36.

Bunker, S. G. (1985) *Underdeveloping the Amazon: Extraction, Unequal Exchange, and the Failure of the Modern State*. Urbana, IL: University of Illinois Press.

Calvo, J. C. (1990) 'The Costa Rican National Conservation Strategy for Sustainable Development: Exploring the Possibility', *Environmental Conservation* 17: 355–8.

Carrière, J. (1991) 'The Crisis in Costa Rica: An Ecological Perspective', in D. Goodman and M. Redclift (eds) *Environment and Development in Latin America: The Politics of Sustainability*, pp. 184–204. Manchester: Manchester University Press.

Carrière, J. (1994) 'Neotropical Investigations: The End of the New World?', *European Review of Latin American and Caribbean Studies* 56: 119–22.

Chambon, A. S. (1995) 'Life History as Dialogical Activity: If you ask me the right questions, I could tell you', *Current Sociology* 43(2–3): 125–35.

Cleary, D. (1993) 'After the Frontier: Problems with Political Economy in the Modern Brazilian Amazon', *Journal of Latin American Studies* 25(2): 331–49.

Duncan, K. and I. Rutledge (eds) (1977) *Land and Labour in Latin America: Essays on the Development of Agrarian Capitalism in the Nineteenth and Twentieth Centuries*. Cambridge: Cambridge University Press.

Edelman, M. (1994) 'Rethinking the Hamburger Thesis: Deforestation and the Crisis of Central America's Beef Exports', in M. Painter and W. H. Durham (eds) *The Social Causes of Environmental Destruction in Latin America*, pp. 25–62. Ann Arbor, MI: The University of Michigan Press.

Escobar, A. (1995) *Encountering Development: The Making and Unmaking of the Third World*. Princeton, NJ: Princeton University Press.

Escobar, A. and S. Alvarez (eds) (1992) *The Making of Social Movements in Latin America*. Boulder, CO: Westview Press.

Fairhead, J. and M. Leach (1994) 'Contested Forests: Modern Conservation and Historical Land Use in Guinea's Ziama Reserve', *African Affairs* 93(373): 481–512.

Fairhead, J. and M. Leach (1995) 'False Forest History, Complicit Social Analysis: Rethinking Some West African Environmental Narratives', *World Development* 23(6): 1023–35.

Fortman, L. (1995) 'Talking Claims: Discursive Strategies in Contesting Property', *World Development* 23(6): 1053–63.

Foweraker, J. (1981) *The Struggle for Land: A Political Economy of the Pioneer Frontier in Brazil, 1930 to the Present*. London: Cambridge University Press.

Foweraker, J. (1982) 'Accumulation and Authoritarianism on the Pioneer Frontier of Brazil', *The Journal of Peasant Studies* 10(1): 95–117.

Gadgil, M. and R. Guha (1994) 'Ecological Conflicts and the Environmental Movement in India', *Development and Change* 25(1): 101–36.

Galli, E. G. (1981) *The Political Economy of Rural Development: Peasants, International Capital, and the State.* Albany, NY: State University of New York Press.
Gandy, M. (1996) 'Crumbling Land: The Postmodernity Debate and the Analysis of Environmental Problems', *Progress in Human Geography* 20(1): 23–40.
Ghimire, K. B. (1994) 'Parks and People: Livelihood Issues in National Parks Management in Thailand and Madagascar', *Development and Change* 25(1): 195–229.
Gudeman, S. and A. Rivera (1990) *Conversations in Colombia: The Domestic Economy of Life and Text.* London: Routledge.
Guess, G. M. (1977) 'The Politics of Agricultural Land Use and Development Contradictions: The Case of Forestry in Costa Rica'. PhD Thesis in Political Science, University of California.
Guha, R. (1990) *The Unquiet Woods: Ecological Change and Peasant Resistance in the Himalaya.* Berkeley, CA: University of California Press.
Hall, C. (1976) *El café y el desarrollo histórico-geográfico de Costa Rica,* transl. J. Murillo. San José: Editorial Costa Rica y Universidad Nacional.
Hass, P. (1993) 'Epistemic Communities and the Dynamics of International Environmental Co-operation', in V. Rittberger (ed.) *Regime Theory and International Relations.* London: Oxford University Press.
Hecht, S. and A. Cockburn (1989) *The Fate of Forests: Developers, Destroyers and Defenders of Amazon.* London: Verso.
Heckadon-Moreno, S. (1984) 'Panama's Expanding Cattle Front: The Santeño Campesinos and the Colonization of the Forests'. PhD thesis in Anthropology, University of Essex.
Holl, K., G. Daily and P. R. Ehrlich (1995) 'Knowledge and Perceptions in Costa Rica Regarding Environment, Population and Biodiversity Issues', *Conservation Biology* 9(6): 1548–58.
Jansen, K. (1994) 'Ecological Degradation in the Production of Food and Export Crops in North-West Honduras'. Paper presented in the 48th International Congress of Americanists (ICA), Stockholm.
Joly, L. (1989) 'The Conversion of Rain Forests to Pastures in Panama', in D. Schumann and W. Partridge (eds) *The Human Ecology of Tropical Land Settlement in Latin America,* pp. 86–130. Boulder, CO: Westview Press.
Jones, J. (1989) 'Human Settlement of Tropical Colonization in Central America', in D. Schumann and W. Partridge (eds) *The Human Ecology of Tropical Land Settlement in Latin America,* pp. 48–85. Boulder, CO: Westview Press.
Leach, M. (1994) *Rainforest Relations: Gender and Resource Use among the Mende of Gola, Sierra Leone.* Washington, DC: Smithsonian Institute.
LeGrand, C. (1984) 'Labor Acquisition and Social Conflict on the Colombian Frontier, 1850–1936', *Journal of Latin American Studies* 19(1): 27–49.
Long, N. (1996) 'Globalization and Localization: New Challenges to Rural Research', in H. Moore (ed.) *The Future of Anthropological Knowledge,* pp. 37–59. London: Routledge.
Maxwell, S. and A. Fernando (1989) 'Cash Crops in Developing Countries: The Issues, the Facts, the Policies', *World Development* 17(11): 1677–708.
Millikan, B. H. (1992) 'Tropical Deforestation, Land Degradation and Society: Lessons from Rondônia, Brazil', *Latin American Perspectives* 72(19): 45–72.
Mitchell, T. (1990) 'Everyday Metaphors of Power', *Theory and Society* 19(5): 545–77.
Moore, D. S. (1993) 'Contesting Terrain in Zimbabwe's Eastern Highlands: Political Ecology, Ethnography and Peasant Resource Struggles', *Economic Geography* 69(4): 380–401.
Morrison, P. C. and J. León (1951) 'Sequent Occupance, Turrialba, Costa Rica', *Revista Interamericana de Ciencias Agrícolas* 1(4): 185–98.
Neumann, R. P. (1997) 'Primitive Ideas: Protected Area Buffer Zones and the Politics of Land in Africa', *Development and Change* 28(4): 559–82.
Nugent, S. (1993) *Amazonian Caboclo Society: An Essay on Invisibility and Peasant Economy.* Oxford: Berg.
Nugent, S. (1997) 'The Coordinates of Identity in Amazonia: At Play in the Fields of Culture', *Critique of Anthropology* 17(1): 33–51.

Nygren, A. (1995a) 'Deforestation in Costa Rica: An Examination of Social and Historical Factors', *Forest and Conservation History* 39(1): 27–35.
Nygren, A. (1995b) *Forest, Power and Development: Costa Rican Peasants in the Changing Environment.* Helsinki: Finnish Anthropological Society.
Nygren, A. (1998) 'Environment as Discourse: Searching for Sustainable Development in Costa Rica', *Environmental Values* 7(2): 201–22.
Painter, M. (1991) 'Re-creating Peasant Economy in Southern Peru', in J. O'Brien and W. Roseberry (eds) *Golden Ages, Dark Ages: Imagining the Past in Anthropology and History*, pp. 81–106. Berkeley, CA: University of California Press.
Painter, M. and W. H. Durham (eds) (1994) *The Social Causes of Environmental Destruction in Latin America.* Ann Arbor, MI: The University of Michigan Press.
Peet, R. and M. Watts (1996) 'Liberation Ecology: Development, Sustainability, and Environment in an Age of Market Triumphalism', in R. Peets and M. Watts (eds) *Liberation Ecologies: Environment, Development, Social Movements*, pp. 1–45. London: Routledge.
Peluso, N. L. (1992a) *Rich Forests, Poor People: Resource Control and Resistance in Java.* Berkeley, CA: University of California Press.
Peluso, N. L. (1992b) 'The Political Ecology of Extraction and Extractive Reserves in East Kalimantan, Indonesia', *Development and Change* 23(4): 49–74.
Peluso, N. L. (1996) 'Fruit Trees and Family Trees in an Anthropogenic Forest: Ethics of Access, Property Zones, and Environment Change in Indonesia', *Comparative Studies in Society and History* 38(3): 510–48.
Pigg, S. L. (1992) 'Inventing Social Categories through Place: Social Representations and Development in Nepal', *Comparative Studies in Society and History* 34(3): 491–513.
Redclift, M. and D. Goodman (1991) 'Introduction', in D. Goodman and M. Redclift (eds) *Environment and Development in Latin America: The Politics of Sustainability*, pp. 1–23. Manchester: Manchester University Press.
Ribot, J. C. (1995) 'From Exclusion to Participation: Turning Senegal's Forestry Policy Around?', *World Development* 23(9): 1587–99.
Rocheleau, D., P. E. Steinberg and P. A. Benjamin (1995) 'Environment, Development, Crisis, and Crusade: Ukambani, Kenya, 1890–1990', *World Development* 23(6): 1037–51.
Roe, E. M. (1991) 'Development Narratives, or Making the Best of Blueprint Development', *World Development* 19(4): 237–300.
Roe, E. M. (1993) 'Global Warming as Analytical Tip', *Critical Review* 6: 411–27.
Roe, E. M. (1995) 'Critical Theory, Sustainable Development and Populism', *Telos* 103: 149–62.
Roseberry, W. (1978) 'Peasants as Proletarians', *Critique of Anthropology* 11(3): 3–18.
Roseberry, W. (1991) 'La Falta de Brazos: Land and Labor in the Coffee Economies of nineteenth-century Latin America', *Theory and Society* 20(3): 351–82.
Sachs, W. (1993) 'Global Ecology and the Shadow of Development', in W. Sachs (ed.) *Global Ecology: A New Arena of Political Conflict*, pp. 3–21. London: Zed Books.
Salas Víquez, J. A. (1985) 'La búsqueda de soluciones al problema de la escasez de tierra en la frontera agrícola: Aproximación al estudio del reformismo agrario en Costa Rica 1880–1940', *Revista de Historia* Número Especial: 97–149.
Sandner, G. (1981/2) 'El concepto espacial y los sistemas funcionales en la colonización espontánea costarricense', *Revista Geográfica de América Central* 15–16: 95–117.
Schmink, M. (1995) 'La matriz socioeconómica de la deforestación', in M. F. Paz (coord) *De bosques y gente: Aspectos sociales de la deforestación en América Latina*, pp. 17–51. México, DF: UNAM.
Schmink, M. and C. H. Wood (1992) *Contested Frontiers in Amazonia.* New York: Columbia University Press.
Segura, O., D. Kaimowitz and J. Rodríquez (1997) *Políticas forestales en Centro América: Análisis de las restricciones para el desarrollo del sector forestal.* San Salvador: EDICPSA.
Sewastynowicz, J. (1986) 'Two-Step Migration and Upward Mobility on the Frontier: The Safety Valve Effect in Pejibaye, Costa Rica', *Economic Development and Cultural Change* 34(4): 731–59.

Taylor, P. J. and F. H. Buttle (1992) 'How do we Know we have Global Environmental Problems? Science and the Globalization of Environmental Discourse', *Geoforum* 23(3): 405–16.

Torres, G. (1992) 'Plunging into the Garlic: Methodological Issues and Challenges', in N. Long and A. Long (eds) *Battlefields of Knowledge: The Interlocking of Theory and Practice in Social Research and Development*, pp. 85–114. London: Routledge.

Touraine, A. (1981) *The Voice and the Eye: An Analysis of Social Movements*, transl. A. Duff. Cambridge: Cambridge University Press.

Townsend, J. G. (1995) *Women's Voices from the Rainforest*. London: Routledge.

Utting, P. (1993) *Trees, People and Power: Social Dimensions of Deforestation and Forest Protection in Central America*. London: Earthscan.

Utting, P (1994) 'Social and Political Dimensions of Environmental Protection in Central America', *Development and Change* 25(1): 231–59.

Valestrand, H. (1991) 'Housewifization of Peasant Women in Costa Rica?', in K. A. Stölen and M. Vaa (eds) *Gender and Change in Developing Countries*, pp. 165–94. Oslo: Norwegian University Press.

Vargas Ulatea, G. (1986) 'La colonización agrícola en la Cuenca del Río San Lorenzo: Desarrollo y problemas ecológicos', *Revista Geográfica* 103: 69–86.

Verschoor, G. (1994) 'Intervenors Intervened: Farmers, Multinationals and the State in the Atlantic Zone of Costa Rica', *European Review of Latin American and Caribbean Studies* 57: 69–87.

Vivian, J. (1994) 'NGOs and Sustainable Development in Zimbabwe: No Magic Bullets', *Development and Change* 25(1): 167–93.

Watts, M. J. (1989) 'The Agrarian Question in Africa: Debating the Crisis', *Progress in Human Geography* 13(1): 1–41.

Williams, R. C. (1986) *Export Agriculture and the Crisis in Central America*. Chapel Hill, NC: University of North Carolina Press.

Zimmerer, K. S. (1993) 'Soil Erosion and Social (Dis)courses in Cochabamba, Bolivia: Perceiving the Nature of Environmental Degradation', *Economic Geography* 69(3): 312–27.

3 Fashioned Forest Pasts, Occluded Histories? International Environmental Analysis in West African Locales

Melissa Leach and James Fairhead

INTRODUCTION

Forests, invoked and evoked, have a curious capacity to structure the ways people describe their histories: they become key motifs in punctuating the past. In West Africa, this is as much the case in histories heard in ministerial and university settings, as it is in discussions shaded by village silk cotton trees. Around the motif it is easy to weave stories of social change and economic transformation with powerful moral connotations. Unsurprisingly, then, histories and their emotive force are frequently contested through forest images. But what does the brute simplicity of the term 'contest' obscure? West Africa offers many cases where the history of a location's forest is presented in highly incompatible ways. This chapter reviews several such cases, and explores in an ethnographic context how these incompatible histories — with all their social and material implications — interrelate.

In our earlier works we have examined the relationships between vegetation management by West African populations, and the forestry and conservation policy applied there (Fairhead and Leach, 1996, 1998). Drawing on ethnographic, archival and comparative documentary sources, largely from the twentieth century, we have framed our analysis of science-policy perspectives in broadly Foucauldian terms, showing the extent to which they can be considered to constitute a discourse of degradation or of deforestation. Framing analysis in terms of discourse has helped us to examine scientific debates concerning forestry and social history within the institutional contexts of colonial and post-colonial West Africa. We have also examined the validity of statements made about West African vegetation change, and traced the role of these statements in conservation policies which have sought to regulate landscapes and people's use of them in particular ways.

In this chapter we reflect further on the value of a Foucauldian approach to discourse in the analysis of West African forestry issues, and also on some of its shortcomings. In this, we engage with a far broader and long-standing set of debates concerning the relationship between agency, structure, knowledge and power, and with the various ways these relationships have been conceptualized in debates concerning science, development knowledge and political institutions.

Extreme interpretations of Foucauldian discourse theory present science and the institutions which generate it as integrated, operating — in more or less monolithic terms — to reproduce the positions of power in which they are historically located (e.g. Escobar, 1995; Ferguson, 1990; Sachs, 1992). Whether attributed to Foucault or his interpreters, this perspective can be challenged on several counts: first, that bureaucracies are far from monolithic; second, that subsuming bureaucratic practice into discourse absolves the actors involved of consciousness, intentionality and responsibility in their deployment of science, at the least obscuring the everyday dilemmas and situations of interaction faced by scientists and administrators, and the ways they respond to them. This leads to a third limitation: the reduction of interactions between administrations and local populations to a confrontation of discourses, falsely casting their interaction as one of assimilation or resistance (cf. Agrawal, 1996; Grillo, 1997; Sivaramakrishnan and Agrawal, 1998).

In contrast, other approaches — notably much actor-orientated sociology in development (such as Long and Long, 1992) and work within actor-network theory in the sociology of science and policy (for example, Callon et al., 1986; Latour, 1993) — give more emphasis to the autonomous intentionality of conscious subjects, their interpersonal interactions, and the ways they may actively strategize to represent issues in certain ways and forge alliances in promoting them. These approaches help identify variations in policy; disputes, tensions and their resolutions within and between departments; and the processes whereby different scientific perspectives and ideas rise or fall. But they also risk, perhaps, over-playing agency and intention at the expense of broader structural features, and losing sight of broader continuities in the ways basic problems are framed.

Hajer (1995), in work on environmental politics in Europe, links the insights of Foucauldian approaches with social-interactive discourse theory. He draws attention to 'argumentative interactions' in which people — provided with subject-positions by discursive fields — creatively marshal ideas and manufacture storylines to secure victory over rival thinkers. He claims that it is through the study of argumentative interactions that one can explain the prevalence of certain discursive constructions (Hajer, 1995: 54). He also introduces the middle range concept of 'discourse coalition', holding that (ibid.: 65):

> in the struggle for discursive hegemony, coalitions are formed among actors (that might perceive their position and interest according to widely different discourses) that, for various reasons, are attracted to a specific set of story-lines ... Discourse coalitions are formed

> if previously independent practices are being actively related to one another, if a common discourse is created in which several practices get meaning in a common political project.

Focusing on argumentative interactions and discursive coalitions, however, risks overplaying the intellectual elements of discourse at the expense of deeper-rooted institutional structures and processes.

A tendency towards different theoretical positions can be considered as partly an artefact of methodology. While some approaches explore science-policy-development issues and debates across countries, and over the long durée, others use fine-grained ethnography to reconstruct interpersonal interactions. The former are perhaps more likely to essentialize 'monolithic discourse' and portray discursive resilience and continuity. The latter are more likely to portray an impression of variability and change, but risk losing sight of broader continuities. These are left in inevitable tension in the multi-sited research strategies which are so necessary for exploring questions of science, knowledge, power and policy (cf. Marcus, 1995) — a tension which we find ourselves grappling with and which this chapter explores.

The two parts of the chapter draw on perspectives rooted in these contrasting methodologies. In the first part we draw on examples from our long-durée analysis of the international and national structuration of perspectives on forestry issues. We argue that there is a sufficiently systematic set of ideas and institutional practices at work for this to be treated as a deforestation discourse. This discourse represents social and population history in relation to forest cover in certain ways — ways which have served to occlude other social and political histories from its vision. If the processes of argumentation identified by Hajer have been at work — and we acknowledge that different methodological lenses would reveal these — they have not, it seems, served to subvert a basic representation of landscape history, and an enduring set of structures and financial/political relations and ideas. These are structuring the deployment of resources, laws and their enforcement in West African conservation, with remarkable similarities in broad forms and effects across a range of countries.

In the second part of the chapter, we pursue a more ethnographic mode, drawing on anthropological fieldwork in the Republic of Guinea. We focus on one locale whose history has been produced in a certain way by deforestation discourse, and examine how the manifestations of the latter are experienced and interpreted by its populations. Here, we find problems where discourse perspectives portray external discourse as confronting local knowledges and frames of reference, thus constructing and reaffirming dichotomies between state administrators and villagers, scientific and 'local' knowledges. Such dichotomies obscure the actual relationships between knowledges produced in engagement with the landscape, and shaped by inhabitants' particular historical and social experiences, and knowledges about landscape shaped by exposure to 'scientific' analysis and its products.

Village school teachers, adult education and development staff, district administrators, forestry officers, as well as urban dwellers and the radio-listening, outward-looking villager would count among those situated very ambiguously amid such dichotomies, leaving virtually no one exempt. Understanding these relationships necessitates a closer focus on interpersonal interactions, their content and context.

Yet as we shall argue, such a focus reveals many situations which seem not to conform to the kind of 'argumentative interaction' described by Hajer (1995). Moreover, observed relationships also bear rather little resemblance to those between 'citizen science' and 'expert institutions' frequently presented in work in the sociology of science and public policy in Europe (such as Irwin, 1995; Wynne, 1996). In the latter, contrasting life worlds and ways of conceiving environmental problems are held to lead to disdain and loss of confidence by citizens in expert institutions (and vice versa). The emphasis is on how intellectual components of discourses become juxtaposed, and resistance is exerted through scientific debate and product.

Rather, we suggest that the instances in which the intellectual content of deforestation discourse become juxtaposed with villagers' alternative ideas about landscape history are rather few and rather insignificant. They tend to be confined to situations which for a variety of reasons preclude intellectual argumentation developing. The dominant analyses of deforestation are very rarely aired or debated, or when they are, in contexts insignificant to political relations. Alternative discourses in which the same phenomena (trees and forests) may figure do not deal with the same questions, so as to make engaged discussion of 'a forestry problem' very unlikely. In these respects, analyses emanating from the discursive field revealed through our long-durée, national and international comparative work on West Africa seem rather irrelevant in interaction with land users. At the same time, the deforestation discourse does exert powerful material effects; but these tend to be interpreted in locality within other frames, reflecting particular social relations and personal histories. In short, people may experience the institutional and material effects of discourse without meaningful engagement with its narratives.

DEFORESTATION DISCOURSE AND THE OCCLUSION OF HISTORY

Moronou, Côte d'Ivoire

The idea of occlusion neatly combines the ideas of illusion of vision and closure encapsulated in the dismissal of oral accounts of landscape history in central Côte d'Ivoire. In 1981 the Ivorian historian Ekanza recorded oral histories among Agni people in the Moronou region (Ekanza, 1981: 59–60). He was astonished to hear elders recounting how, at the time of their occupation of the territory (which Ekanza dated to the eighteenth century), the area had been 'open savanna'. This is because it lies well within what ecologists have always described as the forest zone and was at the time a mosaic of farmland, forest and forest fallow. Once he had recounted these histories, Ekanza drew on prevailing forestry canon to reject their validity as descriptions of landscape pasts, consigning their relevance to political rhetoric and, perhaps, a deep memory of conditions millennia ago.

To have accepted the elders' accounts, Ekanza would have had to challenge a considerable literature documenting the relentless decline of Côte d'Ivoire's forests under population increase and economic transformation (for example, Arnaud and Sournia, 1979; Thulet, 1981), including the major assessments being carried out for the FAO (1981) and those that had been carried out by the Forestry Department (CTFT, 1966) — images which were being popularized in more catastrophic texts (Monnier, 1981). Although reproduced in ongoing forest assessments, these images had deeper roots in early colonial analysis (Aubréville, 1957, 1959; Chevalier, 1909) and in the educational structures in Côte d'Ivoire and France, of which he was a product. However, the elders' version of landscape history actually accorded with certain ecological and historical processes documented by others (including Adjanohoun, 1964; Aubréville, 1962; Spichiger and Blanc-Pamard, 1973). These works describe the northward expansion of forest vegetation in the Baoulé V, whether linked to farming practices or climatic rehumidification. Its credibility would also have been supported by observations made just over the border in Ghana in the 1930s by the forester Vigne (1937), and is subsequently supported by evidence of forest advance into savanna regions elsewhere in West Africa (Fairhead and Leach, 1998). In this case, Ekanza fortunately reproduced the history before occluding it, enabling subsequent reinterpretation, allowing it to be drawn back into a field where its claims about landscape history may find support in further research.

This case is not unique. In many instances, standard analyses of forest loss seem to have served to occlude alternative histories, in the sense of rendering them scientifically invalid or part of a mythic realm. Just as in Côte d'Ivoire, these standard analyses do not simply serve to invalidate other histories, but can frequently themselves be falsified by recourse to other data sets. This underscores the extent to which 'official' forest histories may misconstrue the memory and experience of local populations, in both ecological and political terms. Such incompatibilities between versions of forest cover change can be illustrated further by drawing on several cases from our own primary research and related analyses in the Upper Guinea Coast region of Guinea and Sierra Leone. Importantly, as we shall suggest, the histories thus occluded may support not only different kinds of self imaging, memory and political culture, but also particular resource claims. These alternative histories are of course themselves multiple and debated; limitations of sources and space mean that here we are able only to indicate partial elements of them.

Ziama, Guinea

The Ziama forest in Guinea is, in national and international forestry and conservation circles, considered as a relic of West Africa's diminishing Upper Guinean forest formation. That this is a region where dense humid forest is the natural climatic climax was affirmed by the earliest French colonial botanists (Aubréville, 1939; Chevalier, 1909). Designated a forest reserve in 1932, Ziama was made an international biosphere reserve in 1981 and remains the subject of highly-funded international conservation projects intent on preserving its biodiversity and climatic functions against the supposed depredations of surrounding populations. The image presented in conservation documents is of a pristine forest at risk of clearance for the first time under accelerating population and commercial pressures, requiring conservation-with-development approaches involving radical restructuring of local land tenure and economy. In accordance with the image of a previously little-disturbed forest, socio-economic studies have constructed for the predominantly Toma-speaking farmers living around the reserve a cultural past as a 'forest people' who once lived harmoniously, and in small numbers, with the high forest environment, through hunting and gathering (Baum and Weimer, 1992). Only in the twentieth century, it is claimed, have immigration of Manding farmers and traders, and adjustments in economic and technological orientation towards shifting rice cultivation and tree cash crops, produced the new pressures which threaten the forest. These pressures are seen to reflect the 'cultural degradation' of a once forest-benign indigenous people, and their incapacity to cope with modernity.

Yet villagers in Boo, Koima Tongoro and other reserve-edge villages talk of their past in very different terms, drawing attention to landscape features rendered irrelevant by deforestation discourse (Fairhead and Leach, 1994). They point out groves of silk cotton and kola trees, and extra-dense vegetation within the reserve, overlying the sites of old villages and hamlets, such as the twenty-two small settlements once dependent on Boo. They point out the tall cotton trees in villages which their ancestors planted as look-out posts to see allied settlements across the open savannas — expanses now covered with high forest. Elders recount the handed-down descriptions of the mid-nineteenth century economy: of intensive, short-fallow rice, fonio and cotton farming in savannas and bush fallow, and of vibrant markets where they interacted with the Manding traders long part of the region. These images in oral accounts, so strikingly at odds with the representations of history in deforestation discourse, gain documentary support from the accounts of several Americo-Liberian travellers who visited and described the region between 1857 and 1874 (Fairhead et al., forthcoming).

The painting shown to us on a Boo house wall, of a grandfather killed in wars with the French colonists at the turn of the twentieth century, provides an icon of the intervening events. Decades of protracted war between local polities and then with the French caused depopulation, allowing forest growth over the enriched, farmed soils in climatic conditions which appear to have been becoming more humid. While these wars, and the secondary nature of the forest, were noted by the French botanists who created the reserve in 1932, the archives have remained unread or ignored by modern conservationists. The latter are either unaware of or unconcerned by the ecological status of the forest and the landscape history linked to this. Yet this history remains central to villagers who allude to the present Ziama forest as a place of ancestral and *djinn* spirits, where clearing land (encroaching, in conservationists' terms) is more a political act to restake old land claims and political authority linked to them, than a response to land scarcity elsewhere.

Gola, Sierra Leone

The Gola forest reserves abut the Liberian frontier in the south east of Sierra Leone. Again, this is an area where deforestation discourse has produced powerful representations of landscape history, serving to occlude other accounts of the past. Indeed, Leach's earlier fieldwork among the Gola-Mende in 1987–8 was framed by broad acceptance of the prevalent view that this was a surviving area of mature rain forest amidst pervasive deforestation elsewhere (Leach, 1990). Many internationally-influential analyses hold the country to have lost major forest tracts during the twentieth century. The extreme view of Myers is that 'as much as 5,000,000 ha may still have featured little disturbed forest as recently as the end of World War II. It is a measure of the pervasive impact of human activities that the amount of primary moist forest now believed to remain is officially stated to be no more than 290,000 ha' (Myers, 1980: 164). The analysis of the World Conservation Monitoring Centre is that '50 per cent of the country has climatic conditions suitable for tropical rainforest, but less than 5 per cent is still covered with mature... closed forest. Deforestation is mainly a result of the rapidly increasing human population requiring more agricultural land and fuelwood' (Sayer et al., 1992: 244). In accordance with such analyses, the Gola forest commonly appears on conservation maps as primary rain forest (e.g. Cole, 1980). Sayer et al. (1992: 244) similarly assert that 'the moist evergreen forest in the Gola Reserve, which contains the last large remnant of lowland, closed canopy rain forest in Sierra Leone, is typical of climax Upper Guinean rain forest'.

However, such assumptions were strongly undermined by a reading of early sources describing Sierra Leone's vegetation at the turn of the twentieth century and the insights of Paul Richards and others on this issue (Davies and Richards, 1991; Kandeh and Richards, 1996), coupled to our wider points of critique of the science supporting this style of deducing forest history. Many early accounts note that large parts of the Gola forest reserves had been farmed during the nineteenth century (Government of

Sierra Leone, 1923; Migeod, 1926: 331; Unwin, 1909). In 1908, the forester Unwin collected accounts from inhabitants suggesting that the area had been inhabited and farmed by Gola-speaking populations until around 1850: 'Many years ago, from all accounts, the Mendis attacked the Gola people, and drove them back across the Mano, so that now only the old foundations of the houses, and the Cola trees they planted, mark the sites of their towns, which must have been quite numerous' (Unwin, 1909: 23). In this light, many ethnographic incidents and elements of oral accounts which Leach experienced in Malema chiefdom in 1988, but to which, at the time, she attributed little significance, fall into place. Cotton trees and old village sites pointed out by Mende villagers in the reserve take their place as artefacts in an alternative framing of landscape history.

Kissidougou, Guinea

Further north, Kissidougou prefecture of Guinea lies within the forest-guinea savanna ecotone. The mosaic of forest 'islands' in savanna here has until recently — as we have documented in detail (Fairhead and Leach, 1996) — been interpreted as the result of an ongoing process of deforestation and savannization wrought by the region's Kissi and Kuranko-speaking inhabitants. Scientists since the early colonial period, along with today's national forestry and environmental establishments and a host of internationally-funded conservation and natural resource management projects, seem to accept this view unquestioningly. It accords with widely-repeated images of rural populations and their pasts. These include images of the Kissia as forest people who once lived in harmony with a forest nature that was sacred to them, and of the southwards immigration of Manding-influenced 'savanna people' with a proclivity for fire setting and forest-destructive agricultural technologies (see, for example, Adam, 1948). They also include images of the growing pressures of population increase, commercialization and modernity (Green, 1991; Zerouki, 1993). Kissidougou has been the target of numerous donor-funded environmental rehabilitation projects, as well as government policies to fine people for setting fire and cutting trees.

Yet again, these prevalent histories occlude others told by those living within the forest islands (Fairhead and Leach, 1996). Villagers describe how forest patches, far from being relics of destruction, have been created by themselves or their ancestors in savannas, whether the emphasis is on tree planting, settlement foundation and forests as early war fortresses (as elderly men frequently suggest) or on the gradual vegetation-enhancing effects of gardening, household waste, and the grazing of domestic animals (as others, including many women, imply). Many forest-building and expanding practices are of a highly practical kind, grounded in villagers' everyday ecological knowledge and the fundamental idea in local thought that land is improved through use and work. Villagers consider forest islands as intrinsically linked with settlement and sociality. They consider many local farming and fire-setting practices such as early-burning to have enhanced the progressive expansion of forest into savanna over the last century. Their accounts are confirmed by other data sources. Thus comparison of aerial photographs shows that in 1952 there was less forest, not more, and archives and early travel accounts present an even less forested picture for the turn of the century, although observers then were just as convinced that the few forest islands they saw were relics in terminal decline.

THE STRUCTURING OF DEFORESTATION DISCOURSE

Elsewhere, we elaborate further the similarities between these cases: in their portrayal of an original, natural forest cover progressively destroyed as population-increase and modernity transform cultures which were once more forest-benign. Common analysis is linked to common policy agendas (Fairhead and Leach, 1998). The persistence of such representations, even despite the existence of counter-interpretations and evidence for these, suggests that a certain systematicity is at work in science-policy processes. Here we review and illustrate some of the key arenas for both the production and reproduction of ideas.

International statistics concerning forest cover change, and the various domains in which they are produced and used constitute one such arena. Ample statistics seem to support the view that West Africa has experienced dramatic forest loss during the twentieth century, accelerating during the last few decades. The World Conservation Monitoring Centre (WCMC), for instance, has produced country-by-country figures which aggregate to show that West Africa now has less than 13 per cent of its 'original' forest cover (Balmford and Leader-Williams, 1992: 74), while another internationally-influential survey (Gornitz and NASA, 1985) showed countries to have lost between 69 and 96 per cent of the forest area which they had at the turn of the century. Such statistics draw on and reproduce images of 'original' forest cover as a baseline with which to compare present day assessments derived from techniques such as remote-sensing. They simultaneously produce images of social and demographic history, of a pre-colonial past stretching back indefinitely when minimal populations lived harmoniously with minimally-disturbed forest.

These figures circulate widely in and amongst international organizations, and are put to a variety of uses. They inform assessments of the extent of deforestation important to the elaboration of international agreements, as well as influencing the funding agendas of donor agencies and international conservation organizations. While national forest assessments have an ambiguous relationship with the figures in international circulation, in many circumstances government departments and non-governmental organizations are not averse to drawing on these figures when justifying funding for forest and conservation programmes, especially in front of international donors. Indeed in drawing on national figures which feature in global tables, national assessments can perversely acquire greater authority and rhetorical weight. In particular, as Grainger (1996: 73) notes, 'FAO's forest resource statistics are usually regarded as authoritative and so quickly become established in the literature by default'. Furthermore data from FAO, WCMC, the World Resources Institute and other organizations are often the basis for unsourced tables and figures in secondary and tertiary articles and reports.

The images of 'original' forest cover contained in such statistics are linked to central suppositions in the way ecological science has conventionally treated vegetation change. Ecologists have divided West Africa into a series of bio-climatic zones which, at the scale which informs most national and international statistics, encompass the forest zone of the higher rainfall coastal belt, ceding to drier forest forms and then savannas further north. Many studies of forest cover change assume that at origin, these bio-climatic zones supported their 'climax' vegetation, that is, the ultimate stage in the succession through which vegetation would progress in the absence of disturbance (Clements, 1916). Where the vegetation does not accord with that deemed appropriate to the prevailing bioclimatic conditions, it is commonly construed as a disturbed or degraded form, with the disturbance generally attributed to people.

This set of theoretical tenets within ecological science was elaborated in the West African context early in the twentieth century through the work of colonial scientists, on the basis of specific assumptions current in European and Indian forestry circles prior to African colonization. The botanist Chevalier, for instance, compiled a general description of West Africa's vegetation zones (1911) which formed a basis for subsequent work by vegetation geographers such as Schantz and Marbut (1923) and Mangin (1924). While Chevalier's zones were originally based on descriptions of observed vegetation — or on patches of vegetation thought to be representative — there remained an ambiguity with zonal delineation based on the 'potential' vegetation which could exist under given climatic conditions, leaving room for speculative deduction about vegetation history, by assuming that a zone used once to carry its potential vegetation. Scientific analyses of the savanna areas on the northern margins of the forest zone exemplify such deduction particularly well: early foresters and botanists deemed actual savannas to be bio-climatically capable of supporting forest, and thus assumed that forest had once existed, having since been savannized through inhabitants' farming and fire-setting practices. This was the analysis of Unwin (1909) and Lane-Poole (1911) in Sierra Leone, Chevalier (1912) in Côte d'Ivoire and Benin, and Thompson (1910) and Chipp (1922, 1927) in Ghana. The work of Aubréville (1938, 1949) which correlated specific climates to specific forest types across Africa helped to formalize these speculations, purporting to give a precise delimitation of this ex-forest zone within the savannas — the zone which became popularly known in anglophone circles as 'derived savanna', later appearing as such in the descriptions of Clayton (1961), Hopkins (1965) and Keay (1959a, 1959b).

The notion that there was a balanced, natural vegetation against which present, disturbed vegetation could be compared, in turn supported the development of concepts and methods for defining and measuring departures from it. Scientists identified 'stages of degradation' — for example, those stages supposed to occur in the stepwise degradation of forest to savanna under repeated farming — so that the occurrence of any particular vegetation form came to indicate both a temporal degradation trend, and the stage that such degradation had reached. Thus by examining the species composition, diversity and associations (phyto-sociology) of a vegetation form, it became valid to deduce vegetation history. Such scientifically-authorized representations of history tended simultaneously to invalidate other historical sources. Limited in availability, historical sources were generally rejected because, as Aubréville put it, the term 'forest' was subjective, and the locality referred to too indeterminate, to enable proper comparison of present data with written sources (Aubréville, 1938: 78). Oral history was generally invoked only to illustrate processes already known 'more scientifically'.

Notwithstanding some contemporary debate (such as Aubréville, 1938), it was thus on notions such as climax, equilibrium and succession that the emerging science of forest ecology was built. While the conceptual apparatus which allows forests to be viewed as stable, climax vegetation forms has been strongly challenged in the ecological literature of temperate and dry tropical zones, it still remains the orthodox view among those dealing with tropical humid forests, especially in West Africa. However much certain ecologists working on West Africa have come to reflect critically on this reasoning (Hawthorne, 1996, for example), it has remained unchallenged as a central plank in national and international forest analysis.

Ideas have also infused work in social sciences, such that they come to support statistical and ecological analyses in a mutually-reinforcing way. Many of these works have been authored without direct reference to forest issues and policy, but either draw on prevalent analyses within forest history canon, or are being used by international conservationists and policy makers, or indeed both. They are thus integral to the reproduction of the discourse. The cases which we have presented above indicate how, in broad terms, the idea of recent, intact forest cover has been linked to the presentation of the forest zone as only recently inhabited significantly by agriculturalists. The image is of a zone which once housed only sparse hunter-gatherer or minimal root crop cultivator populations with a benign impact on forest cover, awaiting the introduction of exotic cereal crops and iron technology as enabling conditions for population expansion, beginning gradually around 1500 and awaiting the twentieth century for its major impact. These images, and associated ones linking ethnicity with particular kinds of environmental behaviour, characterized much early analysis by historians and anthropologists of population history, society, economy, ethnicity, and migration. While debate within the social sciences during the past few decades has questioned some of these analyses — for instance, D'Azevedo's influential analysis of Liberian settlement history in terms of a deforestation frontier has been firmly questioned by Jones (1983) — these debates appear to have made relatively little impact on the way social and demographic history is imaged in national and international forestry and conservation circles. At the same time, increasingly sophisticated analyses appear which accept the dominant view of forest cover loss, and offer explanations for it: whether by reference to the history of West Africa's iron industry (Goucher, 1981); veterinary medicine, the spread of trypanosomiasis and the decline of the horse in colonial Freetown (Dorward and Payne, 1975); the dynamics of settlement and kola cultivation in Liberia (Ford, 1992); or the cultural ecology and social dynamics of upland rice swiddens (Nyerges, 1988).

Legal and institutional elements also structure the deforestation discourse, linked to the ways that certain analyses uphold particular types of resource control. The scientific concepts of conventional forest ecology have been concretized in forest policy and law, supporting — and coming to be supported by — the forms that forest reservation and tree permit systems have taken. The assumptions about resource tenure which follow from the notion of 'primary' (climax) forest, for example, have long been used in designating forest reserves, whether by French colonial administrations who explicitly linked the idea of primary forest with a justification for state possession, as under French colonial law the state had rights to 'vacant' land, or in more recent arguments — exemplified in the Ziama case — that such forest should come under the guardianship of the international conservation community. The location of forest reserves and their management embodies ideas concerning both ongoing savannization and the relationship between forest and climate. For instance 'curtains' of reserves were established on the northern margins of the forest zone to defend the forests against ongoing southwards savannization. Tree tenure laws which maintain a distinction between 'natural' trees (claimed by the state) and planted trees (which planters can own) similarly serve to reproduce ideas of natural vegetation, and a reading of many trees in farmers' fields and forest patches as natural, occluding other landscape histories in which they may be artefacts. In short, concepts central to conventional forestry science have gained enduring material expression through conservation practice, while, simultaneously, deforestation discourse occludes competing rights to land and resources.

This aspect of the structuring of deforestation discourse is closely related to economic and administrative structures; and specifically, to the financing mechanisms both for national forestry activities and institutions, and those concerned with development in other fields. In brief, West African forestry services have long derived important revenues from their conserved assets;

from the sale of permits and licences for timber and wildlife exploitation, and from fines for breaking state laws. Revenues — and sometimes institutional survival — have been ensured by a reading of the landscape as deforested and in danger. In many countries, colonial forest codes (which, as we have seen, incorporated particular analyses of forestry problems) have been retained until very recently. Inherited forestry codes brought with them inherited assets, whether in forest reserves or listed timber trees. In their continued presence, these assets, in effect, constantly instantiate the original reasons for their acquisition. Thus the existence of reserve curtains re-invokes the idea that southwards savannization is in progress; watershed reserves continually re-invoke the supposed links between deforestation and hydrological desiccation; and the presence of state-controlled timber trees justified as such by their supposed 'naturalness' continually re-invokes the idea that these trees on farms and near villages are indeed natural relics of a lost forest cover. For forestry administrations to relinquish the analysis linked to these assets would be, in effect, for them to relinquish the resources which they represent. The analysis is thus strongly implicated in the real politics of control over valuable resources and revenues.

International aid flows, and particularly the direction of many funds once directed towards other areas of rural or agricultural development towards 'environmental' issues in the late 1980s and 1990s, has further reinforced deforestation discourse. It is clear that donor–government relations and various forms of 'green conditionality' can influence powerfully the adoption or persistence of particular agendas in national governments. In Guinea, environmental degradation in the forest-savanna transition zone was seen as 'necessary' in a context where aid to environmental rehabilitation had replaced much rural development aid. In effect, having an environmental problem of a certain form was necessary to guarantee donor funding. The discourses and practices of international donors have, of course, themselves undergone many shifts, not least among them increasing emphasis on 'participation' and community-based approaches in conservation and natural resource management. Yet it is striking how little these shifts seem to have affected the broad conceptual framing of forest cover change.

In this part of the chapter, then, we have identified a discourse of deforestation, and have highlighted ways in which it shaped social and demographic history in relation to forest cover, and in particular, how this occludes other social and political histories from its vision. Variation within this policy discourse has tended to debate reasons why the problem matters (timber, biodiversity, livelihood sustainability, and so forth) and approaches to dealing with the problems (repressive, participatory, devolved, etc.), but not the nature of the phenomenon itself. Such debates thus serve to uphold (and hence are part of) the dominant perspective (Foucault, 1980).

The value of this type of discourse approach is to reveal the systematicity in problem framing and policy responses across West Africa, among an array of national and international organizations which formulate and implement forest-related policy. It need not deny the debate, discord and elements of experiential learning present in these institutions, nor subtle regional variations, but it does show the aspects structuring this systematicity, linking scientific and social scientific analysis with institutions, financing and government.

LOCALIZED EFFECTS, LOCALIZED INTERPRETATIONS

In our earlier work in Kissidougou, Guinea, we argued that deforestation discourse had a series of powerful effects. It impoverished people through taxes, fines, resource alienation and the diversion of government activities from more pressing needs. It criminalized many of their everyday activities. It denied the technical validity of existing ecological knowledge and research into developing it. We also suggested that it denied value to their cultural forms, expressions and basis of morality. We argued further that effects were felt in undermining the credibility of outside experts in villagers' eyes; in provoking mutual disdain between villagers and authority, and in imposing on the former images of social malaise and incapacity to respond to modernity (Fairhead and Leach, 1996: 295).

Here, we extend the analysis of the Kissidougou case to reflect further on the manifestations of the broader structuring discourse in specific locales. Our methodological approach turns to a series of ethnographic vignettes. In these, the dichotomy between external discourse and locale does not look so clear, and people do not sit so obviously either side of it. Through the vignettes we explore the extent and types of response to deforestation discourse among different people, and whether, and in what forms, resistance, disdain and loss of confidence in the expert institutions of science and policy can be discerned.

In the north-east of Kissidougou prefecture, European Union-funded environmental rehabilitation programmes are assisting the establishment of village tree plantations. These are intended to exist in perpetuity, and their *raison-d'être* lies in the environmental objectives of watershed protection. To encourage villagers in this environmental, not economic, endeavour, projects are constructing schools and providing wells and other local infrastructure. Although this was supposedly a participatory programme, however, villagers were worried by these plantations and the associated 'gifts'. As young men expressed to our co-researcher, Dominique Millimouno, gifts of such generosity seemed out of proportion to the labour villagers were providing in the tree nurseries and plantations. Was this a sweetener to the ultimate take-over of their lands? What else could the project be after? Could it be attempting to control the gold or diamond deposits rumoured to lie in their land? During that dry season, a number of the project tree nurseries in this area 'burned in wild bush fires', with people either failing to protect them, or — as rumoured — actively causing the incineration.

Villagers in this area had not experienced grave loss of vegetation, and the projects patently did not respond to their environmental understandings or priorities. Nevertheless, this situation cannot be seen merely as a loss in confidence in expert institutions, a lack of credibility of outside experts linked to a distrust in the deforestation discourse that underlay the plantation activities. The fears, worries and political suspicions felt by villagers were real, but these seemed to relate less to readings of the landscape than to broader political and land tenure concerns. They were linked particularly to the expatriate element in the projects, and the French staff members who occasionally visited the villages. Expatriate-led policy and intent aroused suspicion rooted in the harsh experience of French colonial rule. Moreover, it is widely rumoured that white people have the capacity or 'eyes' (iconized in binoculars) to see minerals in the land, and that their primary motivation in life is to accumulate wealth and exert authority. Equally relevant were suspicions rooted in the experience of land nationalization and state socialism under Sekou Toure's First Republic, when large areas of village territory were alienated to create state farms. More recently, land in nearby Banankoro had been alienated from small-scale diamond miners (including Kissidougou migrants) to European/South African controlled multinational corporations. In short, distrust was less of the 'expert institutions' and their deforestation discourse than of certain aspects that the project particulars symbolized, given villagers' experiences of political and economic history. Considering this situation in terms of the intellectual juxtaposition of discourses about landscape history and problematics, and in terms of a dominant discourse and resistance to it, would be to overlook these many other histories and social memories informing the ways people interpret particular interactions.

Dry season fires tend to sweep Kissidougou's savanna lands annually. Living with fire, enhancing its positive effects and minimizing its damage is integral to the art of land management among the prefecture's farmers, herders, hunters and collectors. Yet observing the passage of fire through village territories, forest guards are authorized by legislation grounded in the deforestation discourse to impose fines on villagers. Villagers often accuse forest guards of setting the fires themselves so they can extort either fines, or informal payments in lieu of them.

Again, villagers are experiencing the effects of what we have identified as a deforestation discourse. But again, their interpretations of these experiences are not framed in terms of a juxtaposition of this discourse with their own readings of the place of fire in landscape dynamics. Rather, reflections on the activities of forest guards tended to dwell more on personalized factors concerning the forest guards themselves and their conditions of life than on the *raison d'être* for the institution employing them which analysts such as ourselves might trace to the discourse of deforestation. In Sandaya, the Kuranko-speaking village where we lived, forest guards are in part understood, if not forgiven, for their extorting activities on the grounds that salaries for forest guards are probably recurrently left unpaid (or are appropriated by seniors), so they must by necessity fend for themselves. Several relatives of Sandaya villagers held more senior positions in the national forestry department in Conakry. This was understood more in terms of their position in a good job, as a part of the Conakry bureaucratic élite who might be solicited for the broader influence and brokerage they could provide, than in terms of their mission to 'save the forest'. The relationships between such upper echelons of the forest department and the activities of extorting forest guards was little reflected upon. More personally, another Sandaya relative had become '*fato*' (mad) while working as a forest guard, and was to be seen occasionally, armed as forest guards customarily are, passing through the locality. This created further fields of speculation, orchestrating images of forest guards as wanderers of the bush, and subject to the vindictiveness of its spirits. Through these incidents and relationships the forest department was not so separate from villagers. But their relationship with it was couched not in terms of juxtaposition of discourses, but again in terms of more located experiences and social relationships.

For other reasons, certain villagers appear to be less bothered by forest guards, and less worried by project activities. Take, for example, one of the Imam of Sandaya, a keen farmer and cattle-keeper, but well-linked and well-liked in the Islamic circles centring on Kissidougou town. Respected in the village and town alike, he was unlikely to be targeted by forest guards, and knew it. Moreover through his network he was able to see the possible resource benefits that conservation initiatives might bring and which he might broker for the village; an altruistic act which would reflect well on him too. Such resource benefits would include access to free fruit and other tree seedlings which could be directed to be planted in the vicinity of the village to help form its forest island, and *quid pro quos* — such as rural infrastructure — which could be more advantageous still. For those in a sufficiently strong position, then, conservation projects could be seen as part of useful patronage networks with which positive links could be developed, but without engaging with — or even acknowledging — their broader motives. In this case, the Imam could draw on Islamic notions of coolness and shade, and the benefits which trees could thus bring to village space, echoing, but not articulated with, discourse concerning the rehabilitation of deforested lands and the climate and soil benefits associated with this.

The capacity of the Imam to broker development in this way both exemplifies and contributes to the power of Islam in village politics in the region, frequently in tension with the power of other chiefly lineages. Indeed Islamic circles form a parallel political forum to the state in local governance, dominating Kissidougou's urban politics and having, in many parts of the town, incorporated chiefly lineages. Many environment and development projects have realized the weaknesses of educated state structures in relation to Islam in the area's fragmented political field. The EU environmental rehabilitation programmes have sought to link their analysis of forestry problematics with interpretations of the Koran, sponsoring local Islamic scholars to develop 'green Islam' in citations printed on widely-distributed calendars and booklets. Although politically expedient for the project to secure the 'participation' of influential figures, this activity was also heavily publicized within the national and international development community as an innovative means to build on the supposed complementarities between regionally-rooted culture, and the messages of conservation. Yet this is to misinterpret the relationship between 'high' Koranic interpretation and Islam as rooted in West African village life and its social and ecological practices, which have long been syncretic with non-Islamic beliefs and practices. Arguably it also misinterprets the motives of the Koranic scholars engaged in the activity, who may have been more interested in using it to consolidate politically-advantageous relationships with projects than to promote an environmental discourse which had little meaning to them. It was, after all, in this region that youth expressed the concern that these same projects would alienate their land.

These instances suggest that in many aspects of its everyday operation, the broader *raison d'être* of the forestry service is not brought into play or into the interpretive field of those it affects. In certain circumstances, however, 'expert' readings of forest history and dynamics linked to social issues do engage with villagers' historical perspectives on landscape and vegetation change. In particular, extension agents of the forestry department or development programmes can become aware of the contradictions between the perspectives they have been trained to promote, and the ways villagers talk about and manage land and trees. This is either through their everyday work, their own backgrounds and broader social relations, or through exposure to and use of the participatory methodologies now promoted; some gained such awareness through interactions with us and exposure to our work.

Extension workers have dealt with these contradictions in different ways. One, close to our own work, sought to evaluate what he came to understand from farmers about the effects of certain land use practices on vegetation cover in terms which his project would recognize as suitably scientific. Yet he was unable to pursue this due to other demands on his time, and lack of support — and indeed blockage — from his project employers. Out of this contradiction came a loss of confidence in his employers, and a certain cynicism and critical vision of the discursive structures which shaped his conditions of work.

For other extension workers, encounters with inhabitants' land management practices are met with dismissal. They deem farmers ignorant and unscientific. Farmer perspectives are, they claim, irrelevant to the modern environmental paradigms which their professional education stresses, and which is recapitulated through modern radio media, whether national or international, much of which is itself funded and produced by development organizations. Such dismissal requires a downplaying of the personal experiences of vegetation phenomena which extension workers have gained, justified by asking how a few personal experiences could possibly challenge the canons of science.

The capacity to occlude alternative views of environmental dynamics among extension workers, as well as other professional workers in the locale such as schoolteachers, appears to be linked with the generation gap between high school educated children and their village-living parents, and with the formers' identity as 'modern' citizens. This modern identity is precarious: a precariousness which may be helpful for comprehending the phenomenal arrogance that can be exhibited by certain elements of the urban élite towards village life. In one case a schoolteacher denied that villagers have consciousness. But again, these reactions may be less about contrasting interpretations of landscape change, than part of a broader moralizing discourse about youth and uneducated villagers. Schoolteachers and others are exposed to media which link deforestation with social malaise. In conversation they provide evidence of this malaise in delinquent youth who now hunt with uncontrolled fire, linked to the breakdown of hunters' societies, and the uncontrolled activities of charcoal entrepreneurs, traders and quick-profit loggers. In this, teachers embrace a long tradition relating environmental decay to the malaise of modernity, and environmental quality to a golden age. These stereotypical environmental arguments about modernity are present in the very first French texts which infiltrated into the French educational system in Guinea, and which continue to shape regional perceptions of modernity and its social fallout. In projecting such images on others, many of the urban educated élite support a self image of the opposite: the ability to surf modernity with ease. In this way we can see the seepage of the metaphor of modernity from broader social domains into the ecological, with the ecological providing a vehicle for its reproduction in local social relations.

Nevertheless, even 'traditional' chiefs can become aware of, and engage positively with, modern conservation agendas. This was made clear in statements by a senior elder in the vicinity of the Ziama forest reserve. He described in an interview with conservation project officials how he and his people would take responsibility for conservation were they to be empowered to do so, but that without such empowerment (meaning control over their ancestral domain, now within the reserve) they could do nothing. His comments, asserting the capacity to manage modern issues just as well as the project, were marginalized in state planning reports (Baum and Weimer, 1992; Fairhead and Leach, 1994). It is not that the chief was accepting the deforestation discourse and the image of history that it constructed. He was, however, prepared to accept a part in management of a forest resource which he, just as well as others, could see had acquired global significance, attracting global funding resources.

There are also circumstances in which village residents appear to be engaging with the deforestation discourse on its own intellectual terms. That certain villagers, at least, are able to do this is hardly surprising given that it is embedded in primary and secondary school curricular. The village land chief and elder of Sandaya, for example, told us explicitly that the history of forest in the village was not as foresters — and we ourselves, he presumed in the early stages of our acquaintance — supposed. As he stated in clear rectification of the dominant discourse: 'It is wrong to suggest that cultivators have finished the forest. It is not like that in this country. Everyone has their place. Those who are in the forests found the forests there. Those who are in the savanna, it is not they who have transformed it but God alone'. The analysis did not continue into further critique, or a discussion of alternative perspectives in local land management. It seemed that the context was a polite but firm correction of the tenor of our inquiry.

In another case, a Sandaya farmer who was district political chief, the village primary schoolteacher, and ourselves dominated a discussion in which others were also present, during a visit from the chief's elder brother, resident in France. Here, the discussion turned to the environment and the loss of forest noted by the eminent visitor since his childhood. This conversation in French and Malinke soon passed on. The subject material appeared to create no sense of engagement among other listeners, and raised no particular emotive force either way. The sense was that this had been heard before, but was of little relevance.

It would seem, then, that the deforestation discourse can be acceptably heard and reiterated, and comfortably left unchallenged, within the praxis of certain meetings and social encounters. In neither of these cases did drawing on these perspectives — whether to refute or affirm — have material import, giving any sense of the ramifications which the rejection or acceptance of the discourse could have for control over fire, timber profits or other resources. In effect they took on little more than the connotations of polite speech.

Trees and forest patches do, of course, figure in a range of village affairs which are of deep political and material significance. As we have explored elsewhere, they are important to the ways people consider land tenure, settlement foundation, political authority, fertility, hydrological relations, the lives of animals and spirits, and social and economic transformations and their moral implications (Fairhead and Leach, 1996, 1998). They are drawn on as archives and markers in debates over these issues, thus acquiring a plethora of meanings, without these meanings coinciding with those in the deforestation discourse, and without these local debates being 'about' forest, trees or landscape per se.

DISCUSSION AND CONCLUSIONS

We have argued that a discourse perspective is valuable in capturing the broader historical, international and institutional dimensions of forest histories and problematics, and the ways that these have developed in relation to West Africa. Care needs to be taken, however, in examining the manifestations of such broader discourses in specific locales. If considered in terms merely of juxtaposition of a 'global discourse' of deforestation and 'indigenous perspectives', analysis is forced into certain patterns of representation. People must be described as clearly located (local, national) or ambiguously located (extension agents). They come to be imaged as deceiving (in shifting discourses strategically), as fractal (embodying incommensurable positions without awareness) or as being subject to any particular hegemony (engaged with only one frame of reference).

Without denying that such a vocabulary may sometimes be applicable, and that the juxtaposition of discourses on which it is premised may occur, we have been arguing that these would be rather strong words for issues of much less significance in everyday life. This is for two reasons. First, these characterizations overplay the importance locally of the intellectual (scientific) component of the discourse of deforestation. As we have tried to illustrate, a discourse can operate, and people can experience its material effects, without its analytical components being at the fore. In such circumstances people's interpretations of its material effects are easily diffused into broader interpretive fields drawing on more contextual political and social experiences. Second, where juxtaposition of intellectual discourses such as there is does occur, this takes the form not of 'ever present' disjuncture, but of very particular situated practices. As we have suggested, these may be situations where reflections on landscape change as such are of less significance than the other social or political issues structuring the interaction. Village meetings called by conservation or development agencies need to be considered as such situated practice. As Murphy's (1991) analysis of Mende political meetings emphasizes, in certain contexts it is appropriate or expedient for social reasons to create a front stage appearance of consensus even when backstage opinions over technical issues may be divergent.

In our rendition, then, there would appear to be rather few 'argumentative interactions' in villages of the kind that Hajer (1995) suggests are so crucial to the way that discourses are formed and transformed. At the least, it would be important not to assume such situated practices to be interactions where people voice their opinions about forestry issues and try strategically to influence each others'. Instead, as we have suggested, ideas may pass each other like ships in the night.

Work in the sociology of science — largely in European settings — has explored how the contrasting lifeworlds and ways of understanding environmental problematics between 'citizen science' and 'expert institutions' can lead to mutual disdain and loss of confidence. It shows how citizen-scientific interpretations are fought with those in expert institutions, including through citizen funding of scientific research or lobbying to transform research questions. Yet for such analyses to be applicable presupposes that there is actually a debate and that people are arguing from, at least broadly, the same terms of reference. The examples that we have highlighted do not conform to this condition, beginning to indicate perhaps that engaged scientific debate between citizen science and expert institutions may be highly particular and contingent. In Kissidougou expert institutions have always been quite disengaged from people, linked to the nature of colonial and postcolonial state-building, and the emergence of subject rather than citizen science. Where there is disdain and lack of confidence, it appears to draw less on intellectual analysis than on broader aspects of lifeworlds. Those who have lost confidence in the intellectual components of expert analysis seem to be the more dedicated workers within expert institutions faced with frank incompatibilities. In such circumstances, these 'street level bureaucrats' (cf. Lipsky, 1980) would be expected to be key players in the emergence of effective 'citizen science'.

Villagers' agency to shift the deforestation discourse is inhibited by the limited extent to which they engage with it in its intellectual expression, and the limited traction that their own perspectives have on it. As we have argued, villagers' interpretations concerning the *raison d'être* for institutional structures 'repressing' them, are at the same time both broader (e.g. French and Sekou Toure political history, operation of states) and more specific (e.g. reflections on the working conditions of forest guards). A consequence is that 'resistance' to the broader discourse of deforestation is not manifest. People do have ways of resisting its material effects (for instance, by setting/allowing fires), but these acts tend to be interpreted by those conducting them within broader historical and political frames. They also tend to be accommodated by those expressing deforestation discourse as instances of vandalism.

Our research culminating in *Misreading the African Landscape* (Fairhead and Leach, 1996) brought international and village perspectives on ecology into dialogue in a way that would not normally occur. At one level, our research did formulate Kissidougou practice and perspectives as a citizen science, albeit internally-differentiated and debated. Our work attempted to show the validity of this knowledge within the parameters and vocabularies of ecological debate in scientific circles, pitting its readings of landscape and problematics against that produced by deforestation discourse, and using evidence, such as air photographs, acceptable and comprehensible to the latter. A valid exercise perhaps; but there is no doubt that the picture of citizen science we created was both more integrated and more critically engaged with 'deforestation discourse' than the component elements which we 'researched out'. Nevertheless, it did begin a process of opening up space for shifting frames of reference, and did illustrate some of the parameters around which new discursive coalitions might form.

In work on the sociology of science and public policy, attention is rightly given to the processes by which 'policy space' may open up so that new perspectives and social commitments — such as those linked to citizen science — may feed into policy change. At the same time, there are arguments for more pluralistic approaches to environmental science and policy; for a 'democratization of expertise' (Funtowicz and Ravetz, 1992). However, rather little attention has been given to the necessary prior or simultaneous process which we have suggested would be necessary here: that is, enabling the kinds of syntheses of diverse subject perspectives, and the forging of effective discourse coalitions, which could gain a foothold in policy debates — both within locales, and linking them across regions. This kind of argument becomes more important in an era of globalized environmental science-policy debates, and globalized policy debates more generally; an era which in Guinea, while increasingly important, is already a century old.

REFERENCES

Adam, J. G. (1948) 'Les reliques boisées et les essences des savanes dans la zone préforestiere en Guinee Francaise', *Bulletin de la Société Botanique Française* 98: 22–6.

Adjanohoun, E. (1964) 'Végétation des savanes et des rochers découverts en Côte d'Ivoire centrale', *Mémoire ORSTOM* no.7, Paris.

Agrawal, A. (1996) 'Poststructuralist Approaches to Development: Some Critical Reflections', *Peace and Change* 21(4): 464–77.

Arnaud, J-C. and G. Sournia (1979) 'Les forêts de Côte d'Ivoire: une richesse naturelle en voie de disparition', *Cahiers d'Outre Mer* 127: 281–301.

Aubréville, A. (1938) 'La forêt coloniale: Les forêts de l'Afrique Occidentale Française', *Annales d'Académie des Sciences Coloniales*, IX. Paris: Société d'Editions Géographiques, Maritimes et Coloniales.

Aubréville, A. (1939) 'Forêts reliques en Afrique Occidentale Française', *Revue de Botanique Appliquée et Agronomie Tropicale* 215: 479–84.

Aubréville, A. (1949) *Climats, forêts et désertification de l'Afrique tropicale*. Paris: Société d'Edition de Géographie Maritime et Coloniale.

Aubréville, A. (1957) 'A la recherche de la forêt en Côte d'Ivoire', *Bois et Forêts des Tropiques* 56 & 57, Nogent-sur-Marne (France).

Aubréville, A. (1959) *La flore forestière de la Côte d'Ivoire*. Publ. CTFT no. 15 (3 vols), Nogent-sur-Marne (France)

Aubréville, A. (1962) 'Savanisation tropicale et glaciation quaternaire', *Adansonia* II(1): 233–7.

Balmford, A. and N. Leader Williams (1992) 'The Protected Area System', in J. Sayer, C. S. Harcourt and N. M. Collins (eds) *Conservation Atlas of Tropical Forests: Africa*, pp. 69–80. Cambridge: World Conservation Monitoring Centre and IUCN.

Baum, G. A. and H-J. Weimer (1992) 'Participation et dévéloppement socio-economique comme conditions prealables indispensables d'une implication active des populations riverains dans la conservation de la forêt classee de Ziama'. Conakry: Deutsche Forst-Consult/Neu-Isenburg RFA/KfW.

Callon, M., J. Law and A. Rip (eds) (1986) *Mapping the Dynamics of Science and Technology: Sociology of Science in the Real World*. Basingstoke: Macmillan.

Chevalier, A. (1909) 'Les Vegetaux Utiles de l'Afrique Tropicale Française: Premiere Etude sur les Bois de la Côte d'Ivoire', *Etudes Scientifique et Agronomiques*, Fascicule V.

Chevalier, A. (1911) 'Essai d'une carte botanique forestière et pastorale de l'A.O.F', *C. R. Acad. Sc.* CLII (6) juin.

Chevalier, A. (1912) *Rapport sur une mission scientifique dans l'ouest Africain (1908–1910)*. Paris, 12 janvier. Missions Scientifiques

Chipp, T. F. (1922) *The Forest Officer's Handbook of the Gold Coast, Ashanti and Northern Territories.* London: Crown Agents.

Chipp, T. F. (1927) 'The Gold Coast Forest: A Study in Synecology', *Oxford Forestry Memoires* 7.

Clayton, W. D. (1961) 'Derived Savanna in Kabba Province, Nigeria', *Journal of Ecology* 46: 595–604.

Clements, F. E. (1916) 'Plant Succession: An Analysis of the Development of Vegetation', *Carnegie Institute Washington Publication* 242: 1–512.

Cole, N. H. A. (1980) 'The Gola Forest in Sierra Leone: A Remnant of Tropical Primary Forest in need of Conservation', *Environmental Conservation* 7(1): 33–40.

CTFT (1966) 'Ressources forestières et marché du bois en Côte d'Ivoire'. Abidjan: CTFT/SODEFOR.

Davies, G. and P. Richards (1991) 'Rain Forest in Mende Life: Resources and Subsistence Strategies around the Gola Forest Reserves, Sierra Leone'. London: Report to ESCOR/ODA.

Dorward, D. C. and A. I. Payne (1975) 'Deforestation, the Decline of the Horse and the Spread of the Tsetse Fly and Trypanosomiasis (Nagana) in Nineteenth Century Sierra Leone', *Journal of African History* 16(2): 241–56.

Ekanza, S-P. (1981) 'Le Moronou a l'époque de l'administrateur Marchand: Aspects physiques et économiques', *Annales Univ. Abidjan, Serie I, Histoire* 9: 55–70.

Escobar, A. (1995) *Encountering Development: The Making and Unmaking of the Third World.* Princeton, NJ: Princeton University Press.

Fairhead, J. and M. Leach (1994) 'Contested Forests: Modern Conservation and Historical Land Use in Guinea's Ziama Reserve', *African Affairs* 93: 481–512.

Fairhead, J. and M. Leach (1996) *Misreading the African Landscape: Society and Ecology in a Forest-Savanna Mosaic.* Cambridge: Cambridge University Press

Fairhead, J. and M. Leach (1998) *Reframing Deforestation: Global Analyses and Local Realities — Studies in West Africa.* London: Routledge.

Fairhead, J., T. Geysbeek, S. Holsoe and M. Leach (forthcoming) *Freedom to Roam: African-American Journeys Inland from Liberia in the Nineteenth Century.*

FAO (1981) 'Forest Resources of Tropical Africa, Part I & II' (Country Briefs). Tropical Forest Resources Assessment Project. Rome: FAO

Ferguson, J. (1990) *The Anti-Politics Machine.* Cambridge: Cambridge University Press.

Ford, M. (1992) 'Kola Production and Settlement Mobility among the Dan of Nimba, Liberia', *African Economic History* 20: 51–62.

Foucault, M. (1980) *Power/Knowledge: Selected Interviews and Other Writings 1972–1977,* edited by C. Gordon. Brighton: Harvester Press.

Funtowicz, S. O. and J. R. Ravetz (1992) 'Three Types of Risk Assessment and the Emergence of Post-normal Science', in S. Krimsky and D. Golding (eds) *Social Theories of Risk*, pp. 251–73. Westport, CT: Praeger.

Gornitz, V. and NASA (1985) 'A Survey of Anthropogenic Vegetation Changes in West Africa during the Last Century — Climatic Implications', *Climatic Change* 7: 285–325.

Goucher, C. L. (1981) 'Iron is Iron 'til it is Rust: Trade and Ecology in the Decline of West African Iron-smelting', *Journal of African History* 22: 179–89.

Government of Sierra Leone (1923) *Annual Report of the Lands and Forests Department, 1923.* Freetown: Government Printers.

Grainger, A. (1996) 'An Evaluation of FAO Tropical Forest Resource Assessment 1990', *The Geographical Journal* 162(1): 73–9.

Green, W. (1991) 'Lutte contre les feux de brousse', Report for Projet DERIK, Kissidougou.

Grillo, R. D. (1997) 'Discourses of Development: The View from Anthropology', in R. D. Grillo and R. L. Stirrat (eds) *Discourses of Development: Anthropological Perspectives*, 1–34. Oxford: Berg.

Hajer, M. A. (1995) *The Politics of Environmental Discourse: Ecological Modernization and the Policy Process.* Oxford: Clarendon Press.

Hawthorne, W. D. (1996) 'Holes and the Sums of Parts in Ghanaian Forest: Regeneration, Scale and Sustainable Use', *Proceedings of the Royal Society of Edinburgh* 104B: 75–176. Edinburgh: Royal Society of Edinburgh.

Hopkins, B. (1965) *Forest and Savanna*. London: Heinemann.

Irwin, A. (1995) *Citizen Science: A Study of People, Expertise and Sustainable Development*. London: Routledge.

Jones, A. (1983) *From Slaves to Palm Kernels: A History of the Galinhas Country (West Africa) 1730–1890*. Wiesbaden: Franz Steiner Verlag.

Kandeh, B. and P. Richards (1996) 'Rural People as Conservationists: Querying Neo-Malthusian Assumptions about Biodiversity in Sierra Leone', *Africa* 66(1): 90–103.

Keay, R. W. J. (1959a) *Vegetation Map of Africa south of the Tropic of Cancer*. London: Oxford University Press.

Keay, R. W. J. (1959b) 'Derived Savanna — Derived from What?', *Bulletin IFAN*, Series A2: 427–38.

Lane-Poole, C. E. (1911) *Report on the Forests of Sierra Leone*. Freetown: Government Printing Office.

Latour, B. (1993) *We have never been Modern*. Hemel Hempstead: Harvester.

Leach, M. (1990) 'Images of Propriety: The Reciprocal Constitution of Gender and Resource Use in the Life of a Sierra Leonean Forest Village'. Unpublished PhD thesis, University of London.

Lipsky, M. (1980) *Street Level Bureaucracy: Dilemmas of the Individual in Public Services*. New York: Russell Sage Foundation.

Long, N. and A. Long (eds) (1992) *Battlefields of Knowledge: The Interlocking of Theory and Practice in Social Research and Development*. London: Routledge.

Mangin, M. (1924) 'Une mission forestière en Afrique-Occidentale Française', *La Géographie* 42: 449–628.

Marcus, G. (1995) 'Ethnography in/of the World System: The Emergence of Multi-sited Ethnography', *Annual Review of Anthropology* 24: 95–117.

Migeod, F. W. H. (1926) *A View of Sierra Leone*. London: Kegan Paul, Trench and Trubner.

Monnier, Y. (1981) *La poussière et la cendre: Paysages, dynamique des formations vegetatales et stratégies des sociétés en Afrique de l'Ouest*. Paris: Agence de cooperation culturelle et technique.

Murphy, W. (1991) 'Creating the Appearance of Consensus in Mende Political Discourse', *American Anthropologist* 92(1): 22–41.

Myers, N. (1980) *Conversion of Tropical Moist Forests*. Washington, DC: National Academy of Sciences.

Nyerges, A. E. (1988) 'Swidden Agriculture and the Savannization of Forests in Sierra Leone'. Unpublished PhD thesis, University of Pennsylvania.

Sachs, W. (ed.) (1992) *The Development Dictionary: A Guide to Knowledge as Power*. London: Zed Books.

Sayer, J., C. S. Harcourt and N. M. Collins (1992) *Conservation Atlas of Tropical Forests: Africa*. Cambridge: World Conservation Monitoring Centre and IUCN

Shantz, H. L. and C. F. Marbut (1923) *The Vegetation and Soils of Africa*. American Geographical Society Research Series 13. New York: American Geographical Society and National Research Council.

Sivaramakrishnan, K. and A. Agrawal (1998) 'Regional Modernities in Stories and Practices of Development'. Paper presented at Crossing Borders Initiative conference, Yale University (February).

Spichiger, R. and C. Blanc-Pamard (1973) 'Recherches sur le contact forêt-savane en Côte d'Ivoire: Etude du recru forestier sur des parcelles cultivées en lisière d'un ilôt forestier dans le sud du pays baoulé', *Candollea* 28: 21–37.

Thompson, H. (1910) 'Gold Coast: Report on Forests'. Colonial Reports, Miscellaneous No. 66. London: HMSO.

Thulet, J.-Ch. (1981) 'La disparition de la forêt ivorienne: pertes et profits pour une société', *L'Information Geographique* 45: 153–60.

Vigne, C. (1937) 'Letter to the Editor', *Empire Forestry Journal* 16: 93–4.

Unwin, A. H. (1909) *Report on the Forests and Forestry Problems in Sierra Leone*. London: Waterlow and Sons.

Wynne, B. (1996) 'May the Sheep Safely Graze? A Reflexive View of the Expert-Lay Knowledge Divide', in S. Lash et al. (eds) *Risk, Environment, Modernity*, pp. 44–84. London: Sage.

Zerouki, B. (1993) 'Etude rélative au feu auprès des populations des bassins versants types du Haut Niger', Report for Programme d'Aménagement des Bassins Versants Types du Haut Niger, Conakry.

4 State Sciences and Development Histories: Encoding Local Forestry Knowledge in Bengal

K. Sivaramakrishnan

INTRODUCTION

The scrubby sal *shorea robusta* forests of southwest Bengal have been resuscitated in the last decade. Much of the credit for the dramatic transformation of this rugged landscape has justifiably been given to Joint Forest Management (JFM) by foresters, development experts and villagers involved. However, the gains of JFM have remained precarious. Conflict over its institutionalization has accompanied the spread of JFM to other parts of Bengal and the rest of India. One realm of contention has been the struggle around silvicultural[1] aspects of microplanning.[2] Several development experts, with many years of experience in promoting and evaluating JFM, have noted the scant role of villagers in forest management decisions (Arora and Khare, 1994: 10). Such failure to devolve decision making under JFM particularly afflicts matters of silviculture. This chapter will explore the production of this contested domain of silviculture in JFM, and argue that it has been wrought in several histories — of ecology, politics and scientific discourse.

Scientific forestry was constructed in colonial Bengal by valorizing certain kinds of knowledge and, thereby, privileging attendant modes of forest management. Colonial rule in India was shaped by the flow of information and the categories through which this was absorbed and transformed into what might be lumped together as colonial knowledge (Bayly, 1993). This chapter examines the formation of colonial knowledge (information,

---

1. Silviculture as used in this chapter refers to the art of producing and tending a forest. It comprises, at the very least, the theory and practice of controlling forest establishment, composition, structure and growth. See Smith (1986: 1–28).
2. Microplanning refers to the participatory plans that are prepared by foresters and forest protection committees (set up under JFM) for the jointly managed forests. Defining the realm of silviculture in these plans and conceding that realm to the expert prescriptions of foresters is proving difficult. Such effort recreates the patterns of conflict between local agency and central direction that were etched into the landscape of the managed forest in colonial times.

ideologies, perceptions, and the lessons of practical experience) in the realm of forest regeneration by concentrating on the points of knowledge production, codification and transmission. By considering the conflicted and experimental ways in which sal silviculture was sought to be standardized in colonial working plans, the chapter advances two arguments. First that the scientific and technical discourse surrounding JFM is shaped by a historical legacy. Second that such environmental development discourses are continuously under production. The changing character of scientific discourse is shaped by interlinked changes in structures of resource control and discourses of community.[3]

Through representations of the forest, definition of expertise and manipulation of local structures of authority, colonial foresters created a body of knowledge about natural regeneration for sal forests in eastern India. They worked with, and sometimes against, the vanguard parties charged with establishing the colonial state on its frontiers. In Bengal these frontiers were often the forested fringes of settled cultivation. Managerial knowledge was defined by the colonial state and the landscape was partitioned into discrete jurisdictions, which were then subject to different modes of management.[4]

The discussion which follows thus locates the history of scientific discourse in the related histories of politics and landscapes. A lot of the politics involved was internal to the forestry and larger colonial bureaucracy. The unravelling of this politics is educative because it tells us about enduring mechanisms of state building. For the same reason, I am less concerned with drawing distinctions between local and foreign knowledge, and more interested in the production of knowledge in specific political-ecological-historical settings. In this chapter the term local knowledge is used not to denote the distinct cultural categories of an indigenous people: rather, it is used in a manner which implies that local knowledge is situated practice and, to that extent, is neither a system nor an alternative rationality.[5] I refer more to the intimate acquaintance with a locality — its landscape, social relations of production and environmental management — that people develop as they work to change its appearance. Those who traverse it intensively to make something of the place for their purposes produce such special knowledge of a place, and its spatial history.[6]

---

3. See Li (1996) and Agrawal (1997) for the analogous argument with regard to discourses of community and conservation.
4. A very useful discussion of this historical process as functional territorialization of state resource control may be found in Vandergeest and Peluso (1995).
5. This corrective to 'indigenous knowledge systems' approaches is provided by Agrawal (1995), and in the collection of essays in Hobart (1993).
6. Local knowledge is most usefully recognized by its 'inseparability from a particular place in the sense of embeddedness in a particular labor process' (Kloppenburg, 1991: 522). This formulation is also central to much feminist analysis of science which emphasizes the importance of producing knowledge through sensuous activity, experience that is specifically local. See, for instance, Haraway (1988), Harding (1986) and Smith (1987).

My concern, then, is with such localized production of information, and the processes of translating it into standard terms, that made possible a project like scientific forestry.[7] Historians and sociologists of science have frequently explored the ideological and intellectual processes in which particular branches of science developed.[8] I would suggest that such an approach pays insufficient attention to sites of application, that is, to places where these sciences become technological practices. The terrain of implementation leaves a strong impression on the production and transformation of scientific knowledge.[9] When these sites enter the processes of knowledge production in any specific domain like forestry, they bring with them much else that is going on there. One of the important things that creeps into generating scientific knowledge is the issue of government.

This happens in at least two ways. First, forestry as land management gets entangled in wider issues of land administration — agriculture, revenue, stable local arrangements of production. Second, the pressure on forest departments to develop, standardize and disseminate universal and replicable scientific management models which mesh with larger bureaucratic forms of government influences their selection and codification of procedures. There is then a tension between fitting forestry into a wider universe of managed landscapes of production and identifying it as a distinct, separate, professionalized activity. The effect of this tension suggests a constant production and transformation of science in its applications, the context often being development. We need to track these changes.[10]

In *Representing and Intervening*, Ian Hacking (1983) provided a landmark study that shifted the focus of science studies towards practice, by stressing the 'doing' as much as the 'representing' aspects of science. He later emphasized the multiplicity, patchiness and heterogeneity of the space in which scientists work (Hacking, 1992: 29–64). More research in the history of science has moved in this direction of looking to the social to understand the way rules and practices shape each other (see Pickering, 1992: 1–28). The

---

7. In a recent endorsement and amplification of my proposition, Melissa Leach writes, 'archaeologies of particular knowledges reveal histories of appropriation and interpenetration' that can be traced by the use of heuristic categories like 'local knowledge', 'scientific forestry' and so on. But we cannot ignore the politics of displacement through which local and non-local knowledges emerge and then are perceived as mutually inimical (see Leach, 1998: 5).
8. Scholarly discussions of science as representation have flowered into the sociology of scientific knowledge. Notable examples are Barnes (1977), Bloor (1976), Collins (1992) and Gooding (1990).
9. The relationship between science and practice, and the practitioner debates in which institutionalized 'basic science' is shaped, are well discussed in the context of late nineteenth century American medicine by Warner (1991).
10. For recent work that stresses that scientific ideas were not imported into colonies, but were more often in a process of continuous construction, reconstruction and transformation there, see several essays in Reingold and Rothenberg (1987).

discussion of interests I have suggested allows us to understand how the stabilization of science takes certain routes and not others as the open-ended processes of experimentation unfold. However, when scientists continually explore their way out of a problem, with experience as their guide, their interests intersect with ecological processes.

The institutionalization and professionalization of forest management through its engagement with issues of governance, resource conservation and enhanced productivity also brought scientific forestry into the realm of emerging development discourse in the late nineteenth century. It may be true that 'we do not yet know enough about the global, regional and especially local historical geographies of development — as an idea, discipline, strategy or site of resistance — to say much with any certainty about its complex past' (Crush, 1995: 8). We then need histories of development, to trace the recurrence of ideas, imagery and tropes of development across a range of nineteenth and twentieth century contexts.

FORESTRY AS SCIENCE AND PRACTICE: A DEVELOPMENT REGIME

In 1884 Sir Dietrich Brandis, the first Inspector General of Indian forests, drew a distinction between a modular science that could be transposed from the European laboratory, so to speak, into the Indian field site; and a location-specific forest protection programme where local knowledge was the best asset of the forest guard (Brandis, 1884). But this neat division could not always be sustained in practice: cultural operations were often crucially dependent on local ecological knowledge and understanding the social mechanisms by which scarce labour could be secured for silvicultural tasks. Through the routines of forestry and through their sporting pursuits, forest officers assumed the mantle of authority and scientific expertise that became central to their functioning by the end of the nineteenth century.[11]

At the moment of translating silvicultural prescription into action in any coupe,[12] compartment or block, however, the mantle sometimes slipped, tugged away by a shift in the locus of necessary knowledge to reveal the precarious relationship between assumptions and practice. In the words of a novel written in 1909 and set in the forests of Bengal: 'when it is time to work the forest, *burra sahib* will need Dulall to mark the trees, as no one can tell which can be cut' (Gouldsbury, 1909: 243). These were the travails of moving from lumbering to forest regeneration, which meant the Bengal

---

11. The memoirs of many foresters reveal this; see, for instance, Best (1935), Forsyth (1889) and Stebbing (1920).
12. A coupe is a forest block earmarked for felling or regeneration operations in a particular season.

forest department had to change from building a regime of restrictions around the forests to detailing a set of interventions in them.

Forestry, in this respect, was part of a wider process noticeable in the late nineteenth century when increasing intervention in agricultural production and its justification by appeal to a rhetoric of conservation went hand in hand.[13] Following the Famine Commission reports and the creation of an Agriculture Department in 1885, a new managerial assertiveness was discernible in the whole rural landscape of production.[14] The possibility of transforming the floral and crop composition of this landscape was perceived through the powerful lens of institutionalized science. As the Conservator of Bengal put it, describing the Kurseong forests a few years before the first Working Plan in North Bengal was prepared, 'we should here step in and assist nature . . . and by these means add greatly to the value of the estate'.[15]

Working plans came to symbolize this confidence of the scientific forester; but visualizing a terrain where science could plan unimpeded the manipulation of the forest, compelled a more complete enumeration and disposal of local rights that might obscure the vision. Forest settlements therefore became a necessary prerequisite, and where they could not be concluded, working plans remained an ideal little realized.[16] The lasting irony of the situation was that these plans aspired to define a universal code but their implementation was always caught up in securing and controlling local ecological and political knowledge.[17] The more working plans were reworked in an attempt to render them into modular prescriptions, their successive revisions became disconcertingly aware of specific regional ecological histories, which slowed the drive to modularity. These paradoxes of working plans also reflect the ways in which forest management was caught up in a wider tension in colonial governance between central direction and local autonomy.

---

13. Here the distinction made between conservation and preservation is useful to bear in mind. The former combined utilitarian and developmentalist ideas in environmental management, underpinning soil conservation, water management, sustained yield forestry and so on; while the latter inspired more directly the creation of wilderness areas, parks, and sanctuaries. See Hays (1959) for a discussion of these ideas in the context of American environmentalism. Anderson (1984), Beinart (1984, 1989), Grove (1989) and Peters (1994: 78–80) are among a growing body of scholarship on Africa that suggests the same trends were present there in the late nineteenth and early twentieth centuries.
14. OIOC P/2934 BRP (Agri), Jan 1887, Misc/1/29–30, A progs, no. 510T dated Calcutta 19 October 1886, from M. Finucane, Director Agriculture to Secy GOB Rev.
15. OIOC P/3871 BRP (For), May–July 1891, A progs 98–121, July 1891, Head 4, Coll 2, no. 2669F-G dated Darjeeling 30 Dec 1890, EP Dansey, Offg. CF Bengal to Secy GOB Rev, p. 13.
16. OIOC P/2800 BRP (For), Jan–Feb 1886, A progs 32, Jan 1886, Head I, Coll I, no. 21F dated Simla 31 Aug 1885, resolution by GOI, Home Dept., p. 19.
17. Both Adas (1990: 95–108) and Ludden (1989: 101–130) have pointed out the translation of local knowledge that was basic to the creation of colonial science and its codification as discourses of rule through classification, standardization and textualization.

Forest administration emerged at a time (by 1880) when the colonial state had become a huge investor in India, and was staffed by a large and disciplined bureaucracy that conceived of itself as the custodian of public welfare.[18] This increased governance created a massive documentation project that has now been widely discussed in the literature.[19] As a result, to use David Ludden's words, 'India's development regime evolved on coherent, consistent lines after 1870, the trend being towards more ramified and centralized state power' (Ludden, 1992a: 264). To a limited extent this evolutionary view of the state centralization process is salient. In many ways the post-colonial pattern of democratic socialism and centralized planning carried these tendencies further. To this we must add the redefinition of expertise in terms of high technology, complicated science and environmental uncertainty that privileged metropolitan agency in the identification of environmental problems, outlining solutions and mooting federal legislation even while bringing into sharp relief local conflicts over resources.

According to Mukarji (1989: 468), 'the system inherited from colonial times was in many ways more decentralized than centralized. With the arrival of democratically elected governments ... the balance was tilted heavily towards centralization'. The case of forestry seems to largely exemplify this wider finding: JFM can, therefore, be seen as an effort to regain the greater flexibility that existed in local government during colonial times. However, the argument here is that JFM is fruitfully examined, not as a break from tradition or the recovery of a temporarily abandoned approach, but as the recreation, through a process of engagement and modification, of a complex set of themes in governance that came to the fore in the late nineteenth century. We must re-evaluate epochal transitions in Indian historiography by looking closely at important, if attenuated, continuities in state building and the culture of politics.

Recently the question of whether there was a sharp disjunction between colonial and pre-colonial forest policy, and if there was, when exactly it occurred, has been well discussed (see, for instance, Grove, 1993, 1995; R. Guha, 1983, 1989; S. Guha, 1999; Rangarajan, 1994; Skaria, 1998). There were apparent continuities between pre-colonial and colonial states, notably in forest destruction and monopolization of valuable forest resources. The differences can be seen in ideologies, and concomitant technologies of rule. Arguably, British forest policy as it emerged after 1858, when India was formally integrated into the empire, bore the stamp of forest regulation and management that had been carried on in different provinces as they came

---

18. Ludden (1992a) makes the argument for the process of colonial state formation in general; Brandis (1884) recognizes similar trends in forest administration in particular.
19. Ludden (1992b) has described the process in respect of agricultural surveys and settlements; Dirks (1992) for colonial anthropology; and Sivaramakrishnan (1995) has done the same in the context of forestry. The wider theoretical implications of the process have been discussed in Cohn and Dirks (1988) and Prakash (1990).

under East India Company rule.[20] Systematic forestry could only commence after the stabilization of British rule in the aftermath of 1857.[21] In Bengal, many of the areas that would become reserved forests came under British control only by the 1860s.[22] In other conquered areas the flow of detailed information about the landscape and people had also only quickened in the later part of the nineteenth century.[23] It is in this sense that the late nineteenth century is particularly significant for major aspects of the histories we would seek for scientific forestry as development discourse.

Crafted in this expanding world of empire and information, the development regime for sal forests in Bengal began to take a shape that is discernible in contemporary JFM during the period from 1893 to 1937, when political consolidation converged with technocratic assertion. This chapter, confined to the latter process, dwells on the way expertise in forest management was defined by different elements in the colonial state, in which it was seen to repose, and how it was deployed. The natural regeneration of sal, the most valued tree in Bengal forests, became an elusive goal for scientific forestry by the end of the colonial period, despite tremendous research efforts. During those frustrating years the production of detailed reports, compendia, manuals, calendars, agricultural censuses and of course working plans, created textual bases for fixing and transferring expertise.[24] Such textualization in agriculture was a managerial act that established an instrumental attitude toward farming. From a situation where farm practice was expertise, peasant wisdom was moved to the category of folklore even as it was appropriated and transmogrified into the scientific format of

---

20. The focus during Company Raj remained on individual tree species, leading to protracted efforts to conserve, manage and monopolize teak in south India, western India and Burma: see Anon (1871), Birdwood (1910), Brandis (1860), Bryant (1994), Cleghorn (1861), Falconer (1852), Grove (1995), Rangarajan (1994), Skaria (1998) and Stebbing (1922). In Bengal the first half of the nineteenth century was similarly a period of sporadically knowing and taming the wild. Surveying natural wealth, cataloguing it, and making forested terrain more habitable by 'vermin eradication' were important interventions into a landscape earlier seen as impenetrable and unknowable. This is discussed in chapters three and four of Sivaramakrishnan (1999).
21. Stebbing (1924: 4). He specifically credits the Viceroyalty of John Lawrence (1863–9) as being the period when forest conservancy was systematically introduced in India.
22. The sal forests of the Duars came under British control in 1865 with the conclusion of the Bhutan war and the cession of this area east of the Tista river (Stebbing, 1924: 5).
23. See Anon (1853), Balfour (1862), Ball (1880), Forsyth (1889), Hooker (1854), Hunter (1868), Jackson (1854a, 1854b), Lewin (1869), Ricketts (1854, 1855), Waddell (1899).
24. Commenting upon the discursive strategies whereby expertise is constituted as the exclusive preserve of development agencies, Mitchell (1991: 19) says, 'the discourse of international development constitutes itself ... as an expertise and intelligence that stands completely apart from the country and people it describes'. This theme of objectification and depoliticization recurs in all critiques of development discourse.

statistical tables or maps and reports (Ludden, 1992a: 270).[25] Similar things happened in forestry.

These mechanisms that categorized expertise and directed the forms in which it would be deployed gave control over technologies of improvement to different agencies of the state, and made official research the sole arbiter of what constituted progressive innovation. Thus even in a natural regeneration based scheme, the forest department assumed it would define the scientific modes of such regeneration and hence took upon itself the job of training people in the apparently simple task of protection, enhancement and release of woody growth. This powerful legacy in sal silviculture has carried through into the working of JFM in Bengal today, even though there has been no recent research on natural regeneration and colonial experience in the matter was never satisfactory.

Offering a general explanation of such processes and an agenda for their analysis, Arturo Escobar says, 'the demarcation of fields and their assignment to experts . . . is a significant feature of the rise and consolidation of the modern state. What should be emphasized, however, is how institutions utilize a set of practices in the construction of their problems through which they control policy themes, enforce exclusions and affect social relations' (Escobar, 1988: 435).[26] But the 'modern state' was not as much of a steamroller as projected here. If we look closely within the broad contours provided by Escobar's framework, we find the constitution of expertise was always conditioned by the exigencies of particular contexts. Realized at an uneven pace and in diverse forms, the processes identified by Escobar were constituted through significant regional variations in Bengal.

The emergence of scientific forestry and rational management in late nineteenth century Bengal is best analysed as a product not only of intellectual revolutions and transformations in the organization of knowledge in Europe, but also the practical circumstances of controlling land and labour, and manipulating tree species that became valuable at different points in the history of forest management in India.[27] However, attempts to assign knowledge or material conditions the determining role in policy are likely to prove unsatisfactory. Representations are intimately connected to the production of knowledge, and thereby its codification in government

---

25. Similarly scientific forestry in Europe had already declared the restoration of forests a task beyond mere preventive laws and something that called for scientific expertise; see Harrison (1992: 117) and Lowood (1990).
26. Harvey (1989: 244–6) calls this perspectivism, and points out that it leads to rules of rational practice and the idea that the expert as a creative individual was always capable of a view from the outside — a totalizing vision.
27. Peluso (1992) has done an admirable job of analysing forest policy in terms of these three aspects of control in Java. Rangarajan (1998) has looked at forest management in nineteenth century central India from the same perspective. He attends carefully to changes in the silvicultural agenda — cast in terms of scientific advance — as the political-economic conditions changed.

into legal and policy instruments. At the same time law or policy as instruments of power are not only shaped by this knowledge, but also by practice and experience.[28] The British administrators engaged in framing forest policy clearly brought with them ideas about what was appropriate from European experience. They also encountered and described a pattern of forest use that they displaced only partially, after disparaging it. Grounded thus in regional histories of conquest and ecology, forest management still emerged as a development regime. We therefore have to examine how 'a corpus of knowledge, techniques and scientific discourses is formed and becomes entangled with the practice of power' (Foucault, 1979: 23).

Prefigured by an unprecedented territorialization of forest control after 1860, codified information, standard techniques and rigorous management became central to scientific forestry, through the recognition that these had not been attempted in the past by native states.[29] Reconsidering the nature of state formation in medieval and early modern India, historians are concluding that pre-colonial states in India had little control of production and distribution and what they did have came through social intermediation largely unregulated by 'the State'. Pre-colonial political culture produced multiple overlapping levels and arenas of authority rather than centralized states. Even the Mughal State was more patrimonial than bureaucratic, and its centralization was more ideological than operative (Ludden, 1992a: 266; see also Subrahmanyam, 1992). The British colonial process of state building in India regarded traditional society as something to be both reformed and preserved to retain elements of social hierarchy and stability that would facilitate governance, but at the same time intervened freely in the agrarian economy by treating it as an autonomous domain open to progressive action (Bayly, 1990; Washbrook, 1981). The development of forest management and silviculture in the mixed deciduous forests of Bengal fits within this broad scheme of domination that was put in place in the nineteenth century. There is growing evidence that no previous ruler had such a sustained policy of intrusive exploitation or regulation of forested tracts.[30]

---

28. Thompson (1975: 28–9) gives us an excellent discussion of this complex fusion where forest law takes shape both in agrarian practice and conflicting representations when he says, 'the forest in fact was so by virtue of legal and administrative designation rather than by any unitary organization'.
29. Brandis (1884: 460), a careful and often sympathetic commentator, says, 'in the Central Provinces forests have been preserved by malguzars or landholders on their estates, because the timber or bamboo in them was valuable and could be converted to money. In the drier parts of North-western India, where forests are scarce, there has been a tendency towards the preservation of forests, either as sacred groves or for hunting, or in some cases to provide fuel for iron smelting, to protect the water supply in springs and streams, to provide fuel for towns, or to secure a supply of cattle fodder in times of drought and scarcity. But there has been no organized and effective action to accomplish these objects'.
30. Rangarajan (1996) makes this case specifically for the Central Provinces, R. Guha (1989) for the western Himalaya, Tucker (1989) for eastern and south India, S. Guha (1999), Hardiman (1994) and Skaria (1999) for western India, and Gadgil and Guha (1992) for India as a whole.

An allied feature that converged with the processes of state formation was the urban bias in determining forest value and use. Forest science flourished where central states began to rationalize administration, like in eighteenth century Germany and later in France. This process narrowed the focus of management on wood. As one historian of German forestry notes, 'identifying wood mass as the crucial variable of forestry set the stage for quantitative forest management' (Lowood, 1990: 326). As the measures of wood mass and volume were perfected, the urge to grow the accurately measurable forest increased. But the measurable forest needed more accurate estimation also because its principal products, greatly reduced in diversity, participated in world markets. Plantations of valuable trees were created through an impressive network that supplied seeds and planting material across continents.

Initially, forest conservancy in Bengal, as in the rest of India, commenced with the urge to directly control, systematize and regulate the extraction of timber from what was perceived as the rapidly dwindling hardwood forest wealth of the sub-continent. The rhetoric of conservancy espoused the 'environmental' tones of watershed management, species conservation and wildlife protection; but it also expressed the strident political-economic realities of territorial expansion, the establishment of British rule in strategic regions, and laying down infrastructure for administering empire.[31] While both strains of conservancy ultimately facilitated the disempowering of local communities in the forests, and expedited capital accumulation through forest exploitation, they created, in their discordance, interstitial spaces for the modulation of forest policy.

Protection, reservation, extraction, and marketing of timber were gradually followed by a growing interest in regeneration, particularly by natural methods, of the principal timber species, which were teak and sal. The period broadly coinciding with the first three decades of the twentieth century witnessed burgeoning research on silvics and structural qualities of many hardwoods and a few softwoods like *pinus longifolia* (Champion, 1975; Rodger, 1925). Such research and the dissemination of intensive silvicultural systems through working plans were complementary aspects of scientific forestry that placed new demands on forest management. There was a conflicted expansion of knowledge and a contested growth of managerial arrangements through which scientific forestry was professionalized and institutionalized in the sal forests of Bengal.

The rest of this chapter will, therefore, consider certain programmatic aspects of scientific forestry that illuminate its constitution through demarcation,

---

31. 'Forest Conservancy in Bombay', minutes by Sir Richard Temple, H.E. the Governor of Bombay on the forests of different districts and states of the province; reproduced in *Indian Forester* 5(3): 335–67, 1880, is a good reflection of the shared thinking in different provinces. Temple had been Lt. Governor of Bengal before this assignment (Cleghorn et al., 1852; Clutterbuck, 1927).

inventory, protection, regeneration, working plans and silviculture. In particular, it will examine the vicissitudes of preparing and implementing working plans, and the concomitant reduction of forestry to silvicultural models, to illustrate the two main features through which scientific forestry developed. The first, which can be called *management by demarcation and exclusion* aimed at simplifying land use in state forests by regulating local access. The second, which can be called *management by inventory and controlled regeneration* complicated the earlier regulation of people with the added regulation and transformation of tree growth. In combination, these types of managerial aspirations sought to move the locus of expertise and direction up and out into the higher echelons of the forest service. At their most ambitious, working plans were to serve as powerful instruments that would permit the Inspector General of Forests to dictate what trees were grown where, how, and when they would be harvested. Through the cases of working plan preparation and the tentative, often baffled, silviculture they document, the following sections evaluate the making and unmaking of scientific forestry against the standard of this grand ideal.

WORKING PLANS AS INSTRUMENTS OF REMOTE CONTROL

The first, albeit simple, working plan in India was introduced in 1837 by Munro, the Superintendent of Travancore forests. This was largely an exercise in enumeration of trees by size class and thus an estimate of the harvestible timber in any year. Linear surveys introduced by Brandis in the 1870s to estimate the growing stock were the basis of subsequent early working plans. In 1874 Schlich drew up a preliminary working plan for Buxa reserve, where the problem of water in the dry season had necessitated reconsideration of timber conversion and removal operations.[32] Progress was slow, however, and by 1884–5, only 109 square miles of reserves were under regularly sanctioned working plans. In 1884, Schlich centralized, in the office of Inspector General of Forests, the control and preparation of working plans. This hastened things somewhat and by 1899, 20,000 square miles of government forests were covered by working plans, of which 1900 square miles were in Bengal (Stebbing, 1924: 592–8).

Local governments were to use the working plan as an instrument that would balance the 'reasonable requirements of neighboring populations' with the 'exigencies of sound forest conservancy'. Forest settlements that would ascertain and finally record all admissible rights of village communities and

32. Stebbing (1926: 199); BFAR 1875–6; Hatt (1905: 5). Schlich's plan had proposed an annual removal of 5785 trees over five feet in girth in the next eight years. This turned out to be a rankly optimistic estimate.

private persons in forest lands and their produce were introduced as a necessary corollary of working plans.[33] The plans themselves were declared the most important job of the forest officer as the forest department quickly recognized them as the key to professional continuity in forest management.[34] Struggles to perfect these plans reflect the transformation of forest management into a predominantly scientific question, but such recasting of the terms of argument did not necessarily alter the elements that were the subject of contention. A stable silvicultural system and the locus of its governance remained the most pressing issues.

There are many definitions of working plans, but their important feature was the effort to reduce control into a matter of holding local agencies to plan prescriptions. As one of the last colonial Inspectors General of Forests wrote, 'a working plan is a forecasted framework for the management of a forest over a considerable period, often 120 years, with a detailed plan of what to do in the next 10–15 years to achieve the ultimate result'.[35] These working plans were seen as restoring normalcy to the state forests, which would consist of normal age classes, normal increase and normal growing stock in compartments, blocks and coupes (Schlich, 1876: 104–7). The object of the plan was often limited to showing the quantum of timber and firewood that could be removed without detriment to a continuous output and what works of improvement were desirable.[36] These plans were not, however, introduced as soon as a forest was taken over or reserved.

For any forest area the initial decade after its reservation was one of limited forest management. Operations were confined to forest protection, selection and improvement felling, creeper cutting to release advance growth and conversion of sapling sal into timber. The general pattern in Bengal was that reservation mostly took place in the 1880s, initiation of planning in the 1890s, and approval of short-term (ten to twenty year) plans in the period 1900–1910. During this time, non-timber products like bamboo, grazing fees, lac, and mahua flowers in southwest Bengal were seen as a significant component of forest revenues.[37] Quite often the working plan for a

---

33. OIOC P/2800 BRP Jan–Feb 1886 A progs 32, Jan 1886, Head I (RR), Coll 1, no. 21F dated Simla 31 Aug 1885, GOI Home Res.
34. NAI GOI Rev and Agri (For), A progs 12–16, file 45 of 1901, July 1901, no. 128For. dated Cal 5 Jan 1901, GOB Rev Res, p. 1026.
35. NAI GOI Education, Health and Lands (For), File 13-3/42 — F&L 1942, S. H. Howard, IGF inspection note for the forests of Bengal, December 1941 and January 1942, p. 1.
36. OIOC P/7034 BRP (For), 1905, A progs 71–77, Dec 1905, File 9R/1, no. 271 dated Darjeeling 11 Jan 1905, A. L. McIntire CF Bengal to Secy GOB Rev, p. 122.
37. OIOC P/7033 BRP (For), 1915, A progs 14–16, June 1905, File 3W/9, no. 11WP dated Cal 23 Jan 1905, S. Eardley Wilmot IGF to CF Bengal; no. 1233T-R dated Darjeeling 15 June 1905, M. C. McAlpin US GOB Rev to CF Bengal, pp. 150–3; OIOC P/7034 BRP (For) Oct–Dec 1905, A progs 60–64, Dec 1905, File 3W/4, no. 297 dated Darjeeling 6 Feb 1905, CF Bengal to Secy GOB Rev; no. 527 dated Cal 7 Jan 1905, A. L. McIntire CF Bengal to IGF, pp. 107–9.

particular division was taken up when competing demands grew, like the spread of tea gardens in Darjeeling District which caused a sharp increase in demand for firewood.[38]

Within twenty years of preparing the first working plans in North Bengal, the systems of management prescribed under them came under a cloud, as they were not yielding the desired regeneration of sal. The Inspector General of Forests, after inspecting Jalpaiguri and Buxa forest divisions, observed that the 'forest conditions in Bengal were more diverse than any other province and with more intensive methods of management plans must be revised more frequently'.[39] Yet the revision to the Jalpaiguri plan simplified the landscape classification of its predecessor from eight to four — sal, mixed, evergreen and savannah (Trafford, 1905: 4). Seeking to control these revisions the Inspector General of Forests had deputed the Imperial Superintendent of Working Plans to collect 'necessary local knowledge' from Bengal. The state government refused to organize the tour suggesting that it would impair service discipline and arguing further that providing local knowledge was the task of the regional government, since such knowledge pertained to demand and supply of timber and extraction facilities available.

The real argument soon narrowed onto the silvicultural prescriptions. While the Inspector General of Forests was perturbed by Bengal adopting annual plans of operation that smacked of a lack of professionalism, the Bengal Conservator was adamant that these were appropriate as 'through silvicultural control centralized control of forest management can go too far'.[40] This official went on to forcefully present the case for granting local officers full freedom to experiment with various silvicultural techniques as working plans should not be cluttered with prescriptions not justified by the poor levels of local knowledge.[41] The outcome was a curtailed tour by the Imperial Superintendent of Working Plans. The controversy admirably illustrates the way planning and professionalization, valorized by scientific forestry, repeatedly became the issues around which more important disputes about regional autonomy and central control in forest management were re-enacted.

Another example was the first Singhbhum working plan, prepared in 1903. By this time there were several plans already under operation, or in

---

38. OIOC P/6561 BRP (For), A progs 19–25, July 1903, File 3W/3, no. 14 dated Cal 29 Jan 1903, R. C. Wroughton IGF to Secy GOB Rev; no. 584T-R dated Darjeeling 22 May 1903, A. Earle Offg Rev Secy GOB to CF Bengal, p. 17.
39. NAI GOI Rev and Agri (For), A Progs 28–32, File 162 of 1915, July 1915, Inspection note dated 28 Mar 1915 of Buxa and Jalpaiguri by G. S. Hart IGF, p. 4.
40. WBSRR GOB Rev (For), File 6D/4, B Progs 20–24, Nov 1907, no. 1857/320–7 dated Simla 6 Sep 1907, S. Eardley Wilmot IGF to Chief Secy GOB; no. 2219JR dated Cal 2 Oct 1907, Chief Secy GOB to IGF; no. 11/C dated Simla 11 Oct 1907, IGF to Chief Secy GOB, pp. 1–5.
41. WBSRR GOB Rev (For), File 6D/4, B progs 20–24, Nov 1907, note dated 24 Sep 1907 by A. L. McIntire, CF Bengal in keep with papers, pp. 9–10.

advanced stages of preparation, for different North Bengal areas.[42] When these plans came up for revision, often prior to the end of their initial duration, the changes made in the plan usually incorporated the practical deviations that had been carried out, and sometimes noticed in the annual reports. In Singhbhum, for instance, the plan had been to continue selection felling of sal timber from a 32 square mile area of good valley type forests, half of which had been taken out in 1895–8 to supply sleepers to the East India Railway. For a further 693 square miles of reserves in Singhbhum the plan was fire protection for sal regeneration, something that had already covered 64 per cent of the reserves, increasing from a mere 12 per cent in 1895.[43] At the time the projected revenue from timber operations (major produce) was expected to exceed that from sabai grass (minor produce) sale.[44] But the Inspector General of Forests ruled in 1915 that the poorer quality hill forests being of little value in the production of timber should be given up to sabai grass production. This would entail burning the forest floor and creating large blocks for the production of sabai grass.[45]

At the same time, with the introduction of concentrated regeneration blocks, the working plan was also to be revised to include a regular engineering scheme covering extraction, conversion and transport of timber. The elaboration of such a scheme hinged on the prior completion of procedures allotting specific areas to blocks and determining a sequence of coupes in the first block. Mechanized timber extraction hinged on a prior re-organization of silvicultural practice and thus a re-ordering of the landscape into more systematic and compact blocks of even aged tree crops.[46] As management became more intensive, Singhbhum had been divided into Saranda, Kolhan, Porahat and Chaibassa divisions, with separate plans for each division by 1925. The Uniform System, under which partial clear felling had been prescribed, was introduced under all these plans, but the

---

42. OIOC P/7034 BRP (For), A progs 71–77, Dec 1905, File 9R/1, no. 271 dated Darjeeling 11 Jan 1905, A. L. McIntire CF Bengal to Secy GOB Rev, pp. 122–3; Grieve (1912), Haines (1905), Hatt (1905), Trafford (1905), Tinne (1907).
43. OIOC P/7034 BRP (For), Oct–Dec 1905, A progs 60–64, Dec 1905, File 3W/4, no. 297 dated Darjeeling 6 Feb 1905, CF Bengal to Secy GOB Rev; no. 527 dated Cal 7 Jan 1905, A. L McIntire CF Bengal to IGF, pp. 108–9.
44. The distinction made between major and minor produce indicated the priorities of forest management, especially in the matter of transforming the character of the forest under working plans. The idea was to increase the yield of major produce. These were further classified into valuable and less valuable or jungle trees. I have discussed this schematization emerging as an all-India feature of forest management elsewhere (Sivaramakrishnan, 1995). Prasad (1994: 78–90) has discussed these classifications and their implications for forest management in Central Provinces.
45. NAI GOI Rev and Agri (For), A progs 35, File 136 of 1916, May 1916, no 376/24-WP dated Simla 17 May 1916, G. S. Hart IGF to Secy Rev, GOBO, p. 1.
46. Ibid.: p. 2.

monopoly contractor for timber, the Bengal Timber Trading Company, was generally in arrears with fellings, thereby largely keeping prescriptions at bay (Stebbing, 1926: 524).

During the same period, the Inspector General of Forests and silviculturist visiting Jalpaiguri and Buxa had also recommended taking up artificial regeneration since the evergreen undergrowth in fire protected natural regeneration was suppressing sal.[47] Thus, 300 acres of experimental sowings of sal and 50 acres of other species annually were taken up as approved deviations from the working plans in 1917. The revision of working plans paid great attention to problems of regeneration. Removal and exploitation of both major produce and minor products remained under a regime of permits and concessions to contractors. Shaw Wallace and Company, for instance, had the monopoly for collecting nettle fibre from the reserved forests of Darjeeling, while Burn and Company had the timber and bamboo concessions in Darjeeling and Jalpaiguri forests.[48]

The significant patterns that emerge in working plan revisions are revealed by the case of the Darjeeling plan, which was altered several times in the space of thirty years. Darjeeling forests ranged in elevation from 12,000 feet in Singalila to 600 feet above sea level in the Teesta valley. The ground was steep with deep gorges and rapid mountain torrents. Hill and valley forests covered an area of 115 square miles (Grieve, 1912: 1). The upper hill forests (above 9,000 feet) were mainly silver fir and rhododendron. The middle hills (5,000–9,000 feet) chiefly oak, walnut, toon, laurel, maple, champ and alnus. Sal was found in the valley forests (600–3,000 feet) (ibid.: 5). The middle and upper hill forests had been worked departmentally for three years from 1865 and then under the permit system till they came under Manson's plan of 1892.[49]

This plan had prescribed a 160-year rotation in five 32-year periods, with the first and last block closed to grazing. The first block was to be regenerated by concentrated felling over 1/16th of the area annually under the shelterwood system. Soon after the introduction of the plan, a summary forest settlement extinguished all private rights in all the forests.[50] Osmaston revised this plan in 1902 and shelterwood was removed in all the ten original coupes, but no new regeneration felling was undertaken. Due to inadequate removal of overwood in the Manson plan, plantings had failed to establish (Government of Bengal, 1935: 29).

---

47. NAI GOI Rev and Agri (For), A progs 24–26, File no 311 of 1915, Feb 1916, no. 12319 GOB Rev Res dated Cal 17 Dec 1915, p. 2.
48. NAI GOI Rev and Agri (For), A progs 23–25, File 286 of 1916, Jan 1917, no. 10091-For dated Cal 18 Dec 1916, GOB Rev Res, p. 1.
49. Under the permit system of forest working, the sale of individual stems at a fixed price per tree led to the removal over time of the best trees leaving the defective trees in the forest.
50. No. 1449 dated 26 Mar 1896 from DC Darjeeling to Commr Rajshahi; no. 909G dated 12/14 Sep 1896 from DC Darjeeling to CF Bengal, cited in Grieve (1912).

In 1912 Grieve's plan was introduced to avoid the second felling that destroyed much of the regeneration following the first felling. He excluded all areas open to grazing, put all regenerated areas under a plantation working circle and divided the rest of the forest into High and Coppice working circles. We can see the emerging separation of the managed forest into discrete compartments subject to distinct treatments. More significantly, a clear separation came to mark forests open to grazing and those managed for the production of fuelwood and timber. The plan proposed selection felling in groups, relying on natural regeneration after mature stems had been removed. In areas open to grazing, regeneration was assumed to be impossible and so green felling was confined to the closed areas, from which all stems over two feet in diameter would be removed in fifty years. Overruling the local officers, the Inspector General of Forests changed this plan to one of regeneration felling in groups. This would remove all first and second class trees from closed areas in fifty years and rely on open forests and plantations to yield produce for the second and third fifty-year periods in the 150-year rotation.[51]

By 1920 the plan had been revised again. The failure of natural regeneration and crop improvement methods had led to the adoption of taungya cultivation of sal.[52] Clearfelling and taungya sowings remained the prescriptions in later working plan revisions in North Bengal sal forests of the Duars, though in some cases, as in Buxa, the division of the regeneration areas into sal conversion and softwood working circles brought various blocks under rotations of different duration.[53] Recognizing the local peculiarities of conditions for sal regeneration and a few other desired species undermined the centralizing expertise claims of the working plan, thus hindering the use of these plans as an instrument of control by the Inspector General of Forests. But there were other advantages, and the gains from working plans were sometimes of a different nature. As the

51. NAI GOI Rev and Agri (For), A progs 5–7, File 54 of 1917, Feb 1917, inspection note on the Darjeeling and Kurseong Hill Forests by G. S. Hart IGF, dated 28 Dec 1916, pp. 3–4. According to one official history, this alteration of the plan was based on a misunderstanding. The author (unknown, but probably the CF Bengal of the time, E. O. Shebbeare) notes that the idea of selection felling in groups was unworkable, but Grieve's proposal being confused with the group method of Europe, the IGF approved the modified plan. See Government of Bengal (1935: 31).
52. Taungya was a system of artificial tree regeneration practised by colonial foresters for raising sal and teak plantations. It involved clearing a forest area and planting tree seeds at intervals in rows, between which local cultivators would be allowed to grow rice, corn, tobacco etc. for two to three years. During this period, the cultivators would be required to tend and weed the seedlings. Annual cultivation would cease after the third year, by which time the trees would be established.
53. NAI GOI A progs GOI Education, Health and Lands (For) File 13-3/42 — F&L 1942, inspection note by IGF on the forests of Bengal, December 1941 and January 1942, p. 5; Government of Bengal (1935: 31).

Inspector General of Forests noted in 1942, working plans had become more exact and scientific with increased intensity of management, but chiefly in terms of superior surveillance and estimation of growing stock or merchantable timber. He wrote, 'those parts of India which have the best arrangements for working plans and the best control are just those parts which are providing the most supplies ... because they know what they have and where it is'.[54]

The overall pattern that emerged in working plans placed better quality forests under conversion to the Uniform System, with a rotation or conversion period of sixty to a hundred years. Three or four blocks were formed with roughly equal areas of teak and sal, the area allotted to the first block being regenerated under shelterwoods, while selection felling and thinning were continued in the third and fourth blocks. Forests deemed to be of poorer quality were placed under a coppice working circle to produce small timber and firewood on a thirty to forty year rotation.[55] By making conversion to the Uniform System the ultimate aspiration of any working plan, the entire forest management regime in Bengal was placed on a scale of approximation to scientific forestry, with very little being close to the top of the scale. This then served both as a powerful push to achieve the ideal and as a way of accommodating deviations or distortions.

SILVICULTURAL MODELS AND ELUSIVE PARTICULARITIES

From the beginning, natural regeneration was the preferred method of crop development. Brandis therefore favoured the training of recruits to the Indian forest service in Europe, where this method was successfully used for high timber production (Stebbing, 1924: 47). This was not by any means a recent innovation since natural regeneration through coppice for fuelwood and selection felling for timber had emerged by the fourteenth century as standard forestry practices (Fernow, 1907: 38). The main crop-improvement activities were creeper cutting and improvement felling. Cleanliness of crops (absence of creepers) became a measure of whether foresters and guards were working hard on their beats. The main idea underlying improvement felling was to 'favour the valuable species and eliminate the less valuable and those interfering with the growth of the former' (Stebbing, 1924: 576–8). Subsidiary work would encourage the younger age classes by removing weeds and low growth that choked them, and lighten the cover overhead by

54. NAI GOI Education, Health and Lands (For), File 13-3/42 — F&L 1942, inspection note by the IGF on the forests of Bengal, December 1941 and January 1942, p. 2.
55. See, for example, Phillips (1924), Sinha (1962). This does not apply to sal forests of North Bengal where taungya had been introduced or other forests of Bengal, like those in Chittagong and Sundarbans, where sal was not the valuable species.

girdling or removing inferior species of trees. Thinning aimed to achieve similar ends of improving light for valuable species regeneration and producing even-aged crops.[56] Although experienced foresters often confessed to their diaries that it was far from being an exact science,[57] a model forest for each species was envisioned through thinning alone.

The ideal was light and frequent thinning but lack of men and money made this unattainable in Bengal, where rapid forest growth made under-thinning the main problem (Homfray, 1936: 4–6). The chief purpose of thinning was to attain the largest possible timber trees per acre with straight, well-formed boles, an ideal as aesthetic as it was commercial. Noting the reluctance of marking officers to take out trees of large girth that were forked or crooked, Homfray wrote, 'it cannot be too carefully impressed on thinning officers that the chief point about the tree is its shape and not its size' (ibid.: 9).[58] Work progressed better where valued species appeared in unmixed stands, a condition that crop improvement aspired to create, imitate, or stimulate. By 1915, prescriptions for intensification in sal management included moving into the regeneration block system. Here crop improvement was introduced into every felling series on a continuing basis as improvement felling became a management practice in forests of all age classes instead of a forest harvesting technique allied to selection felling.[59]

By 1925, selection and improvement fellings were being replaced in Bengal by concentrated regeneration under taungya or various coppice systems, aspiring to the shelterwood compartments or uniform system. With the exception of North Bengal taungyas, where clear felling and artificial regeneration were essential to securing sal growth, in other parts the transition remained mostly an ideal. Movement toward the ideal was interrupted by wars that tended to increase timber demands in sharp peaks, recurrent labour problems in forest working, and financial stringency imposed on departmental working. The most significant thing about incomplete, unattained, and constantly revised working plans and silvicultural arrangements, however, was the repudiation by local governments of central direction and control, after a point. Typically the local officials would emphasize the need for short-term and revisable schemes that were trans-

---

56. OIOC P/10122 BRP (For), 1917, A progs 4–6, Mar 1917, File 3I/1, inspection note by IGF G. S. Hart on Darjeeling forests, dated 28 Dec 1916; no. 85F/54-1 dated Simla 14 Feb 1917, A. E. Gilliat US GOI Rev and Agri to Secy GOB Rev, pp. 16–17.
57. One forester was candid enough to assert that 'no two officers would mark exactly the same tree in a given area'. CSACL, Wimbush Papers, undated typescript entitled 'Life in the Indian Forest Service, 1907–1935', pp. 88–9.
58. Since the marking of trees was often left to village mandals, forest guards and even intelligent coolies, omission of crooked trees may have been deliberate as these were needed for making ploughs.
59. NAI GOI Rev and Agri (For), A progs 28–32, File 162 of 1915, July 1915, Inspection note of Buxa and Jalpaiguri, dated 28 March 1915, by G. S. Hart IGF, p. 17.

formed gradually into long-term plans. Such rejection or resistance that challenged notions of absolute expertise residing at the centre emerged as part of the development regime in the forests of Bengal.

Changes in silvicultural practices necessitated the modification of all access rules to production and protection forests. With the shift to regeneration by blocks, which was considered permanent, provisional arrangements like selection and improvement felling were displaced and this change required closing protected forests for longer periods (twenty to thirty years).[60] As closer attention was paid to tree crops that maximized valuable species in even-aged stands, concomitant changes in silvicultural practice altered labour requirements. Subordinate departmental staff who were skilled in the organization of tasks like thinning and cleaning became necessary and were sought through the pool of trained personnel recruited in the Kurseong Forest School.[61] Silvicultural tasks like thinning always posed a challenge where the benefits of scientific prescription were realized only through local and intimate knowledge of the managed forest. This displaced the balance of expertise down the forester hierarchy, making the guards and rangers key personnel in identifying trees to be marked or in determining the intensity of thinning in any season.[62] Periodic thinnings thus became not only a management operation: they served as the performative moment when expertise, like ritual knowledge, was passed on to younger generations of specialists in the making.

Until 1910, all the sal forests in Bengal were worked under the selection method with some improvement felling, a system necessitated because the supply of large sal trees was scant and dispersed through the forests.[63] In the decade following the First World War, radical changes were introduced in the silviculture of sal. By the late 1920s, Shebbeare, the Conservator of Forests in Bengal who fifteen years earlier had pioneered taungya experiments for sal in North Bengal, could report that taungya had become the single most important means to regenerate sal forests. The taungya method was labour-intensive and required land preparation by clear felling and firing of scrub, but its advantage was that the outcome was easily managed

60. NAI GOI Rev and Agri (For), A progs 43–46, File 124 of 1916, May 1916, no 174For. dated 3 May 1916, Offg Secy Revenue Punjab to GOI.
61. NAI GOI Rev and Agri (For), A progs 5–7, File 54 of 1917, Feb 1917, inspection note on the Darjeeling and Kurseong Hill Forests by G. S. Hart IGF, dated 28 December 1916, pp. 2–3.
62. The respect that senior foresters had for the local knowledge of the venerable guard of long standing is well conveyed in Gouldsbury (1909). Homfray (1936: 11) writes in his authoritative manual on thinnings that 'village mandals, guards and especially intelligent coolies . . . know a good deal about silviculture'. Stebbing (1926: 460) notes that in France too the Inspector would carry out thinnings after assembling all guards and rangers and the oldest forest guard would be the acknowledged expert.
63. McIntire (1909: 6). For selection felling the minimum prescribed diameter was two feet, that is, a girth of six feet.

for timber. Taungya working plans were simple. They prescribed clear-felling and restocking 1/r of the total area every year, where r was the length of rotation, usually eighty years.[64] One consequence of taungya and the switch to concentrated regeneration of any other sort was that trees like mahua, valued mainly for their non-timber forest product yields to villagers, were also removed under silvicultural prescriptions. Under earlier selection systems they were left alone in recognition of their utility to local people.

From 1926 to 1939 most sal forests were brought under the Uniform System, but within the following ten years this was being abandoned, as it was found that natural regeneration of sal could not be obtained through canopy manipulation.[65] As the sal study tour had noted, conversion to the Uniform System was predicated on the belief that regeneration *de novo* could be established in sal forests through the management of edaphic and light factors, but the highly mixed results had revealed that success was limited to regions with existing adequate advance growth, which was 'due to the past history of the forest, not the purposeful action of the forest officer' (Anon, 1934; see also De, 1941; Griffith and Gupta, 1948; Raynor, 1940; Smythies, 1940; Warren, 1940, 1941). Ironically, this was only possible under conditions not subject to the criteria of scientific management, namely, financial prudence and standardized silviculture. Field foresters in the late 1920s were finding that sal seedlings established in situations of varying light, possible only under an uneven-aged canopy. Felling to simulate these conditions would be both uneconomical and violate the ideal of concentrated regeneration blocks (Osmaston, 1928). Conversion to the Uniform System had failed on several counts: first preparatory felling did not induce regeneration; second, final removal of overwood damaged established crops; third, the debris of concentrated felling posed a pest and fire hazard; fourth, in damp areas weed infestation became rampant (ibid.: 643).

Thus the Uniform System came under critical scrutiny. On the eve of the Second World War, the national silvicultural conference was still calling for research on sal regeneration and the problems of pure teak plantations, pointing to the enduring problems in the silviculture of the two most valued species of Indian forests.[66] The same issues were revived as Indian foresters convened for sal and teak regeneration planning after India's first

64. OIOC P/11712 BRP(For), 1928, A progs 20–24, April 1928, File 9R/19 of 1927, no. 5191/R-53 dated 26 Sep 1927, E. O. Shebbeare CF Bengal to Secy GOB Rev, p. 39.
65. The work of R. S. Hole and E. A. Smythies initially, and the supplementary research of a host of other field foresters, had been the basis for moving into the shelterwood compartment system; notably, Bailey (1924), Ford-Robertson (1927), Hole (1919), Makins (1920), Osmaston (1928), Sen and Ghose (1925), Smythies (1926).
66. NAI GOI Education, Health and Lands (For), A progs, File 22-4/41 — F&L, 1941, no. 14803/40-IV-130 dated Nov 14, 1940, S. H. Howard President FRI to Secy GOI, pp. 6–7.

post-independence National Forest Policy was announced in 1952 (Forest Research Institute, 1953). In short, scientific forestry remained a development discourse under production.

DEVELOPMENT REGIMES AND HISTORICAL PROCESSES

Forest guards (often assigned to lead off tricky silvicultural operations in their beat), mahouts, and the toils of the field forester all generated information that was schematized to assemble scientific forestry in Bengal as a complex of knowledge and technologies of power.[67] In this way they contributed to processes redefining forest conservancy and providing it with several distinct elements. These were: a clear delimitation of its domain (demarcation); identification of indigenous vegetation and its valuation (inventory); enumerating, simplifying and circumscribing the bundle of usufructuary and other rights of local people (reservation/protection); devising a scheme of management that will produce the most desirable woods in the quickest time (regeneration/plantation); formalizing the arrangements through codes, manuals and division of responsibility among forestry officials, to ensure the establishment of routines (working plan). These elements were and still are crucial to the definition of scientific expertise and the privileging of its role in forest management, but they also contain traces of what has now widely come to be called development discourse.[68]

I am not speaking of development here as a received doctrine or cultural schema.[69] Dogged by uncertainty, conflict within its agencies, and attrition-causing resistance from local political configurations that it sought to co-opt, scientific forestry remained a discourse continually under production. It is therefore necessary to examine the historical processes in which such production occurs. One benefit of such an approach is that 'by tracking development historically, one can appreciate the complex origins of what came to be the unitary meaning of development that seemed to surface in the late colonial period in and around the Second World War' (Watts, 1995: 49). Furthermore, identifying the locus of production for any particular variant of a discourse should be integral to any such historical inquiry. The constellations of debates and expertise that surrounded scientific forestry a

---

67. For instance, in thinning operations forest guards were required to work ahead of the officer to present a clear view (see Homfray, 1936: 29).
68. See, for definitions of discourse of development, several essays in Sachs (1992), also Escobar (1995), Ferguson (1994), Mitchell (1991). The characteristic of importance to us is the ability of development discourse to comprehend any situation requiring 'improvement' or 'development' through a non-local technical expertise that can then offer modular and generalized solutions.
69. My usage of the term follows Ortner (1989).

hundred years ago remain salient through forest management institutions that emerged then and exist today. Hence the importance of an historical approach that takes us into the making and unmaking of colonial forestry science as a development regime.

The discourse of development has a long history precisely because it is not a received doctrine.[70] This chapter has tried to demonstrate its continuous production and negotiation through processes of state building in the forests of Bengal, spanning the colonial and entering the post-colonial period. The argument advanced here recognizes that recent scholarship on development, such as that of Arturo Escobar, acknowledges comparability between colonial discourse and development. James Ferguson does the same with notions of improvement, conservation, and development. Yet both these powerful analysts of development firmly locate the production of a distinct development discourse in the decolonization period after 1945, and in the corridors of western development agencies, principally the World Bank (see Escobar, 1995: 9, 23; Ferguson, 1994: 67–8, 86, 264). They also focus on the effects of development discourse working as a system of knowledges. They thus conform to the currently dominant mode of analysis where development is treated as a schematic representation of the third world and thereby the agent of certain political consequences, including those in which local 'targeted populations' absorb and manipulate the discourses of development.[71] In contrast, I propose a deeper history for development in general and suggest that the particular case of conservation and development reveals a conflicted and contested production of development discourse.

Scientific forestry, with its focus on the efficient and systematic production of timber for the 'public interest', clearly enunciated the productionist agendas that we take as characterizing developmentalist state policies in the twentieth century. On the other hand, the authors and propagators of scientific forestry launched a sustained critique of shifting cultivation, deplored private forest management, and inquired into the relations between deforestation and desiccation. This aspect of their work helped formulate the discourse of conservation so carefully identified and traced back to the seventeenth century in *Green Imperialism* (Grove, 1995). By conflating conservation with colonialism's civilizing mission, at the end of the nineteenth century, scientific forestry took on another important feature of development discourse — the notion of progress. Emerging thus as a devel-

---

70. I am extending here Ludden's (1992a) argument that post-colonial Indian development discourse was shaped by the colonial governmental forms that emerged in the later part of the nineteenth century. More recently Cowen and Shenton (1995: 29–33) have presented a longer European history of development. In this context, see also Hettne (1990).
71. Appfel-Marglin and Marglin (1990), Dubois (1991), Esteva (1992: 6–25), Parajuli (1991), are some of the general writings promoting this approach. Brow (1996), Gupta (1998), Pigg (1992) and Woost (1993) are instances where development as a received and systematic discourse engages other discourses in a local setting in South Asia.

opment regime, colonial scientific forestry became a complex of changing institutions and ideas informing post-colonial forestry.

Acknowledgements

The research on which this paper is based was carried out as part of a larger project entitled 'Revising Laws of the Jungle: Changing Peasant-State Relations in the Forests of Bengal'. This research was assisted by a grant from the Joint Committee on South Asia of the Social Science Research Council and the American Council of Learned Societies with funds provided by the Andrew W. Mellon Foundation and the Ford Foundation. Financial support for this project was also provided by the Wenner-Gren Foundation for Anthropological Research, New York; Center for International and Area Studies, Yale University; and the Program in Agrarian Studies, Yale University. I thank Mark Ashton, Indrani Chatterjee, John Cinnamon, Renald Clerisme, Susan Darlington, Margaret Everett, Vinay Gidwani, Ramachandra Guha, Sumit Guha, William Kelly, David Ludden, Patricia Mathews, Nancy Peluso, Saroj Sivaramakrishnan and Heinzpeter Znoj for reading and commenting on earlier versions. I am especially grateful to Jim Scott for patiently going through several drafts and always finding something new and insightful to say about them. Discussion of a previous, and much longer, version of this essay with Paul Greenough, Anna Tsing and others at the SSRC conference on Environmental Discourses and Human Welfare, 28–30 December 1995, was invaluable for revising the paper. Residual shortcomings are entirely my responsibility.

Abbreviations Used (archive references)

Agri = Agriculture; BFAR = Bengal Forest Administration Report; BORP = Bihar and Orissa Revenue Proceedings; BRC = Bengal Revenue Consultations; BRP = Bengal Revenue Proceedings; Cal = Calcutta; CF = Conservator of Forests; Coll = Collection; Commr = Commissioner; DC = Deputy Commissioner; DCF = Deputy Conservator of Forests; For = Forests; GOB = Government of Bengal; GOBO = Government of Bihar and Orissa; GOI = Government of India; IGF = Inspector General of Forests; LP = Lower Provinces; NAI = National Archives of India, New Delhi; Offg = Officiating; OIOC = Oriental and India Office Collections, London; Progs = Proceedings; Res = Resolution; Rev = Revenue; Secy = Secretary; US = Under Secretary; WBSA = West Bengal State Archives, Calcutta; WBSRR = West Bengal Secretariat Record Room, Calcutta; WP = Working Plan.

REFERENCES

Adas, M. (1990) *Machines as the Measure of Men: Science, Technology and Ideologies of Western Dominance*. Delhi: Oxford University Press.

Agrawal, A. (1995) 'Dismantling the Divide Between Indigenous and Scientific Knowledge', *Development and Change* 26(3): 413–39.

Agrawal, A. (1997) 'Community in Conservation: Beyond Enchantment and Disenchantment'. CDF Discussion Paper. Gainesville, FL: Conservation and Development Forum.

Anderson, D. (1984) 'Depression, Dust Bowl, Demography and Drought: The Colonial State and Soil Conservation in East Africa During the 1930s', *African Affairs* 83: 321–43.

Anon. (1853) *Report on the Political States of Southwest Frontier Agency, Revenue Administration of Assam and Wild Tribes Bordering Chittagong*. Selections from the Record of the Bengal Government, No. 11. Calcutta: Bengal Secretariat Press.

Anon. (1871) *Papers Relating to East India Forest Conservancy,* Part II, Madras. London: House of Commons.
Anon. (1934) 'The Report of the Sal Study Tour', *Indian Forest Records* XIX (3).
Appfel-Marglin, F. and S. Marglin (eds) (1990) *Dominating Knowledge: Development, Culture and Resistance*. Oxford: Clarendon.
Arora, H. and A. Khare (1994) 'Experience with the Recent Joint Forest Management Approach'. Paper presented to International Workshop on India's Forest Management and Ecological Revival. New Delhi (10–12 February).
Bailey, W. A. (1924) 'Moribund Forests in United Provinces', *Indian Forester* 50: 188–91.
Balfour, E. (1862) *The Timber Trees, Timber and Fancy Woods as also the Forests of India and of Eastern and Southern Asia*. Madras: Union Press, Cookson and Co.
Ball, V. (1880) *Jungle Life in India, or the Journeys and Journals of an Indian Geologist*. London: Thomas de la Rue and Co.
Barnes, B. (1977) *Interests and the Growth of Knowledge*. London: Routledge and Kegan Paul.
Bayly, C. (1990) *Indian Society and the Making of the British Empire*. Cambridge: Cambridge University Press.
Bayly, C. (1993) 'Knowing the Country: Empire and Information in India', *Modern Asian Studies* 27(1): 3–43.
Beinart, W. (1984) 'Soil Erosion, Conservationism and Ideas about Development: A Southern African Exploration, 1900–1960', *Journal of Southern African Studies* 11(1): 52–83.
Beinart, W. (1989) 'The Politics of Colonial Conservation', *Journal of Southern African Studies* 15(2): 143–62.
Best, J. W. (1935) *Forest Life in India*. London: John Murray.
Birdwood, H. M. (1910) *Indian Timbers: The Hill Forests of Western India*. London: The Journal of Indian Arts and Industry.
Bloor, D. (1976) *Knowledge and Social Imagery*. Chicago, IL: University of Chicago Press.
Brandis, D. (1860) *Report on Teak Forests of Pegu*. London: HMSO.
Brandis, D. (1884) 'The Progress of Forestry in India', *Indian Forester* 10(9): 399–410; 10(10): 452–62; 10(11): 501–10.
Brow, J. (1996) *Demons and Development: The Struggle for Community in a Sri Lankan Village*. Tucson, AZ: University of Arizona Press.
Bryant, J. (1994) 'From Laissez Faire to Scientific Forestry: Forest Management in Early Colonial Burma', *Forest and Conservation History* 38(4):160–70.
Champion, H. (1975) 'Indian Silviculture and Research over the Century', *Indian Forester* 101(1): 3–8.
Cleghorn, H. (1861) *The Forests and Gardens of South India*. London: W. H. Allen.
Cleghorn, H., R. Forbes Royle, R. Baird Smith and R. Strachey (1852) 'Report of the Committee appointed by the British Association to Consider the Probable Effects in an Economical and Physical Point of View of the Destruction of Tropical Forests', *Journal of the Agricultural and Horticultural Society of India*, 8: 118–49.
Clutterbuck, P. (1927) 'Forestry and the Empire', *Empire Forestry Journal* 6: 184–92.
Cohn, B. S. and N. B. Dirks (1988) 'Beyond the Fringe: The Nation-State, Colonialism and the Technologies of Power', *Journal of Historical Sociology* 1(2): 224–9.
Collins, H. M. (1992) *Changing Order: Replication and Induction in Scientific Practice*. Chicago, IL: University of Chicago Press.
Cowen, M. and R. Shenton (1995) 'The Invention of Development', in J. Crush (ed.) *Power of Development*, pp. 27–43. London: Routledge.
Crush, J. (1995) 'Introduction: Imagining Development', in J. Crush (ed.) *Power of Development*, pp. 1–23. London: Routledge.
De, R. N. (1941) 'Sal Regeneration de novo', *Indian Forester* 67(6): 283–91.
Dirks, N. B. (1992) 'Castes of the Mind', *Representations* 37: 56–78.
Dubois, M. (1991) 'The Governance of the Third World: A Foucauldian Perspective on Power Relations in Development', *Alternatives* 16(1): 1–30.

Escobar, A. (1988) 'Power and Visibility: Development and the Invention and Management of the Third World', *Cultural Anthropology* 3(4): 428–43.
Escobar, A. (1995) *Encountering Development: The Making and Unmaking of the Third World.* Princeton, NJ: Princeton University Press.
Esteva, G. (1992) 'Development', in W. Sachs (ed.) *The Development Dictionary*, pp. 6–25. London: Zed.
Falconer, H. (1852) *Selections from Records of Bengal Government no. 9, Report on the Teak Forests of the Tenasserim Provinces, With Other Papers on Teak of India.* Calcutta: Military Orphan Press.
Ferguson, J. (1994) *The Anti-Politics Machine: 'Development', Depoliticisation and Bureaucratic Power in Lesotho.* Minneapolis, MN: University of Minnesota Press.
Fernow, B. (1907) *A Brief History of Forestry in Europe, the United States and Other Countries.* Toronto: University of Toronto Press.
Ford-Robertson, F. C. (1927) 'The Problem of Sal Regeneration with Special Reference to the "Moist" Forests of the United Provinces', *Indian Forester* 53(9): 500–11, 53(10): 560–76.
Forest Research Institute (1953) *Proceedings of the All-India Sal Study Tour and Symposium.* Dehradoon: Forest Research Institute.
Forsyth, J. (1889) *The Highlands of Central India: Notes on their Forests and Wild Tribes, Natural History and Sports.* London: Chapman and Hall.
Foucault, M. (1979) *Discipline and Punish: The Birth of the Prison*, translated by Alan Sheridan. New York: Vintage.
Gadgil, M. and R. Guha. (1992) *This Fissured Land: An Ecological History of India.* Delhi: Oxford University Press.
Gooding, D. (1990) *Experiment and the Making of Meaning.* Dordrecht: Kluwer Academic.
Gouldsbury, C. E. (1909) *Dulall the Forest Guard: A Tale of Sport and Adventure in the Forests of Bengal.* London: Gibbings and Co.
Government of Bengal (1935) *The Forests of Bengal.* Calcutta: Superintendent of Government Printing.
Grieve, J. W. A. (1912) *Working Plan for the Darjeeling Forests.* Darjeeling: Darjeeling Branch Press.
Griffith, A. L. and R. S. Gupta (1948) 'The Determination of the Characteristics of Soil Suitable for Sal', *Indian Forest Bulletin*, Silviculture (new series) no. 138.
Grove, R. (1989) 'Early Themes in African Conservation: The Cape in the Nineteenth Century', in D. Anderson and R. Grove (eds) *Conservation in Africa: People, Policies and Practice*, pp. 21–39. Cambridge: Cambridge University Press.
Grove, R. (1993) 'Conserving Eden: The (European) East India Companies and Their Environmental Policies on St. Helena, Mauritius and in Western India, 1660 to 1854', *Comparative Studies in Society and History* 36: 318–51.
Grove, R. (1995) *Green Imperialism: Colonial Expansion, Tropical Island Edens and the Origins of Environmentalism, 1660–1860.* Cambridge: Cambridge University Press.
Guha, R. (1983) 'Forestry in British and Post-British India: An Historical Analysis', *Economic and Political Weekly* 17: 1882–96
Guha, R. (1989) *Unquiet Woods: The Ecological Bases of Peasant Resistance in the Himalaya.* Delhi: Oxford University Press.
Guha, S. (1999) *Environment and Ethnicity in India, 1200–1991.* Cambridge: Cambridge University Press.
Gupta, A. (1998) *Postcolonial Developments: Agriculture in the Making of Modern India.* Durham, NC: Duke University Press.
Hacking, I. (1983) *Representing and Intervening.* Cambridge: Cambridge University Press.
Hacking, I. (1992) 'The Self-Vindication of the Laboratory Sciences', in A. Pickering (ed.) *Science as Practice and Culture*, pp. 29–64. Chicago, IL: The University of Chicago Press.
Haines, H. H. (1905) *Working Plan for Reserved Forests of Singhbhum, 1903–1918.* Calcutta: Bengal Secretariat Press.

Haraway, D. (1988) 'Situated Knowledges: The Science Question in Feminism and the Privilege of Partial Perspective', *Feminist Studies* 14(3): 575–99.
Hardiman, D. (1994) 'Power in the Forests: The Dangs, 1820–1940', in G. Pandey and P. Chatterjee (eds) *Subaltern Studies, Volume VIII: Essay in Honor of Ranajit Guha*, pp. 89–147. Delhi: Oxford University Press.
Harding, S. (1986) *The Science Question in Feminism*. Ithaca, NY: Cornell University Press.
Harrison, R. P. (1992) *Forests: The Shadow of Civilization*. Chicago, IL: The University of Chicago Press.
Harvey, D. (1989) *The Condition of Postmodernity: An Enquiry into the Origin of Cultural Change*. Oxford: Basil Blackwell.
Hatt, C. C. (1905) *Working Plan for the Reserved Forests of the Buxa Division*. Calcutta, Bengal Secretariat Book Depot.
Hays, S. P. (1959) *Conservation and the Gospel of Efficiency: The Progressive Conservation Movement, 1890–1920*. Cambridge, MA: Harvard University Press.
Hettne, B. (1990) *Development Theory and the Three Worlds*. London: Methuen.
Hobart, M. (1993) *An Anthropological Critique of Development: The Growth of Ignorance?* London: Routledge.
Hole, R. S. (1919) 'Regeneration of Sal Forests', *Indian Forester* 45: 119–32.
Homfray, C. K. (1936) *Notes on Thinnings in Plantations*. Bengal Forest Bulletin, no. 1, Silviculture Series. Alipore: Bengal Government Press.
Hooker, J. D. (1854) *Himalayan Journals: In Bengal, the Sikkim and Nepal Himalayas, and the Khasia Mountains*. London: Ward, Lock and Bowden.
Hunter, W. W. (1868) *The Annals of Rural Bengal*. New York: Leypoldt and Holt.
Jackson, W. B. (1854a) *General Report of a Tour of Inspection*. Selections from the Record of the Bengal Government, No. 16. Calcutta: Bengal Secretariat Press.
Jackson, W. B. (1854b) *Report on Darjeeling*. Selections from the Record of the Bengal Government, No. 17. Calcutta: Bengal Secretariat Press.
Kloppenburg, J. (1991) 'Social Theory and the De/Reconstruction of Agricultural Science: Local Knowledge for an Alternative Agriculture', *Rural Sociology* 56(4): 519–48.
Leach, M. (1998) 'Challenging the Relations Between Global and Local Knowledges in Environment and Development'. Paper presented at the Conference on Globalization, Modernity, and Locality in Stories of Development, Yale University (1 March).
Lewin, T. H. (1869) *The Hill Tracts of Chittagong and the Dwellers Therein*. Selections from the Records of the Bengal Government, No. 43A. Calcutta: Bengal Secretariat Press.
Li, T. M. (1996) 'Images of Community: Discourse and Strategy in Property Relations', *Development and Change* 27(3): 501–28.
Lowood, H. E. (1990) 'The Calculating Forester: Quantification, Cameral Science, and the Emergence of Scientific Forestry Management in Germany', in T. Frangsmyr, J. L. Heilbron and R. E. Rider (eds) *The Quantifying Spirit in the Eighteenth Century*, pp. 315–42. Berkeley, CA: University of California Press.
Ludden, D. (1989) *Peasant History in South India*. Delhi: Oxford University Press.
Ludden, D. (1992a) 'India's Development Regime', in N. B. Dirks (ed.) *Colonialism and Culture*, pp. 247–88. Ann Arbor, MI: University of Michigan Press.
Ludden, D. (1992b) 'Anglo-Indian Empire', in B. Stein (ed.) *The Making of Agrarian Policy in British India, 1770–1900*, pp. 150–86. Delhi: Oxford University Press.
Makins, F. K. (1920) 'Natural Regeneration of Sal in Singhbhum', *Indian Forester* 46: 292–6.
McIntire, A. L. (1909) *Notes on Sal in Bengal, Forest Pamphlet no. 5*. Calcutta: Superintendent of Government Printing.
Mitchell, T. (1991) 'America's Egypt: Discourse of the Development Industry', *Middle East Report* March–April: 18–34.
Mukarji, N. (1989) 'Decentralization Below the State Level: Need for a New System of Governance', *Economic and Political Weekly* 23: 467–72.
Ortner, S. (1989) *High Religion: A Cultural and Political History of Sherpa Buddhism*. Princeton, NJ: Princeton University Press.

Osmaston, F. C. (1928) 'Sal and its Regeneration', *Indian Forester* 54(12): 639–55.
Parajuli, P. (1991) 'Power and Knowledge in Development Discourse', *International Social Science Journal* 127: 173–90.
Peluso, N. (1992) *Rich Forests, Poor People: Resource Control and Resistance in Java*. Berkeley, CA: University of California Press.
Peters, P. (1994) *Dividing the Commons: Politics, Policy and Culture in Botswana*. Charlottesville, VA: University of Virginia Press.
Phillips, P. J. (1924) *Revised Working Plan for the Reserved Forests of the Saranda and Kolhan Division in the Singhbhum District, Bihar and Orissa Circle*. Patna: Superintendent of Government Printing.
Pickering, A. (1992) 'From Science as Knowledge to Science as Practice', in A. Pickering (ed.) *Science as Practice and Culture*, pp. 1–28. Chicago, IL: The University of Chicago Press.
Pigg, S. L. (1992) 'Constructing Social Categories through Place: Social Representations and Development in Nepal', *Comparative Studies in Society and History* 34(3): 491–513.
Prakash, G. (1990) 'Writing Post-Orientalist Histories of the Third World: Perspectives from Indian Historiography', *Comparative Studies in Society and History* 32(2): 383–408.
Prasad, A. (1994) 'Forests and Subsistence in Colonial India: A Study of the Central Provinces'. PhD Thesis. New Delhi: Jawaharlal Nehru University.
Rangarajan, M. (1994) 'Imperial Agendas and India's Forests: The Early History of Indian Forestry, 1800–1878', *Indian Economic and Social History Review* 31(2): 147–67.
Rangarajan, M. (1996) *Fencing the Forest: Conservation and Ecological Change in India's Central Provinces, 1860–1914*. Delhi: Oxford University Press.
Rangarajan, M. (1998) 'Production, Desiccation and Forest Management in the Central Provinces, 1850–1930', in R. Grove, V. Damodaran, and S. Sangwan (eds) *Nature and the Orient: The Environmental History of South and Southeast Asia*, pp. 575–95. Delhi: Oxford University Press.
Raynor, E. W. (1940) 'Sal Regeneration de novo', *Indian Forester* 66(9): 525–9.
Reingold, N. and M. Rothenberg (eds) (1987) *Scientific Colonialism: A Cross-Cultural Comparison*. Washington, DC: Smithsonian Institution Press.
Ricketts, H. (1854) *Report on the District of Singhbhum*. Selections from the Record of the Bengal Government, No. 16. Calcutta: Bengal Secretariat Press
Ricketts, H. (1855) *Papers Relating to the Southwest Frontier Comprising Reports on Purulia or Manbhum and Chota Nagpur*. Selections from the Record of the Bengal Government, No.20. Calcutta: Bengal Secretariat Press.
Rodger, A. (1925) 'Research in Forestry in India', *Empire Forestry Journal* 4: 45–53.
Sachs, W. (ed.) (1992) *The Development Dictionary*. London: Zed.
Schlich, W. (1876) 'Notes on Preliminary Working Plans', in D. Brandis and A. Smythies *Report of the Proceedings of the Forest Conference held at Simla October 1875*, pp. 104–7. Calcutta: Superintendent of Government Printing.
Sen, J. N. and T. P. Ghose (1925) 'Soil Conditions Under Sal', *Indian Forester* 51(6): 243–53.
Skaria, A. (1998) 'Timber Conservancy, Desiccationism and Scientific Forestry: The Dangs, 1840s–1920s', in R. Grove, V. Damodaran, and S. Sangwan (eds) *Nature and the Orient: Essays in Environmental History of South and Southeast Asia*, pp. 596–635. New Delhi: Oxford University Press.
Skaria, A. (1999) *Hybrid Histories: Forests, Frontiers, and Wildness in Western India*. Delhi: Oxford University Press.
Sinha, J.N. (1962) *Working Plan for the Reserved, Protected and Private Protected Forests of Manbhum Division*. Patna: Superintendent of Government Printing.
Sivaramakrishnan, K. (1995) 'Imagining the Past in Present Politics: Colonialism and Forestry in India', *Comparative Studies in Society and History* 37(1): 3–40.
Sivaramakrishnan, K. (1999) *Modern Forests: Statemaking and Environmental Change in Colonial Eastern India*. Stanford, CA: Stanford University Press.
Smith, D. (1986) *The Practice of Silviculture*. New York: John Wiley.

Smith, D. (1987) *The Everyday World as Problematic: A Feminist Sociology*. Boston, MA: Northeastern University Press.
Smythies, E. A. (1926) 'The Problem of Sal Regeneration', *Indian Forester* 52: 395–400.
Smythies, E. A. (1940) 'Sal Regeneration de novo', *Indian Forester* 66(4): 193–9.
Stebbing, E. P. (1920) *The Diary of a Sportsman Naturalist*. London: John Lane.
Stebbing, E. P. (1922) *The Forests of India*, Volume I. London: John Lane.
Stebbing, E. P. (1924) *The Forests of India*, Volume II. London: John Lane.
Stebbing, E. P. (1926) *The Forests of India*, Volume III. London: John Lane.
Subrahmanyam, S. (1992) 'The Mughal State — Structure or Process? Reflections on Some Recent Western Historiography', *Indian Economic and Social History Review* 29(3): 291–322.
Thompson, E. P. (1975) *Whigs and Hunters: The Origins of the Black Act*. New York: Pantheon.
Tinne, P. (1907) *Second Working Plan for the Reserved Forests of the Tista Division*. Calcutta: Bengal Secretariat Press.
Trafford, F. (1905) *Working Plan for the Reserved Forests of the Jalpaiguri Division*. Darjeeling: Bengal Secretariat Tour Press.
Tucker, R. P. (1989) 'The Depletion of India's Forests under British Imperialism: Planters, Foresters and Peasants in Assam and Kerala', in D. Worster (ed.) *The Ends of the Earth: Perspectives on Modern Environmental History*, pp. 118–40. Cambridge: Cambridge University Press.
Vandergeest, P. and N. Peluso (1995) 'Territorialization and State Power in Thailand', *Theory and Society* 24: 385–426.
Waddell, L. A. (1899) *Among the Himalayas*. London: Archibald Constable.
Warner, J. H. (1991) 'Ideals of Science and their Discontents in Late Nineteenth-Century American Medicine', *Isis* 82(313): 454–78.
Warren, W. D. M. (1940) 'Sal Regeneration de novo', *Indian Forester* 66(6): 334–40.
Warren, W. D. M. (1941) 'Sal Regeneration de novo in B-3 Sal', *Indian Forester* 67(3): 116–23.
Washbrook, D. (1981) 'Law, State and Society in Colonial India', *Modern Asian Studies* 15(3): 649–721.
Watts, M. (1995) '"A New Deal in Emotions": Theory and Practice and the Crisis of Development', in J. Crush (ed.) *Power of Development*, pp. 44–62. London: Routledge.
Woost, M. (1993) 'Nationalising the Local Past in Sri Lanka: Histories of Nation and Development in a Sinhalese Village', *American Ethnologist* 20(3): 502–21.

5 The Changing Regime: Forest Property and *Reformasi* in Indonesia

John F. McCarthy

INTRODUCTION

During 1997 and 1998 some of the most extensive forest fires this century ravaged the Indonesian islands of Kalimantan and Sumatra, burning some seven to ten million hectares of forest and agricultural land.[1] As political turmoil erupted in the wake of a currency meltdown, the environmental devastation fused into a wider economic and political crisis. As in other sectors of the economy, during the New Order (1966–98) powerful conglomerates and politico-business families had been able to dominate the forestry sector at the expense of sound environmental management. Critics now directed the same charges levelled against the regime's economic management against its environmental management: precarious policies, 'the capricious exercise of power, cronyism and incompetent systems of governance' (Robison and Rosser, 1998: 1593).

There are a variety of social, economic, demographic and ecological factors involved in deforestation in Indonesia.[2] Yet, as Bromley and Cernea

---

1. Estimates of the area burnt vary widely. Conservative government figures put the total area burned at 383,870 ha (KLH and UNDP, 1998a). Using satellite data, GTZ Integrated Forest Fire Management project (IFFM) estimated that 4 million ha burned in East-Kalimantan alone in 1998 (IFFM/GTZ, 1998). The EU-sponsored fire project in south Sumatra estimated that roughly 2 million ha of land had burned in that province during 1997 (Schindler, 1998). Spot Asia, a satellite operator, estimates that 10 million ha of forest and agricultural lands were burnt. WWF and EEPSEA estimated that the total area burned in 1997 amounted to 5 million ha (Reuters, 25/11/98; WWF, 29/5/98). In December 1998, a report quoted in the Jakarta Post estimated that the fires had consumed some 10 million ha of Sumatra and Kalimantan (*Jakarta Post*, 18/12/98). In comparison, the large fires of 1982/83 destroyed an estimated 3.6 million ha of rainforest (Achmadi, 1988). For more detail on the 1997–8 fires, and the contrasting discourses in their explanations, see Harwell (this volume).
2. Sunderlin and Resosudarmo (1997) distinguish between the agents involved in changing land use, the immediate causes influencing the decisions of agents and the underlying causes of changes to forest cover. In this framework, the underlying causes include overarching national, regional or international forces that influence the 'decision parameters' of agents (Sunderlin and Resosudarmo, 1997: 15).

have observed, 'an institutional crisis typically underlies a tangible environmental problem' (Bromley and Cernea quoted in Southgate and Runge, nd). This chapter therefore focuses on the social structures and power relations underlying the problems facing the management of Indonesia's forests, particularly in connection with the changing structure of property rights operating in the vast forest domain. I will argue that the official discourse of the New Order period tended to ascribe the failings of the state forest institutions either to the managerial and logistical difficulties facing a state forestry agency in managing such an extensive forest area with few resources, or to the difficulties facing law enforcement. In this view, management problems could be remedied by making available greater resources and by a more thoroughgoing implementation of the law. However, while these two factors were clearly at play, this policy discourse avoided discussion of a third issue — the allocation of property rights over forest territories to politico-business interests.

Considering environmental policy as the imposition of a particular structure of property rights, this chapter will focus on the way competing property claims are linked to overlapping property regimes and their attendant notions of rights. In constructing a regulatory order for resource extraction, following the colonial example, the New Order overlaid indigenous notions of tenure and territoriality with the concepts and regulatory institutions of a state property regime. However, in timber-rich rainforests, the state allocated *de jure* use rights to commercial interests. By failing to institute a state property regime in anything more than name, a *de jure* state property regime slipped into the *de facto* control of private interests. These consisted of the conglomerates and politico-business families at the apex of the state, together with the local networks of power and interest dominating illegal timber operations at the district level. The end result was a complex layering of competing property claims covered with the normative discourse of the state forestry. The fallout of these practices was seen in the recent forest fires.

Now, following the demise of Suharto, the possibility emerges of setting Indonesian forest management on a more sustainable footing. Advocates of reform have argued for a dismantling of the state forest regime as it developed under Suharto and a redirection of the flow of benefits derived from the forests to local communities. This chapter will consider this demand as the re-emergence of a debate over the structure of property rights that gives protection to particular uses of environmental resources.

The following section will consider the policy framework constructed by the New Order for managing 143 million ha of land classified as state forest.[3] A case study from West Kalimantan will illustrate how this forestry

3. While the Government of Indonesia (GOI), working with international agencies, has carried out forest inventories since the 1980s, the exact area of remaining forest remains

discourse obscured underlying property rights issues that are embedded in Indonesian political history. After considering the impact on local communities of the property rights regime that emerged under Suharto, I will discuss its consequences: the devastating forest fires of 1997–8. The chapter ends by considering the prospect for reform of the forestry regime following the fall of Suharto.

POLICY FRAMEWORK FOR FOREST MANAGEMENT

As political commentators have noted, the New Order sought to rest its legitimacy upon constitutional and legal processes (Langenberg, 1990: 132). The state forestry discourse drew its authority from the clause in the Indonesian Constitution that grants control of designated forest lands and resources to the state. During the Suharto period, the government interpreted this to mean that the state had exclusive authority over all aspects of human activity within any territories classified as 'forest area' *(kawasan hutan)* (Barber et al., 1994: 67). This was expressed in a large body of laws, decrees and administrative regulations that governed access to and use of forest resources. The cornerstone of this legal edifice was the Basic Forestry Provisions (No.5/1967), a law that set out the framework for forest management over thirty years.

The Basic Forestry Provisions divided the 'forest area' into distinct land-use categories, deploying separate policy objectives in areas set aside for timber production, conversion to agriculture, and conservation. To advance state control over the 'forest area' the government embarked on a series of forest mapping exercises (Peluso, 1995). To overcome intersectoral planning problems that resulted from these initial efforts, and to facilitate the development of the forestry industry — especially the allocation of access and use rights to the benefits to be derived from this 'forest area' — in 1980 the Minister of Agriculture (then responsible for forestry) asked each provincial governor in areas outside Java to prepare a Consensus Forest Land Use Plan (*Tata Guna Hutan Kesepakatan* or TGHK).[4] The TGHK

---

subject to debate, not least because GOI uses the term 'forest area' to denote the lands under the jurisdiction of the Ministry of Forestry, rather than the area actually covered by forest (KLH and UNDP, 1997: 364). Using satellite data from 1986–91, the National Forest Inventory estimated in 1996 that there are 120.6 million ha of forested land covering 69 per cent of the land area outside Java. Of this, the government estimated that 92.4 million ha were intact in 1993 (KLH and UNDP, 1997: 364, 377; Sunderlin and Resosudarmo, 1997).

4. After intersectoral discussions at the regional level, regional TGHKs were prepared and recorded on provincial maps known as *peta Rencana Pengukuhan dan Penatagunaan Hutan* (RPPH). After ministerial agreement at the national level, ministers then signed these maps. The Ministry of Forestry then used the TGHK as the basis for allocating HPH (long-term logging leases) and HTI (concessions for timber plantations).

classified 143.8 million ha (approximately 75 per cent of the nation's land area) as 'forest land'. This was subsequently placed under the jurisdiction of the Ministry of Forestry (Departemen Kehutanan, 1992; KLH and UNDP, 1997).

As a spatial planning and mapping process, the TGHK formed part of a large-scale accumulation strategy that worked to the disadvantage of forest dwellers (Peluso, 1995: 383). On paper at least, the forest authorities marked out the acceptable boundaries for different forest land uses within the boundaries of the vast 'forest area' without taking into account local communities and cultivation areas or local notions of territoriality. The TGHK merely classified areas for land use according to six categories involving either direct or indirect control. In the areas mapped as 'protection forests' (30.8 million ha) and nature reserves/conservation areas (18.8 million ha), the operational section of the Forestry Department maintains direct control, at least in theory. In 64.3 million ha mapped as 'production forest' or 'limited production forest', the forest authorities maintain indirect control, granting long-term leases for logging the forest (HPH) to concessionaires under Forestry Department supervision (KLH and UNDP, 1997: 364).[5] In areas mapped as 'conversion forest' (26.6 million ha), land is set aside for 'planned deforestation'; here the Forestry Department receives requests for conversion of forest to plantation agriculture, granting use rights (HGU) to leases over what remains state land. The land later reverts to the leaseholder under the statutes of other state agencies.[6]

In 1992 the Spatial Use Management Act No. 24 established a new process of spatial planning. While strongly influenced by the TGHK spatial plans, the act established significant differences from previous land use allocations. The new act gave the National Development Planning Agency (BAPPENAS), the Ministry of Home Affairs and the Ministry of State for Environment (KLH) the key roles in implementing spatial planning. Since the act threatened the power of the Ministry of Forestry as the main decision maker over land use in the 'forest area', it has led to a struggle within the bureaucracy over the power to allocate property rights. The Ministry of Forestry has tried to protect the TGHK system by claiming that the Forestry Act 'should have legal priority over the new Act' (Gunarso and Davie, forthcoming). While this dispute over sectoral responsibility remains unresolved, many of the basic assumptions guiding the allocation of resources persist. For instance, under the new system, outside areas set aside for cultivation or timber production (*kawasan budaya*), the Act determines

---

5. In 1971 the state forest authorities began leasing these areas to private companies who were entitled to extract timber according to the Selective Cutting and Replanting System or TPTI (*Terbang Pilih Tanam Indonesia*). Due to the inefficiencies of this system, in the late 1990s GOI began developing the Production Forest Management Unit (KPHP) concept to replace TPTI (KLH and UNDP, 1997: 368).
6. The sixth category consisted of unclassified state forest land (*hutan negara bebas*).

that remaining areas should be maintained under natural forest cover as protected areas (*kawasan lindung*) for environmental protection.[7] While the sum of protected areas set aside for their extensive biodiversity and conservation values or ecological functions amounts to an impressive 25 per cent of the total land area of Indonesia, the forest authorities are unable to defend such an extensive area and in many cases these areas are protected on paper only.[8] Together with the widespread misuse of production forests this means that, despite the stated policy aims to maintain the 'permanent forest', deforestation rates increased over the 1990s to somewhere around 1 million ha per year.[9]

In attempting to come to terms with the deforestation problem, recent government policy documents, such as the 'Biodiversity Action Plan' and 'Agenda 21-Indonesia', recognize a complex cocktail of problems.[10] These documents highlight unsustainable logging, forest conversion for mining and plantation development, disregard for traditional land tenure and resource utilization rights, transmigration, shifting agriculture and other changes in land uses that occur as an increasing population struggles to meet its needs (BAPPENAS, 1993: iv; KLH and UNDP, 1997: 365). Each class of the 'permanent forest' faces its own distinct problems. For instance, 'Agenda 21-Indonesia' identifies the major shortcoming in the management of production forests, including current concession policies and logging practices, the low price of logs and low economic rents for forest exploita-

---

7. It is not altogether clear how this classification system can be integrated with the TGHK spatial plans that set aside large areas of 'permanent forest' (*kawasan hutan tetap*), either for timber production and timber plantations or for biodiversity conservation and protection of ecological functions. This lack of integration between TGHK and the new spatial plans has meant that in West Kalimantan the TGHK land allocations still persist, preventing the implementation of the new spatial planning act and retaining the power of the forestry department land allocations to HPH and HGU (Steering Committee, 1998).
8. There are very few resources available for forest protection. A KLH/UNDP study noted that 'of the 30,000 civil servants in the Ministry of Forestry, only 477 are employed in the office of the central PHPA (Directorate General of Forest Protection and Nature Conservation) which is in charge of protected areas'. There are 8,500 forest guards (*jagawana*) responsible for policing over 100 million ha of 'forest area' (Suara Pembaruan, 10/4/98; KLH and UNDP, 1997: 378).
9. Figures for the rate of deforestation vary 'depending on the sources and methods of analysis', the most accepted estimates ranging between less than 600,000 ha and 1.3 million ha per year (KLH and UNDP, 1997: 364). According to Sunderlin and Resosudarmo, this wide discrepancy is due to: (1) the lack of reliable primary data regarding forest cover change; (2) lack of clarity over what constitutes 'deforestation', leading to severe distortions of the issue. This chapter follows suggestions based on an FAO study: 'forest' refers to 'natural forest', and 'deforestation' is 'the sum of all areas' transitions from natural forest classes to other classes' (i.e. net deforestation) (Sunderlin and Resosudarmo, 1997).
10. The 'Biodiversity Action Plan' was prepared by BAPPENAS in the run-up to the UNCED conference in 1993. Following UNCED, 'Agenda 21-Indonesia' was drawn up in 1997 by the KLH and the United Nations Development Program as a long-term strategy for sustainable development.

tion, and conflicts with local communities (KLH and UNDP, 1997: 368). The management of protected areas also faces problems, including 'lack of public participation, lack of management framework, the need for regional income, insufficient funding and lack of law enforcement' (ibid.: 378).

As Dove (1993a: 23) has written, 'it is widely understood that state elites seek to control valuable forest resources; it is less widely understood that an important means to this end is the control of resource-related discourse'. Accordingly, it is useful to explore how the official forestry discourse deals with a specific problem, in this case illegal logging. The following case study offers insights into how, in attempting to account for failings of the state forestry regime, the prevalent forestry discourse of the New Order period concealed the property rights issues at the centre of the problem. It also reveals the networks of patronage and collusion that undermined the aims of state forest policy, and the underlying legitimacy problem facing this policy.

TIMBER POACHING IN STATE FORESTS

Under Indonesian law, logging interests can extract timber from native forests through two legal channels. First, logging companies can obtain long-term logging concessions or HPH (*Hak Pengusaha Hutan*) to selectively log production forests. Second, those wishing to open timber plantations or agricultural plantations can obtain leases over state forest-land.[11] If the land to be cleared still contains productive stands of timber, they can obtain Timber Harvest Permits (*Ijin Pemanfaatan Kehutanan* or IPK) to remove and process valuable logs. Beside widespread abuses of the laws, decrees and administrative regulations relating to legal concessions,[12] companies and individuals with (and without) HPH and IPK permits also poach timber in forest areas across Indonesia, usually with the support of local government and military officials (KLH and UNDP, 1998b: 83). The Minister of Forestry has frequently made statements of concern regarding the deforestation resulting from this uncontrolled logging (*penebangan liar*) of state forests. Moreover, the issue has featured prominently in local and national newspapers. In February 1996, for instance, *Forum Keadilan,* a national weekly, ran a special feature under the headline 'Hunting Invisible Wood in Kalimantan'. The article describes the operations of a special integrated forest security team (*Tim Pengamanan Hutan Terpadu*) involving members of the military, police and forestry apparatus which was set up to tackle uncontrolled logging in the province:

11. Those wishing to open palm oil plantations obtain HGU (*Hak Guna Usaha*) leases over conversion forest areas while those wishing to open timber plantations (*Hutan Tanaman Industri* or HTI) obtain HPHTI concessions.
12. These problems are widely discussed in the literature; see Barber et al. (1994); Dauvergne (1994, 1997a, 1997b, 1997c, 1998); Potter (1991, 1996).

> In the Jungle of West Kalimantan, one intense day last November, the flow of the Padu Ampat was unusual. Around five hundred freshly cut logs appeared in the form of a long raft floating slowly down the river. A few moments later, around twenty forest police known as *jagawana*, emerged from downstream. That month, the Regional Forestry Office (*Dinas Kehutanan*) was carrying out a 'functional operation' against forest thieves.
>
> Arriving at the location armed with machetes, *jagawana* adeptly jump onto the floating wood, while others continue to stand on the back of the speedboat. They are alert. 'We must watch out because sometimes those who transport the wood are ready to act without thinking about the consequences,' says Syamsul, a *jagawana* who has already worked for more than 10 years in the West Kalimantan forest.
>
> Fortunately, there was no fight on the water. Not because those transporting the wood didn't fight, but because the *jagawana* failed to meet a soul. They concluded that this was stolen wood. 'If it was not stolen wood, it is impossible to think that it would be abandoned just like that', says the Head of the Batusumpar Forest Functionary Unit (*Kesatuan Pemangku Hutan*) Gst. M. Saleh. (Baskoro and Sihotang, 1996: 51)

Over several months of paramilitary operations, the forest police confiscated 2,600 logs on this one river, Sungai Padu Ampat.[13] The report indicated that this kind of illegal activity represented a loss to state tax receipts and a breach of the authority of the Forestry Department.

In discussing the illegal logging problem, the article identifies two levels of failure. The first relates to the technical and managerial capacity of the forestry apparatus more generally, and the forest police of West Kalimantan specifically. In this case fifty *jagawana*, many of them desk-bound administrators, police 1.4 million ha of state forest in West Kalimantan. The *jagawana* do not possess a single speedboat, and have to rent boats for their operations. They are poorly paid, and, while they face potentially violent resistance from illegal loggers, they are only armed with machetes, although the Ministry of Forestry had asked the Indonesian army to supply rifles with plastic bullets. To ensure the working of the state forestry regime, the article suggests that the Forestry Department should increase the wages and improve the facilities provided for the forest police.

Beyond these managerial and technical issues, the journalists identify a second class of problems facing policy implementation: the law enforcement issue resulting from the widespread collusion between loggers and local government officials.[14] A wide range of people participate both directly and indirectly in a wood trade operating outside the formal forestry rules. A large number of saw mills, many operating illegally, create a demand for

---

13. The article estimated that this amounted to 1,200 cubic metres, worth over 15 million rupiah at the time (approximately US$6000).
14. In Aceh and other resource-rich regions, local people including officials resent the fact that HPH issued from the centre extract local forest resources, leading to environmental damage (such as deforestation-induced flooding) without sufficiently contributing to local development. Local feeling is that Jakarta-based commercial interests and the central government coffers have captured an inequitable proportion of profits generated from local resources (Cohen, 2/7/98). This has helped stimulate the emergence of locally-based illegal logging networks.

wood that greatly outstrips the amount of wood that can be produced from legal logging concessions operating in the area.[15] Concessionaires or *Cukong* (financial backers) increase supply by providing capital and chainsaws to local people who are prepared to take to the forest for weeks. The loggers cut wood, carry it to the river and construct rafts to carry it downstream. To ensure the smooth working of their logging operations, the *Cukong* form relationships with powerful local officials; for instance, to obtain a fig-leaf of legality for transporting the wood, they buy wood hauling permits from forestry personnel.[16] *Cukong* also obtain legal immunity through seeking the patronage of key figures in local government. They make payments and contribute to development projects favoured by local government officials. When the forestry officials are ordered to crack down on illegal logging, these patron–client ties between local business interests and local officials come into play: it is the woodcutters rather than the *Cukong* who face jail. In the one case when the police arrested two *Cukong*, they were later freed because of 'insufficient evidence' (Baskoro and Sihotang, 1996: 55).

During the New Order period this pattern was widespread beyond West Kalimantan. In the Sumatran province of Aceh, for instance, provincial newspapers have carried stories for many years describing the involvement of local government officials in illegal logging. In August 1995 the head of the Golkar faction in the provincial parliament, retired Major General H. T. Djohan, told a reporter that many officials support (*membeking*) logging or directly act as bosses (*tauke*).

> Every week there are community figures that report the involvement of government figures in wood theft ... Many officials (*aparat*) have chainsaws. Observe the scores of illegal timber trucks that are on the roads each day. See also the scores of illegal fee collection posts (*pos pungutan*) along the whole length of the highway... The wood business also involves many government agencies and the provincial government. Regional forestry offices, the army and even TKPH (special anti-logging teams) are involved. Observing the potential and power of the agencies involved it is difficult to believe that the wood and forest problem can be handled with justice, and this affronts the community who see this taking place before their eyes (*Serambi Indonesia*, 7/8/95).

At the regional level, policy failure directly proceeds from the patronage links between state officials and their local business clients. According to Dauvergne, those charged with implementing forest regulations 'ignore state

---

15. The prevalence of illegal logging relates to the rapid growth in the capacity of the wood processing industry under Suharto. In 1998 KLH/UNDP quoted a 1993 report that, if all timber-processing industries were to work to capacity, they would require 54 million cubic metres of timber per annum. The current legal supply amounts to only 30 million cubic metres per years, meaning that HPH are unable to meet even 80 per cent of capacity (44 million cubic metres), creating an incentive for illegal logging (KLH and UNDP, 1998b: 79).
16. There are two kinds of permits for transporting logs: those for unprocessed logs (*Surat Angkutan Kayu Bulat* or SAKB), and those for transporting sawn timber (*Surat Angkutan Kayu Olahan* or SAKO).

rules in exchange for bribes, gifts and career stability ... the result is rampant illegal logging, timber smuggling, tax and royalty evasion, flagrant violations of logging rules, and avoidance of reforestation duties' (Dauvergne, 1997b: 72).

Beyond these two problems, the contents of the *Forum* article, rather than the explicit argument, point to a third, more deeply rooted problem:

> Of course the lower class (*klas bawah*) does not consider the consequences or the amount [of revenue] lost by the country on the wood that they cut. They just form their link [in the wood supply chain]. Because, if they don't cut wood, in other words without permits, so many stomachs around them will stay empty. 'According to my father, my family used to work cutting wood. And no one forbade this', states Mahmud, a man twenty-five years of age. (Baskoro and Sihotang, 1996: 54)

As already noted, the Indonesian land system classifies some 143 million ha of the nation's land surface as 'forest land' and places it under the jurisdiction of the Ministry of Forestry. However, *de jure* status is one thing, *de facto* control another. This classification of 'forest land' means that Ministry of Forestry jurisdiction extends over areas inhabited or used by some 60 million people.[17] As many case studies have shown, in spite of legal sanctions, local communities continue to cut wood and open land for agriculture according to local customary rights. The state has held the activities of shifting cultivators and 'forest squatters' responsible for a large amount of deforestation (Angelsen, 1995; Colfer and Dudley, 1993; Dove, 1983, 1993a, 1993b; Guha, 1994; Hayes, 1997; Myers, 1980). Yet, as Peluso has noted, when local groups contest the state system of resource extraction, production and conservation, and a state has to resort to violent means to protect its claims over natural resources, this indicates that the state's claim to legitimacy for forest policy has partially or wholly failed (Peluso, 1993: 47).

However, in proposing a solution to the illegal logging, the *Forum* article, rather like the official forestry discourse, does not directly discuss this legitimacy problem. In 1996, the Minister of Forestry at the time, Djamaludin, candidly admitted that illegal logging practices were widespread and intractable: they were occurring in all classes of state forest, including officially designated 'protection forest' and national parks. Due to the involvement of local officials, he acknowledged, law enforcement was proving too difficult. Yet, once again, he suggested that the answer lay in

---

17. Estimates of the number of 'forest squatters' vary. A 1998 article in *Asian Economic News* placed the figure at 1.7 million. According to the RePPProt study team, in 1991 there were 1,199,970 families of swidden agriculturalists using 11,402,300 ha of forest land. Poffenburger has estimated that 60 million people live in or near the forest lands of the outer islands (Moniaga, 1993: 135), while the Ministry of Forestry has estimated that 16.5 million people directly depend on the forest (KLH and UNDP, 1998b: 79). In 1997 Fox and Atok carried out a study in West Kalimantan to help clear up this controversy. Results suggested that 20 to 30 per cent of the population of West Kalimantan (between 650,000 and 1,000,000 people) lived in areas classified as state forest land (Fox and Atok, 1997).

improving state control. By 'taking firm action in accordance with the law' the Ministry of Forestry would be able to successfully carry out its mandate — wood production, sustainable forest management and protection of ecological functions and nature conservation (*Kompas*, 12/4/96). One critic noted that this reaction 'represented an official language that was now routine and clichéd. Only what a pity that it is never realized; it always remains a promise for the future' (*Detektip*, 20/12/96: 6–7).[18]

Two years later we find a similar response to the forest fires (Muryadi et al., 25/8/97; Muryono, 10/1/98). In April 1998, an article in a national daily, *Kompas*, discussed the role of forest plantations (HTI) in the fires. HTI companies were widely accused of deliberately starting fires to open forested lands for timber planting. The Minister for the Environment, Juwono Sudarsono, admitted that these companies were intentionally starting fires. 'The problem is that this is neither controlled nor well planned', he said. 'Everywhere there is a problem with legal organisation. Here we must look carefully, because the main problem is implementation in the field. The personnel, funds, equipment and provisions that are available are very inadequate with the result that local government or even government departments are overwhelmed in the face of it' (*Kompas*, 3/4/98).

According to the official forestry discourse, policy failure is ascribed to managerial and technical problems facing the forestry agency, and to obstacles standing in the way of law enforcement. By defining the issue in this narrow fashion, the official forestry discourse sees policy failure in terms of 'poor implementation'. By making a rigid distinction between 'good' policy and 'poor' implementation, forest policy can retain the ideal of state forest management — of experts making, interpreting and carrying out laws and regulations within a clearly defined mandate — in essence the legitimacy of the state forestry apparatus (Barber et al., 1994: 69). An article in the *Ecologist* provides a model of policy failure, suggesting that often when policy makers are failing to achieve stated goals, 'they may be achieving other, unstated goals' (*Ecologist*, 1995: 218). Often, powerful groups have agendas that differ from those of policy makers, and are able to shape policy outcomes. Where the implementation of policy may upset the latters' agendas, environmental policy makers may be powerless to cajole implementing agencies into executing policy. Or, by attempting to do so, decision makers might court failure, exposing their powerlessness. To the extent that decisions can be seen to spring from the existing way in which problems are defined, therefore, other problems are prevented from being considered. By limiting the range of issues taken into account while making decisions, a policy discourse enables policy makers to avoid considering sensitive political difficulties in the design of policy (*Ecologist*, 1995).

---

18. Original: 'Pernyataan 'akan ditindak tegas' memang merupakan bahasa pejabat yang telah lazim dan klise. Cuma sayangnya, jarang berbukti jadi kenyataan, kecuali tetap akan . . .'

In the present international climate, most states 'now employ the language of sustainable development, environmental protection and biodiversity conservation' (Dauvergne, 1998: 14). Public sector directors within the Ministry of State for Environment, the National Development Planning Board and the Ministry of Finance have been troubled by the declining state of Indonesia's forests, the lack of economic rents extracted by the state from the timber industry and the social inequalities, conflicts and environmental destruction resulting from the system of forest management (Barber et al., 1994: 70–3).[19] Yet, powerful networks of patronage close to Suharto had vested interests in property arrangements that gave them privileged access to the state 'forest area'. These interests ensured that, while the official forestry policy discourse seemed to deal with a range of problems, it actually failed to face sensitive political issues. Even for public sector directors with a genuine interest in reform, it was not possible to question the political underpinning of a regulatory framework that allocated property rights over vast areas of forest to vested interests with ties at the apex of the state. This reality was reflected in the bureaucratic culture of the policy élite. Here designing plans is a high status activity, and policy documents contain impressive principles. However, progress is 'impressive until the moment when implementation is required' (Emerson quoted in Dauvergne, 1997b: 221 fn 65). Even when policies are implemented, they have to work their way down from 'insulated policy makers in Jakarta' through the chains of power and interest to the forests of distant provinces. This ensures that 'the actual effect of policies often bear little resemblance to the original content outlined in Jakarta' (Dauvergne, 1997a: 69).

In essence, then, the official forestry discourse of the New Order period tended to obscure a range of issues. These issues become explicit from reading the history of forest management in Indonesia — a history reconstructed in terms of the changing property regimes operating in Indonesia's forests.

THE CHANGING REGIME

In pre-colonial times, forest farmers and peasants across the Archipelago used forests to meet basic needs. Over time communities developed complex ranges of rules that were woven around the socio-cultural features as well as the economic activities of local communities (Arora, 1994). Villages maintained notions of territoriality over neighbouring forests, a type of *lebenraum* where, under specified conditions, members of a community had the right (later known as *hak ulayat* or the 'right of disposal') to open plots of land, to harvest forest products, including timber for housing, firewood,

---

19. Klinken would also include Jamaludin Suryohadikusumo, the Minister of Forestry from 1993 to 1998, in this reformist group (Klinken, 1998).

food, fibres, medicines and other forest products (Parlindungan, 1997: 218). The self-governing authority systems of local communities enforced compliance to various types of communal land management systems. Land cleared for swidden was often held in common by residential or kin-based groups, and 'cultivators were given temporary use rights extending through a rotation cycle or cycles' (Poffenberger, 1990: 9). However, in the more intensely farmed areas, such as the lowlands of Java, agriculture 'required a great deal of labour to build such permanent structures as walls, terraces, and irrigation systems'. In the wet rice regions of Java, 'the concept of individual ownership and land sales became more common' (ibid.). In these areas, local kingdoms became more organized, and rulers concentrated on controlling the most valuable resources: people, and the products of agriculture and forests. Pre-colonial states were less successful in controlling land use, particularly in the vast forested regions far removed from the seat of government (Peluso, 1990: 29).

When the Dutch first came to Java, local resources were under the control of local nobility, and initially the Dutch used the existing system to extract teak from the forests. Later, as colonial power extended, they sought to secure a monopoly on teak, forest labour and shipbuilding, trying to assert control over access to and use of the forests. However, in 1808, in the face of widespread forest destruction on Java, the Dutch Governor General declared all forests to be the domain of the state and set up a government forest service. Henceforth, a branch of the colonial bureaucracy would manage land, trees and labour. The Governor General also passed the first law punishing all uses of the forest not authorized by the state, in effect criminalizing many customary uses of the forest (Peluso, 1990: 32).

Thus, the early nineteenth century marks the beginning of a shift in the structure of property relations in Indonesia's vast forests — the emergence of a state forestry regime. Henceforth, through law, the state would claim the right to be the ultimate arbiter of property relations. As Daniel Lev has noted, 'law (literally) records the structure of the state and reflects (virtually) the distribution of political, social and economic advantages' (Lev, 1985: 57). The colonial law, Lev concludes, 'which established the genetic pattern of the Indonesian State, was intended primarily to make exploitation efficient' (ibid.).

In this context, the Agrarian Acts of 1870 were a crucial point. Over the previous century, land had already been made available for rent or granted as private land for sugar, tea, rubber and coffee production. In 1870, the colonial authorities wanted to encourage further private investments in plantation agriculture. To facilitate this, the Agrarian Acts declared that all lands without certified ownership belonged to the state. Under this declaration, all land not under constant cultivation, including swidden land lying fallow and forest used for hunting and gathering, became the 'free' domain of the state (Kano, 1996). Thereafter, commercial interests could obtain seventy-five year leases for opening plantations (Suhendar, 1994: 7).

With the implementation of this policy in Java, the colonial government in effect 'nationalized' large areas of forest and then gave use rights over areas of this land to foreign commercial interests. In this way the Dutch colonial government generated revenue to support economic development in the Netherlands (Geertz, 1963). In the colony, as foreign interests gained tenurial rights, large areas of forest were cleared. As the population increased along with this commercial land use, farming communities lost their land reserves and increasing landlessness and poverty emerged in Java.

In claiming control over the forested areas, on paper the Dutch colonial system had overridden the tenurial systems of traditional communities in forest lands that were not under constant cultivation. However, outside Java, this policy had limited effect. Given the size of the area concerned, the Forestry Service had difficulty controlling access and use of forests. Moreover, in mapping out state forest reserves, colonial foresters sometimes became involved in protracted conflicts with local communities. In 1928–9, for instance, local populations in West Sumatra resented the establishment of a state forest reserve that limited their land use possibilities, especially at a time when the colonial forest authorities were granting concession rights for wood exploitation to foreign enterprises. In response, some colonial foresters sought ways to accommodate local communities (Goor, 1982: 436–7), but in many cases, especially in the extensive forest areas outside Java, the traditional management systems were able to continue, albeit without any recognized legal status (Goor, 1982; Poffenberger, 1990). As in other parts of the world, the colonial regime set up a regulatory order that overlaid a pre-existing customary regime with its own concepts of property rights (Bruce et al., 1993; Spiertz and Wiber, 1998). The scene was set for conflict in the post-colonial period between élites using national law to justify access to local resources, and local people seeking to preserve their own tenure systems.

With the final victory of the Indonesian Republic in 1950, land reform became a key priority for the nationalist leaders, particularly the need to overcome the inequities of the colonial pattern of land use, to narrow the gap between the colonial legal heritage and customary law, as well as to recognize the land rights of those working the land (Tjondronegoro, 1991). Finally, after extended discussion, legislators drafted the new Basic Agrarian Law (UUPA 5/1960) in 1960. This law attempted to distribute land more justly. To ensure greater access to land for small farmers and for the landless, the law set down the principles of maximum landholdings and called for the registering of land for redistribution. While it recognized the existence of customary (*adat*) rights, it specified that state sovereignty should be superior to *adat*: *adat* tenure systems would not be allowed to interfere with national or regional economic development (SKEPHI and Kiddell-Monroe, 1993). The Agrarian Law became associated with PKI 'unilateral actions' — conflicts between landlords and village officials on one side and farmers, the Communist Party (PKI) and the Peasants' Front on the other. However, the killings of 1965–6 that accompanied the downfall of Sukarno saw the destruction of the PKI and the Peasants' Front. Land reform and the agrarian law were now seen as a product of the PKI, and the New Order regime that emerged turned its back on all questions pertaining to land. While the New Order never rescinded this law, the agrarian reform was never fully implemented (Tjondronegoro, 1991).

FOREST PROPERTY RIGHTS AND THE NEW ORDER STATE

In defining legal and illegal activities and in deciding which property claims it will support in the field, a state can allocate and enforce property rights, or let them lapse. As Bromley has pointed out, 'environmental policy is nothing, if not a dispute over the putative rights structure that provides protection to mutually exclusive uses of certain environmental resources' (Bromley, 1991: 3). A key policy question for the New Order regime was thus: whose claims and interests would the state act to protect?

The New Order that emerged under Suharto's tutelage has been characterized as a comparatively strong state (Dauvergne, 1997c). As MacIntyre has written, under Suharto, the determination of policy remained 'largely unfettered by societal interests' (MacIntyre, 1990: 17). This meant that the New Order was able to allocate *de jure* property rights over Indonesia's vast forests largely according to its own priorities.[20] At first, in 1967–70, the New Order faced an economic crisis. As the state treasury could gain large economic rents from logging, forestry policy shifted towards bringing in urgently needed revenue (Tjondronegoro, 1991). As had occurred under the colonial regime, this process was facilitated by new laws: the Basic Forestry Provisions (Act No. 5/1967) extended the *de jure* sovereignty the state claimed over the 'forest area' under the Indonesian constitution. This new law stated that 'all forests within the territory of the Republic of Indonesia, including the natural resources they contain, are taken charge of by the state' (Act No. 5, 1967). Despite the property claims of isolated communities over surrounding territory, in the name of revenue generation, the state allocated twenty-year timber harvesting rights over large areas of state forests to urban élites with strong political connections.[21]

As local partners joined large multinational corporations to facilitate timber operations, logging became an opportunity for financing the political

20. In some respects, as Dauvergne has argued, the New Order state was 'strong': it was able to limit foreign control of the timber industry and guarantee access to commercial interests. However, from another perspective, due to the networks of power and patronage that involved state officials in timber extraction, in other ways the state was 'weak' — unable to enforce environmental policy for timber management (Dauvergne, 1997b, 1997c).
21. This process was reminiscent of the granting of leases over what was nominally state land for plantation agriculture a century earlier (Mubyarto, 1992).

and military élites. To this end, Suharto handed out logging concessions 'to benefit loyal military officers, appease potential opponents and bolster the military budget' (Dauvergne, 1997c: 7). Many concessions were linked to military organizations, and the central and regional military commands controlled over a dozen timber companies. Meanwhile, political and bureaucratic power holders within the central, provincial and local bureaucracy and military apparatus also formed partnerships with private business interests creating networks of social power and interest, supporting illegal logging and other lucrative activities at odds with state forest policy. In the course of the 1980s the government banned the export of logs, and the multinationals withdrew. Predominantly Chinese owned conglomerates as well as business groups owned by powerful political families, some linked to the Suharto family, consolidated their interests in the heavily protected forestry sector.

In 1998, after the resignation of Suharto, the Ministry of Forestry and Estate Crops revealed that 422 concessionaires (HPH) were still active, controlling some 51.5 million ha of production forest. However, with cross-ownership, hidden deals and silent partnerships, ownership of forest concessions was anything but transparent.[22] In late 1998 the Ministry of Forestry and Estate Crops released a report showing that just twelve companies controlled 'virtually all' of Indonesia's forest concessions, with three groups controlling more than 2 million ha each (*Republika*, 28/12/98; Timber and Wood Products, 28/11/98).

The best known of the timber tycoons were Prajogo Pangestu and Bob Hasan. Prajogo Pangestu's Barito Pacific Group's holdings stretched over 5.5 million ha of forest, an area larger than the province of West Java: Prajogo controlled 'more of the world's rainforest than any other individual' (Barber et al., 1994: 72). He was said to employ 55,000 people in timber enterprises that included the world's largest tropical plywood export business with annual sales of about US$600 million (Dauvergne, 1997c). With close ties to Suharto and interests in several Suharto family ventures, Prajogo enjoyed extensive presidential patronage. In the 1990s, for instance, many timber interests discovered lucrative opportunities in moving into plywood manufacture and export. When Prajogo wanted to move into the plywood industry, Suharto wrote a memo to the Minister of Forestry to facilitate his plans, making Barito Pacific the number one company in timber plantations overnight (Barber et al., 1994: 72; Schwarz and Fredland,

---

22. While a company might be formally registered as controlling 1 million ha, by buying up smaller concessions, it might control a much more extensive area (Mubyarto, 1992: 80). Following the resignation of Suharto, public pressure mounted to make a clean slate of the intricate networks of patronage that had flourished under the former president. In December 1998, the Minister of Forestry and Crop Production, Muslimin Nasution, announced that the Ministry had discovered that Suharto's family (*keluarga Cendana*) had shares in concessions operating over 4.22 million ha of state forests (*Republika*, 28/12/98).

12/3/92). A second group under Mohammad 'Bob' Hasan also held logging rights over more than 2 million ha of forest. As a close friend of Suharto, Hasan gained attention as Suharto's last Trade and Industry Minister. However, earlier, as 'the strongest player in setting Indonesia's forest policies', he was said to be the unofficial Minister of Forestry. As head of the Indonesian Wood Panel Association (Apkindo), Hasan told mill-owners how much plywood they should produce, where they would export and at what price (Barber et al., 1994: 71; Schwarz and Fredland, 12/3/92).

In the course of their rise to commercial prominence, the politico-business families and timber tycoons used their privileged position to gain access to state facilities and economic opportunities, and to push aside public sector directors charged with managing the national resource sector (Robison and Rosser, 1998). Suharto's use of logging royalties earmarked for reforestation demonstrated this. In accordance with several presidential decrees (*Keppres*), this 'Reforestation Fund' was reallocated to projects now held to be connected with 'corruption, cronyism and nepotism', unrelated to reafforestation and even damaging the environment. These included clearing peat swamp forest for a million ha agricultural project in East Kalimantan, cheap loans to commercial timber plantations implicated in the forest fires, and a 250 billion rupiah transfer to Kiani Kertas, a pulp company controlled by Hasan (Dauvergne, 1997c; Klinken, 1998; *Republika*, 30/12/98)

As ownership and control over forest use and management rest in the hands of the state and are administered by the state forestry agency, on one level this constitutes a *de jure* state property system. Individuals and groups may use natural resources, but only with the approval of the Ministry of Forestry, 'the administrative agency responsible for carrying out the wishes of the larger political community' (Bromley, 1997: 6). However, during the New Order, private interests were able to capture resource management nominally under the control of the state forestry apparatus. By co-opting or marginalizing state managers, the press, forestry experts, NGOs and local communities, these private interests ensured that forest management slipped beyond public control. Due to the intervention of ruling power interests and large-scale industries, these forest management institutions disregarded the *de jure* rules. Ties at the pinnacle of the state distorted policies and weakened 'the supervision of middle- and lower-level state implementors' leading to a 'culture of corruption, cronyism and nepotism' at the district level (Simon, 1998).

At the same time, property rights allocated to commercial interests overlapped with pre-existing layers of local property claims. The state allocated 'forest area' under the *de jure* disposal of the Forestry Department based on a mapping exercise that failed to take into account *de facto* local property regimes — the informal concepts of territoriality and land tenure in use among the local communities surrounding the forest.[23] However, despite

23. Regarding the forest land use agreement (TGHK) mapping, see footnote 4.

the formal property rights that the state agencies have allocated, commercial interests wishing to control forest areas needed to come to terms with local communities and their *de facto* property systems. Where the Ministry of Forestry has allocated land considered by local communities to be subject to 'right of avail', there has often been conflict with local communities. This was demonstrated by a Forestry Department report which states that one of the main obstacles for allocation of rights to use the 'forest area' by the Forestry Department to commercial interests, is that *Hak Ulayat* are still operative in such areas even though the Ministry of Forestry had not suspected that they existed.

> Not infrequently the land allocation [for plantations] has to be moved because no agreement can be reached with the local *adat* community. The result is that many requests [for plantations lands] ask for dense forest to be made available for conversion to agricultural lands. (Departemen Kehutanan, 1992: 22)

In order to illustrate the impact on indigenous resource and forest management, it is useful to consider a 1992 study carried out by a team of Indonesian researchers from Gadjah Mada University (Mubyarto, 1992). Peluso has noted that changes in patterns of forest use management and control are complex and require analysis from various angles, 'starting from the activities of the forest users themselves and extending to national and international political economy' (Peluso, 1992: 210). In many cases depletion of forest resources is associated with changing resource use patterns linked with the arrival of timber companies together with the advent of new ways and means of exploiting the forest (Colfer, 1987; Peluso, 1992). As in this case, there is a clear link between the granting of property rights by the state to logging companies and agricultural plantations, and land shortages, resource scarcity, the decline of local livelihood systems and indigenous management structures.

IMPACT ON INDIGENOUS RESOURCE MANAGEMENT: A JAMBI CASE

For centuries, subsistence farmers on many of the outer islands of Indonesia have utilized a system of agriculture known as 'swidden' or 'shifting agriculture' (Colfer, 1993; Dove, 1986; Hayes, 1997). As elsewhere in Southeast Asia, in Jambi province, Sumatra, this forest farming system was 'a practical and successful way of utilising land where poor soils, steep gradients, and heavy rainfall make conventional farming methods unproductive or impossible' (Myers, 1980: 1). As described by a team of researchers from Gadjah Mada University, this form of agriculture is primarily orientated to producing food crops, including rice (Mubyarto, 1992). Customary rules governing resource use determine that farmers primarily open swidden in secondary forest, and that to prevent erosion, the swidden (*ladang*) area should not be too steeply sloped. As the fertility of the *ladang* declines after use, in accordance with *adat* practices, farmers leave swidden areas fallow for seven or eight years. In this way, the *adat* system aims to maximize livelihoods by ensuring fertility and protecting the ecological processes upon which agriculture depends. Consequently, to operate effectively on traditional lines, swidden agriculture requires extensive areas of forest.[24]

Community-based tenure systems operating in Jambi involve a complex bundle of group and individual property rights under the control of a customary authority system. Forest, land and dispersed resources such as minor forest products are ultimately the property of an interdependent community of users. However, individual use rights over resources that are more concentrated tend to be obtained in various ways. Villagers who want to work an unused swidden area previously worked by another forest farmer may do so with the consent of those who previously worked the area or through the arbitration of the *adat* authorities.[25] As in many other areas of the world, for all practical purposes, control of trees confers control of the land on which they stand. Consequently farmers gain *de facto* control and long-term tenurial rights over land by planting trees (Bruce and Fortmann, 1991: 473). To mark their continued property rights over a piece of land, before abandoning a swidden, forest farmers often plant fruit trees. After harvesting a swidden, farmers also may plant the area with rubber trees, thereby turning the *ladang* into a permanent forest garden (*kebun*). While property rights over rubber *kebun* are abiding, property rights in *ladang* areas are not enduring. However, besides marking permanent property rights, rubber has the additional advantages of growing on less fertile land and being productive without intensive maintenance. Moreover, as Dove (1996) has noted, the combination of market-orientated rubber production with extensive, subsistence-orientated agriculture offers several advantages: rubber cultivation generates cash for essential needs while enabling forest farmers to avoid occasional failures of swidden; at the same time, swidden cultivation helps farmers avoid over-dependence on rubber cultivation, a product subject to price fluctuations.

However, the researchers found variability in land tenure systems within Jambi. In Sarko district, for example, there is a category of village land known as *Pungko* land, an area set aside for food crops and made available to needy villagers each year according to customary law and with the

24. According to one estimate from the 1940s, swidden-based agriculture could support populations of up to about 50 per square kilometre. This means that increasing population density is a critical factor, as fallow periods grow shorter and forest resources become over used. At a certain point the swidden system then becomes non-sustainable (Hayes, 1997).
25. It is worth noting that the *adat* system described here is by no means a static one. Historically farmers in Jambi also cultivated wet rice fields (*sawah*); villagers only began to cultivate rubber around 1910. As rubber produced greater profits compared to sawah, forest farmers began to turn areas previously used for *sawah* as well as newly opened *ladang* areas into rubber plots (Mubyarto, 1992 #194: 40–1).

consent of *adat* leaders. In this district, villages also set aside *Hutan Rimbo Sejati*, areas of primary forest protected according to *adat* connected with traditional belief systems. In other areas, villages simply divide their lands into two areas — permanent areas for rubber cultivation and areas reserved for swidden agriculture. By ensuring that there is always sufficient forest reserve available for opening swiddens, this land tenure system ensures that the village produces sufficient food. This system depends either on customary laws enforced with sanctions, including fines; or on the social cohesiveness which flows from the reality that many tasks, such as opening a *ladang*, depend on collective labour.

In the 1970s, logging companies and agricultural plantations began to flood the area. By 1990 there were thirty HPH operating in Jambi, controlling 2,662,000 ha of forest (Mubyarto, 1992: 39). According to the Indonesian Selective Cutting System or TPTI (*Terbang Pilih Tanam Indonesia*) regulations, which are meant to ensure forest regeneration, logging companies could extract specified amounts of timber from their logging concessions each year.[26] However, as the *Forum* journalists had reported for West Kalimantan, the Gadjah Mada University researchers found that Forestry Department policy to control logging failed in Jambi — once again, apparently, due to technical and managerial difficulties and law enforcement problems. First, the Forestry Department lacked the resources and personnel to monitor logging, a problem exacerbated by the huge extent of the forest concession controlled by each concessionaire and the lack of involvement of local government in the process (Mubyarto, 1992). Meanwhile forest concessionaires sub-contracted forest areas to smaller enterprises, and these sub-contractors harvested as many trees as possible, in violation of government selective logging regulations. Second, as a 1998 article in *Suara Pembaruan* notes, widespread collusion between forest personnel and the concessionaires ensured that criminal sanctions for infringements of regulations regarding forest exploitation were not dealt with according to the law. For instance, although Jambi forest police confiscated almost 2 million cubic metres of illegal wood connected with HPH operations in 1996–97, not one case was brought to court. The reason given was lack of evidence, despite the mountain of confiscated wood. Even in those cases when logging companies were fined, the fines were smaller than the logging taxes which companies would have been obliged to pay had they obtained the timber legally (*Suara Pembaruan*, 27/11/98b).

As in other studies (Fox, 1993; Peluso, 1992), the researchers also discovered a link between the allocation of property rights to concessionaires

---

26. TPTI provides that only trees with diameters above 50 cm may be taken. Loggers must also leave a certain number of 'mother trees' in each ha to ensure the regeneration of the forest within thirty-five years. Due to the inefficiencies of this system, in the late 1990s GOI began developing the Production Forest Management Unit (KPHP) concept to replace TPTI (KLH and UNDP, 1997: 368).

and the decline of indigenous land management practices. Concessionaires and plantations secured all possible titles and certificates from relevant government agencies. Within the local regime, land was inalienable from the community; community members obtained rights by investing labour and time into the land. Within this system, trees marked tenurial rights, and local farmers had no need for formal evidence of their property rights. Yet concessions enclosed land previously available for opening swiddens, and land shortages ensued, a problem exacerbated by population growth and in-migration.[27] Under these conditions local people found it difficult to locate land that met *adat* criteria for opening *ladang*. In areas close to logging concessions, rather than the stipulated seven to eight years, forest farmers had to re-work *ladang* areas after only two or three, or sometimes even one-year fallow periods. This led to increased erosion, falling fertility, smaller harvests and increased poverty. As local people lost property rights over nearby areas, taking wood now became 'theft', while those opening land in forest under *de jure* state control were now classified as 'forest squatters' (*perambah hutan*). In this context, according to official figures, cases of illegal clearing, uncontrolled logging and obtaining wood without valid documents increased by a third over the period 1987–9 (Mubyarto, 1992: 40–1). While at times logging concessionaires attempted to control access to their areas, officially-classified protection forests under primary forest cover were poorly protected by forestry officials: local people opened land in these areas despite the state's statutory claims over land and trees. Between 1982 and 1990, for example, the area of protection forest in Jambi fell from 1,147,500 ha to 181,000 ha (Mubyarto, 1992: 89).

Other research from Sumatra and Kalimantan has indicated that, besides the forest clearing carried out by plantation, logging or transmigration, government-sponsored land claims by external users in some instances led to increased clearing of forest by local communities. By failing to recognize local tenurial rights, such as those existing over swidden areas, large land claims have created uncertainty of tenure. Expectations of future land shortages have encouraged local farmers to race to clear new areas of forest. As the state regards fallow land as unused, and unused land reverts to the state, farmers have opened up swidden areas and then planted tree crops such as rubber to establish more secure rights over the land (Angelsen, 1995; Colfer, 1997: 79).

As in other areas, state intervention to increase central authority also contributed to the decline of *adat* systems of authority controlling resource use (Nababan, 1996). Before 1979, *adat* regulation and village government

---

27. Population growth caused by in-migration is clearly a serious threat to forests: Colfer found that in-migration was the principal source of East Kalimantan's population increase, which in turn was one of the principal threats to its forests (Colfer, 1993: 79). For a discussion of the effect of migration on forest management in Kalimantan, see also Potter (1996).

were in the same hands: assisted by a community council (*Dewan Marga*), the *Pasirah* leader was both the *adat* head and the primary government official in the village. Village resources belonged to the village based kinship group (*marga*), and the *Pasirah* had the authority to manage the land and the forest, allowing individuals to open areas of forest and ensuring they protect key ecological functions. However, following the implementation of the Village Government Law (Act No. 5/1979), the institutional linkages between *adat* and village government were cut. The *Pasirah* lost authority over the forests and was reduced to carrying out traditional ceremonies. At the same time, the village head became the official primarily responsible for implementing policy 'from above', at times without the opportunity to consider whether new regulations or instructions were appropriate or in conflict with local *adat*. At this time villages were also rearranged into administrative units containing a standard 100 families. While this innovation was intended to facilitate administration, in some places it meant joining geographically-isolated hamlets into one village. As a consequence, some hamlets were left without effective connections with the village apparatus.

The researchers of the Jambi case study noted that HPH activity opened new roads to isolated communities and gave villages new access to markets; the HPHs also increased government taxation revenue and contributed to development budgets, generating new work opportunities in the timber industry. At the same time, these changes contributed to the transformation of local ways of life. Generally, with greater access to the outside, forest farmers become more readily able to provide cash crops and forest products to the world market. To a larger extent, villagers have become consumers of imported goods, including motorbikes, fuel, radios, televisions, mechanical pumps, pesticides and fertilizers. Where these changes have been associated with a greater emphasis on producing an agricultural surplus for sale, integration into the wider cultural and market relations also affected the tenacity of older traditions. In many cases the old culture that sustained a local ecological order has begun to weaken. With the arrival of modern technology, the rapid conversion of natural resources into cash has become possible (McNeely and Wachtel, 1991).

The Gadjah Mada researchers described the ecological impact of these changes and their effect on local communities. Uncontrolled logging caused erosion, the silting up of rivers and loss of hydrological functions. This in turn led to falling water levels with droughts in the dry season, floods in the wet season, decreasing fishing stocks and the devastation of wet rice agriculture. Local communities lost income previously gained from the collection of non-timber forest products including rattan and bamboo. Meanwhile, the decline in wildlife populations, including natural predators such as tigers, led to crop infestations by wild pig. These changes cumulatively made local communities more susceptible to food scarcity.

While HPH activities have such a significant impact on forests and local livelihoods, the government persistently blames shifting cultivators for forest destruction (Dove, 1996; Potter, 1996). The government has recently stated that indigenous forest dwellers form part of an estimated 1.7 million 'forest squatters' whom it holds responsible for the systematic destruction of forests (*Asian Economic News*, 1998). However, in line with other researchers, the Gadjah Mada researchers argue that the activities of traditional shifting agriculturalists should be distinguished from those of 'forest pioneers' or 'forest squatters'. In Jambi, those practising shifting agriculture open small plots of secondary forest according to *adat* principles; they plant primarily subsistence crops to support their families, returning to the same area after a fallow period. It is therefore difficult to argue, the report maintains, that their activities are the main cause of forest destruction. On the other hand, pioneer agriculturalists, usually truck farmers and transmigrants, move up logging roads into areas of forest to open new plots and to plant cash crops. Exploiting the forest for short-term commercial purposes, they use the land continuously, without fallow periods, so that it is soon exhausted. They then abandon the land and move into new areas of forest (Dove, 1996; Mubyarto, 1992: xxiv; Potter, 1996; Sunderlin and Resosudarmo, 1997). However, drawing on other research, Sunderlin and Resosudarmo warn against creating a simple polarity between shifting cultivators and forest pioneers. As 'traditional' farmers modernize, there is actually 'a continuum of farming systems running from traditional "shifting cultivation" (involving long fallows and long-term forest conservation) at one extreme, and "forest pioneer" cultivation (often involving long term degradation and deforestation) at the opposite extreme' (Sunderlin and Resosudarmo, 1997). A number of 'different and rapidly evolving agricultural systems' are being grouped together under the label of shifting agriculture (Potter, 1996: 27). Due to this confusion over definitions, there are widely diverging views on the relative responsibility of shifting cultivators, forest pioneers, smallholders and transmigrants for deforestation.

As this case study attests, forest policy entails a transfer in the flow of benefits arising from the forests away from local communities. The state allocates areas of forest under *de jure* state sovereignty to private interests, usually urban élites with close ties to key political figures. Where these claims overlap with customary land rights, local communities become 'squatters' in local forest territory. Afraid of future land shortages, farmers try to gain *de facto* control of available land by planting trees, including in poorly-protected protection forests and national parks. Meanwhile, underpaid local officials join in networks of power and interest that benefit from uncontrolled logging. As a consequence the forest authorities fail to implement the strictures of the state forest regime.

In many places, then, we find the co-existence of overlapping property regimes with local communities, commercial interests and other actors engaging in struggles over property rights. Such conflicts are best understood 'as conflicts over resources among different actors (the state, national, and local élites, and peasant), who appeal to different legal systems to substantiate their claims' (Bruce et al., 1993: 628). At times, especially recently, this has lead to open conflict (Anon., 1998; Barber, 1998; Dove, 1997; Human Rights Watch Asia, 1997). However, in situations where the balance of power is stacked against them, local communities prefer to avoid open conflict. This leads to what Scott calls 'everyday forms of resistance' — unheralded sabotage, clandestine theft and passive resistance — the disorganized, unplanned guerilla-style activities through which peasants make their resistance felt (Scott, 1985).

If a playwright were to fit this story into the dramatic form of a Greek tragedy, surely the denouement would be the forest fires of 1997–8. The fires were directly linked to environmental policies that allocated large concessions and conversion areas to conglomerates and politico-business families, as well as to overlapping and conflicting property claims.

DENOUEMENT: THE FOREST FIRES

Between May 1997 and May 1998, forest fires raged across large areas of Kalimantan and Sumatra and in other pockets of the archipelago. Producing a sickening pall of smog that hung over Southeast Asia for several months, the conflagration poisoned the air of an estimated 70 million people. Besides causing widespread health problems, the fires shut airports, disturbed navigation and led to large losses to the tourist industry (WWF, 1998). Extensive areas of agricultural and forest land were burnt: according to one appraisal, fire consumed 5 million ha in 1997 and a further 3 million ha in 1998, causing estimated total damages of more than US$4.4 bn[28] (Reuters, 25/11/98; WWF, 29/5/98).

Once the smoke had cleared, KLH and UNDP produced a report that demonstrates how forestry practices and land clearing had created an environment vulnerable to fire (KLH and UNDP, 1998b). While a natural rainforest is a humid environment that leaves very little dry litter on the forest floor, logging leads to the accumulation of dead biomass on the forest floor; by opening the forest canopy and leaving the forest exposed to intense tropical light, logging also dries out the forest. If conditions are dry, such as a severe drought caused by an El Niño event, logged or degraded forests burn easily. Areas converted into timber estates and plantations also represent increased fire risks. Many timber estates replace natural forests with acacia monoculture, a species that drops excessive amounts of litter, while the use of herbicides by plantations results in large areas of dead grass and weeds; both of these create a readily inflammable environment during prolonged dry periods.

While there is some debate about the rate and extent of deforestation, the KLH/UNDP report tried to estimate the scale of environmental change that

---

28. See footnote 1 for estimates of the extent of land burnt.

has occurred over recent years.[29] Using data from the national forest inventory, the report estimated that, of the 109.5 million ha of 'permanent forest' (*kawasan hutan tetap*), by 1996 some 19.4 million ha were no longer covered in forest or had experienced changes in the structure of vegetation. Outside these 'forest areas', the report estimated that another 30.5 million ha of forest classified for other land uses had also been degraded (KLH and UNDP, 1998b: 87). Logically enough, these increased rates of deforestation and land conversion correlate with increased frequency of forest fires: there have been three major fires in less than eight years: 1991, 1994 and 1997–8. As the largest deforested areas lie in Kalimantan and Sumatra, the most extensive forest fires occurred there.

The report found that, besides fires started by accident or by arson, most fires were deliberately set to clear forest areas. Even in areas not particularly affected by the El Niño drought, this burning-off led to fires spreading out of control over extensive areas. Three agents are commonly held responsible for starting fires: commercial interests opening forest for industrial timber estates; plantation companies; and smallholder farmers.

The government first promoted the timber estates (*hutan tanaman industri* or HTI) under the former Minister of Forestry, Hasjrul Harahap (1988–93). The idea was to increase forest cover by replanting degraded forest areas — with a monoculture of quick growing exotic species — and to provide alternative timber sources to the native forests. The development of HTI fitted Indonesia's wish to become the world's largest pulp and paper producer (Gellert, 1998: 76). The government provided incentives to make HTI very attractive to investors. Using money earmarked for reafforestation (the 'Reforestation Fund'), the government provided a range of generous credit schemes, subsidies, capital grants and even low interest or interest free loans; these schemes provided companies with up to 65 per cent of the capital required to open a timber plantation (*Suara Pembaruan*, 12/1/98). While companies were meant to establish HTI on 'empty land' (*tanah kosong*) or degraded production forests, many of these areas granted to HTI were not 'empty'. They included fallow areas and other local territories under local property regimes (KLH and UNDP, 1998b: 78). Moreover, companies that obtained forested areas for developing HTI acquired timber harvest permits (IPK) that enabled them to harvest productive stands of timber. Even before companies accepted the generous credit offered by the government to develop HTI, they could gain huge profits from cutting timber. They therefore preferred to open HTI in areas still containing valuable tree stands, including areas covered by primary forest, moderately logged forests or even community forested lands (KLH and UNDP, 1997: 378). In theory, after logging, companies would clear the remaining degraded forest and

29. FAO has estimated that the rate of deforestation in the 1970s was 300,000 ha per year, while by the 1990s deforestation rates were estimated to be between 700,000 and 1.2 million ha per year (KLH and UNDP, 1998b: 86).

replant the area as a timber plantation, but many were happy just to reap windfall profits from harvesting the wood. For those wishing to proceed to the timber plantation stage, the cost of land clearing without fire is four times that of burning off the remaining vegetation; burning was therefore the method of choice (KLH and UNDP, 1998b: 76–9).[30]

In addition to opening timber estates, commercial interests in the 1990s moved into the booming palm oil sector. The area under palm oil plantations more than quadrupled between 1985 and 1995, and the government has plans to turn Indonesia into the world's largest producer of palm oil (Gellert, 1998: 77). For this to occur, Indonesia needs to increase production of palm oil from 4.1 million tons per year in 1994, to 7 million tons by the year 2000. The KLH/UNDP report noted that there was considerable evidence that, to meet ambitious planting targets, companies intentionally used fire in a planned fashion to clear and prepare land for planting (KLH and UNDP, 1998b: 76–9).

Local forest farmers traditionally use fire to open up to a hectare of land for growing food crops. Indigenous forest management practice is to burn off at the end of the dry season, in the weeks before the first rains, with farmers carefully establishing fire breaks to ensure that the fire does not spread. According to one report from Kutai sub-district, East Kalimantan:

> During the whole fire period no fires originated from swiddens, not a single farmer had dared to burn before the first rain came in early November. Traditionally, clean corridors are made around swiddens before burning in dry years. But 1997 was much too dry to prevent a fire from escaping into the forest. Since the origin of such a fire would be obvious and *adat* fines for burning forest gardens are high, farmers waited for rain to come. (Gonner, 1998: 3)

However, the KLH/UNDP report argues that in some areas social change, in-migration, loss of property rights over surrounding areas, decreasing social cohesiveness, the effect of government policies towards indigenous forest dwellers (*masyarakat terasing*) and village government had undermined *adat* authority systems. In many places forest pioneers, including unemployed migrants from other areas of Indonesia — sponsored and spontaneous transmigrants — had taken up farming. These migrants were either unfamiliar with traditional management practices or not subject to *adat* authority structures. This meant that some farmers used fires without taking precautions, leading to uncontrolled fires spreading to surrounding properties (KLH and UNDP, 1998b: 72–5; Schindler, 1998).

Consequently, on the one hand, there is a strong link between the fires and the policy of allocating concession rights to private interests in areas set

---

30. Gellert noted a further incentive for going through the motions of developing HTI: 'it has been argued that the owners of plantations can use the capital [borrowed for developing HTI] for their own purposes if they make only minimal efforts in those subsidised plantations' (Gellert, 1998: 76). In the absence of a real interest in developing an HTI, a company might burn the land to satisfy government monitors that an initial planting effort had been made.

aside by government agencies for conversion to plantations and timber estates. Driven by short-term economic incentives, and in the absence of effective control from the state management regime, these commercial interests then purposely set fire to areas of forest under hazardous conditions. On the other hand, the report shows how in areas subject to smallholder agriculture, in-migration leading to increasing local population in outer areas of Indonesia, together with a range of social and cultural changes, also meant that some forest farmers were now opening areas of land without taking precautions.[31] As we have seen, in many areas, these two phenomena overlap, and there are competing authority structures and opposing property claims. In the dry situation of the El Niño event this also led to fires.

Several reports reveal that property disputes were associated with widespread arson. While these property disputes had a wide variety of causes, they were ultimately connected to a land-use practice that created a conflict between co-existing customary tenure regimes and property rights based in national law. This occurred in a context of legal uncertainty and in the absence of clear dispute resolution mechanisms. In many cases, commercial interests had obtained *de jure* property rights over land previously under customary tenure regimes: the new owner may have obtained rights through state agencies without the consent of traditional owners or without paying adequate compensation. Reports from Kalimantan mention that some community members who had never accepted this transfer of ownership retaliated by setting fire to plantations. In other instances companies or their representatives with a legal claim over local land were known to burn that land to facilitate access. By reducing productive gardens to ashes, the agents of companies decreased the economic value of the land, minimizing the amount of compensation they would have to pay to the traditional owners, and allowing those responsible for negotiations to pocket the difference (Gonner, 1998; KLH and UNDP, 1998b: 72). In other cases the arrival of plantation companies offering compensation led to disputes within communities over land titles. According to an account from Kutai sub-district, East Kalimantan (Gonner, 1998), an oil palm company's 'way of practice' led to conflict within the community. The company had ignored traditional land rights and existing resource management systems. 'Most villages were divided between those who wanted to join the oil palm project and those who rejected it'. Disputes broke out even within families, and large sums of compensation led to jealousy, causing 'revenge and envy fires'. 'Too much changed too fast, too little remained under the control of villages'; in this

---

31. Summarizing earlier research, Sunderlin and Resosudarmo observe that data on population densities by province show a strong inverse relationship with forest cover. However, while population growth is clearly linked to deforestation, due to a range of other variables involved, it is not possible to construct a simple causal link between population growth and deforestation (Sunderlin and Resosudarmo, 1997).

situation, traditional dispute resolution methods failed, and disputes were 'solved' by fire (ibid.).

In this context the policy failures that have long dogged the official forestry discourse surfaced: efforts to deal with the fires faced large 'bureaucratic obstacles' (Schindler, 1998: 1), including the highly centralized command structure. The long chain of the bureaucracy meant that, while agencies at the operations level were following 'complicated and tedious' reporting procedures, requesting equipment and arranging logistics, the fires were spreading (KLH and UNDP, 1998a: 47). Moreover, according to the report of a German fire prevention project (IFFM/GTZ),[32] law enforcement was 'either weak or almost non-existent' (Schindler, 1998: 1–2). Although there were sanctions stipulating that uncontrolled burning was a criminal act, before the fires government regulations did not strictly prohibit burning. A 1997 forestry decree required that those wishing to burn off areas of land needed to have a licence from authorized officials and fulfil certain conditions before starting controlled burning (KLH and UNDP, 1998b: 84, 148): however, logging companies and plantations usually lacked an approach to fire management and were unprepared for fire emergencies. When the fires intensified, in October 1997, controlled burning was also prohibited. Yet regulations were not enforced, according to one account because 'too often public servants have to top-up their miserable salaries by turning a blind eye on unlawful activities in the forests' (Schindler, 1998: 1–2).

Meanwhile, a highly politicized debate over the causes of the fires broke out at the national level. NGOs blamed timber estates and plantation companies for deliberately starting fires to open new areas and called for legal sanctions to be brought against them (*SINAR*, 4/10/97). Reformist ministers also made statements blaming the irresponsible behaviour of the logging and plantations firms. In reply, the spokesman for the timber interests, Bob Hasan, argued that the big plantation firms had no interest in burning down the forests. 'Why should we burn [the forests]', he said, 'We need the raw materials. It does not make sense'. Hasan claimed that the fires were a result of the irresponsible activities of 'people's plantations, shifting agriculture and wood thieves' (*SCMP*, 30/9/97). This view gained presidential support: after quietly telling his ministers to tone down their statements, ministerial comments became more conciliatory. The presidential line held that the fires were a natural disaster — a consequence of the severe drought caused by the El Niño effect (Klinken, 1998; Muryono, 10/1/98). Once again, although perhaps for the last time under Suharto, these clients were able to derail state regulation. The Ministry of Forestry investigated companies involved in forest conversion. In October 1998 the Ministry revoked the forest clearance licences of sixty companies suspected of caus-

32. Gesellschaft fur Technische Zusammenarbeit's (GTZ) Integrated Forest Fire Management Project (IFFM).

ing fires, but shortly afterwards the decision was overturned. By the time Suharto was forced from office (Forrester and May, 1998; Vatikiotis and Schwarz, 4/6/98), the Ministry had failed to bring a single company to trial for breaking regulations (KLH and UNDP, 1998a: 59).

REFORMASI

The 1997–8 forest fires brought the whole system of resource management and the patronage system in which it was embedded into disrepute. At a time of economic and political crisis, the balance of power that had determined access and use of forest resources at the national level shifted away from Suharto and the patronage network around him. In 1993 the World Bank had attempted to make a US$120 m loan conditional on reform to the forestry sector, but had failed because the changes threatened business interests close to Suharto (Ross, 1996). As Suharto fell from power, reformist public sector managers became more influential. In exchange for a US$43 bn rescue package, the International Monetary Fund was now able to obtain government agreement for a raft of changes similar to those previously advocated by the World Bank. These included creating new rent taxes on timber resources, changing the way concessions were allocated, dismantling the paper and plywood cartels, allowing the export of unprocessed logs, and incorporating the reforestation fund in the state budget (Sunderlin, 1998).

Following Suharto's departure, it was no longer possible to defend the earlier forestry discourse that had helped stave off reform. Critics condemned the state forestry regime on two grounds. First, the system had failed in terms of equity: private interests and networks of 'corruption, cronyism and nepotism' (*Korrupsi, Kollusi dan Nepotisme* or KKN) had enjoyed the benefits while the costs had been borne by local communities. Second, in terms of ecological outcomes, the cumulative effects of the previous three decades had led to acute ecological problems culminating in the forest fires. In due course NGOs, student groups and academics, now freed from the tight restrictions of the Suharto period, openly called for a 'new forestry paradigm' that provided for the people's welfare (Simon, 1998; *Sylva Indonesia*, 1998). These groups also demanded a range of other reforms, including repealing the Basic Forestry Act (1967/No.5) and its implementing legislation (PP No.21/1970) and replacing them with new laws that facilitated community-based forest management. Another demand was for the restructuring of the (renamed) Ministry of Forestry and Estate Crops (Dephutbun) and the elimination of KKN from its ranks. One meeting suggested that forest management be surrendered to regional governments, turning Dephutbun into little more than a co-ordinating body. There was also a call for the cancellation of forest concessions (HPH), agricultural and timber plantation (HTI) leases and timber use permits (IPK) (FKKM, 22/9/98; Latin, 1998; Simon, 1998; *Sylva Indonesia*, 1998).

As we have seen, a pattern of environmental degradation was linked to a system where élites had captured formal rights over areas of forest. In the absence of a state authority system with either the capacity or political will to regulate resource use, under the New Order access and use of forest resources in many areas had been captured by commercial interests and local networks of power and interest. Now, with the demise of Suharto, there would, on the one hand, be new possibilities for reforming state rules that privileged élite control over large areas of the forest estate; on the other hand, there might be opportunities for overcoming the conflicts over forest resources by finding ways to nest customary forest regimes within the wider state property regime.

In response to calls for reform (*reformasi*) and the conditions set by the IMF, the new Minister for Forestry and Estate Crops, Mr Muslimin, produced a raft of legislative changes. He moved to break up the concentration of ownership that had developed in the forestry sector; for instance, one new decree (Kepmen No.728/Kpts-II/1998) proclaimed that forestry companies would no longer be able to hold concessions of more than 100,000 ha in a single province or more than 400,000 ha in the whole country, while another decree (Kepmen No.728/Kpts-II/1998) created a new system to auction concessions. In February 1999, the Ministry began to auction the rights to HPH licences that had expired or been withdrawn for gross mismanagement, including several owned by the timber tycoons or politico-business families close to Suharto (*Jakarta Post*, 15/1/99). In accordance with the populist emphasis associated with calls for a 'People's Economy' (*Ekonomi Rakyat*), the Minister decided to allocate 3 million ha of forest to co-operatives and small or medium-sized businesses. He announced that logging companies wishing to extend their concession rights would also have to allocate 20 per cent of their shares to co-operatives (*Suara Pembaruan*, 27/11/98a)

However, the success of forestry reform will depend on whether reformist state sector managers, NGOs and other critical elements of civil society will be able to prevail against commercial logging interests and their partners within the local, provincial and central bureaucracy. In the Philippines under Aquino, reform was possible because loggers tainted by association with Marcos were now excluded from the government coalition (Ross, 1996: 191). The changes in Indonesia seem to signal the eclipse of many of the interests associated with Suharto. However, as Dauvergne (1998) has pointed out, there are also key differences between the Philippines and Indonesia. When Marcos fell from power, the timber industry was already in sharp decline. As forests were largely depleted, patronage links between political leaders and loggers could no longer have the same potency. Moreover, a more vocal NGO sector, greater environmental concern and the strengthening of democratic practices also supported reform (Dauvergne, 1998: 5–6). In Indonesia, by contrast, the move towards democracy is still fragile and there are still significant timber stocks to fuel patronage systems.

While conclusions are perhaps premature, sceptical observers have speculated that the new system of allocating property rights over 'forest areas' does not represent a clean break with the past. For instance, while the new system of auctioning concessions was meant to be open and transparent, the parties allowed to take part in the auction of concessions are selected in advance (Down to Earth, 1999: 9). The new opportunities for co-operatives also seem open to abuse: one critic noted that an HPH could manipulate the co-operative scheme, making an 'ad hoc co-operative' that is actually just an extension of the logging company (*Suara Pembaruan*, 27/11/98a). Others doubt whether co-operatives can be separated from the existing systems of privilege or really represent a reallocating of property rights to local communities:

> Cooperatives ... have become an extension of a vast corrupt bureaucracy which permeates every village ... As few indigenous forest communities have the formal education, business skills or finance to establish a cooperative, it is likely that the initiative will be of greater benefit to entrepreneurs from urban areas and other outsiders who may have little or no interest in or experience of sustainable forest management. (Down to Earth, 1999: 9)

Wary of the rise of a new class of '*pribumi* cronies'[33] close to the new decision makers, some observers see the promotion of thousands of co-operatives and small businesses as a part of a process of distributing 'business opportunities' to chosen beneficiaries to drum up political support for the 1999 elections (Khanna, 15/12/98; McCawley, 18/12/98). In this view the Habibie government is moving the flow of benefits derived from State forests away from Suharto's circle and towards a new group of clients.

Meanwhile, the basic framework that allows commercial interests to gain access to 'state forests' endures: at a time of economic crisis, the government needs to increase revenues from the forestry and plantation sector. While there is a ban on opening up new areas for logging, Dephutbun continues to reallocate leases over forest areas. Although the allocation process has changed, the aim remains the same: 'exploiting Indonesia's forests for maximum commercial gain' (Down to Earth, 1999: 9). As the booming agribusiness sector is seen as the silver lining on a dark economic horizon, the Ministry has continued to allocate large new areas for the development of oil palms.[34]

The existence of overlapping property claims and competing authority systems in forest areas has been connected to a pattern of deforestation, resource conflict and forest fires. Yet there have been few clear resolutions

---

33. The term *pribumi* is used to refer to the Indonesians of ethnic origins other than Chinese descent.
34. Once again, statistics are inconsistent. The *Far Eastern Economic Review* reported that there are plans afoot to expand the plantations from 2.4 million ha to 3.9 million ha by the end of 1999 (Gilley, 14/1/99). Meanwhile an Indonesian NGO estimated that there were plans to develop another 3.2 million ha of new palm oil plantations by the year 2000 (Alwy, 1998: 4). According to one report, 'some 5.5 million hectares of Sumatra and Kalimantan were earmarked for oil plantations and another 24.5 million hectares of East Indonesian forests earmarked for conversion' (*Jakarta Post*, 18/12/98).

to the problem aside from a series of legislative reforms. Here we will consider two of the new initiatives.

The Minister of Forestry and Estate Crops has created new legislative initiatives that in theory move a step closer to recognizing the overlap between state claims over areas and local notions of territoriality. A new regulation on Forest Utilization and the Harvesting of Forest Products from Production Forest (PP No.6/1999) granted *adat* communities the right to take forest products for their daily needs within concession areas. Concession holders must also allow the widest possible participation of communities within their areas, informing them of planned activities and providing opportunities to take part in forest activities. Moreover, concessionaires were now obliged to develop the capacities of co-operatives and small enterprises within their concession areas. According to a new ministerial decision on 'Community Forestry' (Kepmen 677/Kpts-II/1998), communities could gain the right to manage areas of forest based on community practices and *adat* law.[35] The Ministry would permit community groups to form co-operatives to obtain thirty-five year 'community forestry leases' (*Hak Pengusahaan Hutan Kemasyarakatan*) over production and protection forests as well as in specific conservation zones. The decree allows communities to utilize traditional forest management systems as long as they do not conflict with 'forest sustainability' (*kelestarian hutan*). Moving beyond an earlier community forestry decree, the new initiative also allows for community rights to harvest timber (Menteri Kehutanan, 1995; *Suara Pembaruan*, 3/11/98, 11/14/98).

By early 1999, in areas that have been sites of long-standing community-based conservation programmes, two communities in South Aceh and West Kalimantan had processed authorizations to manage areas of forest in accordance with *adat* principles (Dephutbun, 1988; *Suara Pembaruan*, 26/12/98). The procedures for gaining these community forestry concession rights were somewhat onerous. Under the new decree, based on the recommendation of local and provincial authorities, community groups can apply for leases to the head of the regional forestry office. If their application is successful, the forestry authorities will then grant a lease according to a management plan drawn up with the guidance of NGOs, university experts and forestry staff. While the new initiatives facilitate co-management strategies advocated by NGOs, it is unlikely that forest communities will gain these rights on a wide scale.

While these initiatives represent several steps forward from recent policies, there seems to be some difficulty breaking with the legislative regime for

35. This seems to be a realization of earlier statements of reformist state sector managers. As early as 1993, the Biodiversity Action Plan recognized that an estimated 40 million people were living in or dependent upon resources in the 'public forest estate' and that these people were the '*de facto* forest managers' (BAPPENAS, 1993).

forest management developed under the New Order. Within this framework, Dephutbun has no authority to legally recognize *adat* land rights over state forest land. However, the Ministry can issue use rights in the form of licences and leases to communities already inhabiting the land.

> This type of agreement implies that the community is relinquishing its claims to fundamental land rights for limited usufruct rights. By doing so, the community is unwittingly acknowledging the land's status as state land and admitting that they only have specified rights of use, rather than traditional tenure. (Haverfield, 1998: 63)

A more radical legislative programme would be required to substantially overcome a situation where élites rely on the law to substantiate their claims over resources, while local people seek to preserve their tenure system — which has such a weak basis under the national law. This means that the disjunction between *de jure* rights and *de facto* structure of property relations associated with earlier conflicts over resources is likely to continue.

A draft version of a new Basic Forestry Act — to replace the much maligned Basic Forest Act (No.5/1967) — which was circulating in late 1998 indicated that law makers within the state bureaucracy were reluctant to move away from the underlying assumptions of the state property regime as it existed under Suharto. While the draft Act recognizes the existence of *adat* forest (*hutan adat*), it maintains the concept of the 'forest areas' (*kawasan hutan*) as falling under the dominion of the state (RUUPK, 1998). Under this draft, the Minister retains the managerial responsibility for these 'forest areas' — some 70 per cent of the nation's land surface. According to Section 31, special state corporations will be set up to represent the government in allocating rights to private interests, corporations, community groups and *adat* communities. Critics noted that the draft law retains the power of a centralized forestry bureaucracy, leaving forestry management in the hands of poorly performing state enterprises. In the words of one observer, 'the bill also provided a big opportunity for officials of the appointed state forestry companies to practice collusion with big timber companies' (*Jakarta Post*, 10/12/98). What is more, in retaining the power of the centralized forestry bureaucracy, the new draft Basic Forestry Act has not met the demands for more regional autonomy (*Republika*, 10/12/98; *Suara Pembaruan*, 26/11/98).

While reforms are debated at the centre, the political and economic crisis has clearly affected the management of resources in the regions. With many areas of the formal economy in crisis, desperate people are forced to secure a meagre subsistence harvesting forest resources for short-term benefit and cutting down forest areas to open plots to plant subsistence crops. The currency crisis has led to booming prices in some cash crops, at least in local currency terms, and this has also spurred many to open new areas of forest (*Suara Pembaruan*, 23/10/98; Sunderlin, 1998; Waldman, 27/10/98). In the best of times forestry officials often needed to use paramilitary means to apply unpopular regulations. Now, with accusations of corruption levelled against local officials in many places, and with local communities under such intense economic stress, local officials are hardly in a strong position to enforce state regulations governing forest resources. Moreover, there have been calls for a loosening of central control over the provinces and a distribution of a greater share of profits generated from local natural resources back to the regions. However, if experience in the Philippines is any guide, decentralization without greater political reform at the regional level may give greater scope for local political figures to increase their personal control over logging (Dauvergne, 1998: 8).

CONCLUSION

Official forestry policy in Indonesia has been based on a normative ideal that state institutions can effectively govern the huge area classified as state forest land. Accordingly, the failure of the state forest regime has been seen to lie in the lack of state capacity to manage state lands and implement the law effectively. Now forest fires, economic and political crises have exposed the link between the shortcomings of previous practices and the property regimes operating in the nation's vast forest estate. Given the range of other issues involved in deforestation, policy initiatives dealing with the property rights issue will not be a 'silver bullet' for Indonesia's ecological crisis. However, reform in this area is needed if Indonesia is to move beyond the inequities and ecological destruction associated with the New Order period.

Reforming the structure of property rights operating in Indonesia's vast forest estates is a highly political issue. In the past land reform was an element of the polarized politics of the revolutionary period and the killings of 1965–6. Nowadays, as Dove has noted, while so many interests covet the rich land and forest resources, the forest-dwelling people themselves are too weak and too far from power to insist on a more just and rational management of local resources (Dove, 1996). Initiatives to nest local institutions within the state forestry regime face other serious obstacles, including the need for foreign revenue, resistance to change amongst the state forest bureaucracy, and the influence of corporate interests at the centre wishing to retain their control of land and forests. Moreover, if past experience is any guide, existing networks of power and interest clustered around local government are likely to tacitly resist the implementation of forestry initiatives that they see as interfering with local development plans or their access to the spoils of uncontrolled logging (Barber et al., 1995; Perbatakusuma et al., 1997).

The forest authorities are only beginning to experiment with ways to establish local institutional arrangements governing resource use within the state system. Recent legislative reforms open up new possibilities for co-management and amount to a tacit recognition that this state regime overlaps pre-existing customary regimes. Yet, as the subtitle of a recent publication on the forestry reforms — 'Reform without Change' (*Reformasi Tanpa Perubahan*) — indicated, *Reformasi* has not brought the changes hoped for by critics (Haryanto, 1998). Up to now, the reforms have failed to confront the property rights issue underlying the social inequity and the ecological destruction associated with the state forestry regime as it operated under Suharto: there is still some distance to go before emerging from the morass.

REFERENCES

Achmadi, H. (1988) 'Forest Fires: Logging is to Blame', *Inside Indonesia* 14: 16–17.

Alwy, M. (1998) 'Palm-Oil Plantation and the World Bank/IMF: Its Relation and Impacts in Indonesia'. http://www.latin.or.id/palm_oil.htm

Angelsen, A. (1995) 'Shifting Cultivation and "Deforestation": A Study from Indonesia', *World Development* 23(10): 1713–29.

Anon. (1998) 'Local Conflicts over Forest Resource Access (Opposition to Indonesian Government's Forest Resource Activities)', *Environment* 40(4): 32–3.

Arora, D. (1994) 'From State Regulation to People's Participation: Case of Forest Management in India', *Economic and Political Weekly* March 19: 691–8.

Asian Economic News (1998) 'Indonesia mulls Indigenous Rights to Forest Resources', 7 September.

BAPPENAS (1993) *Biodiversity Action Plan for Indonesia.* Jakarta: Ministry of National Development Planning/National Development Planning Agency.

Barber, C. V. (1998) 'Forest Resource Scarcity and Social Conflict in Indonesia', *Environment* 40(4): 4–20.

Barber, C. V., N. C. Johnson and E. Hafild (1994) *Breaking the Logjam: Obstacles to Forest Policy Reform in Indonesia and the United States.* Washington, DC: World Resources Institute.

Barber, C. V., S. Afiff and A. Purnomo (1995) *Tiger by the Tail? Reorienting Biodiversity Conservation and Development in Indonesia.* Washington, DC: World Resources Institute.

Baskoro, L. R. and L. F. Sihotang (1996) 'Memburu Kayu Siluman di Bumi Khatulistiwa', *Forum Keadilan* 12 February: 50–5.

Bromley, D. W. (1991) *Environment and Economy: Property Rights and Public Policy.* Cambridge, MA: Blackwell.

Bromley, D. W. (1997) 'Environmental Problems in Southeast Asia: Property Regimes as Cause and Solution'. http://www.idrc.org.sg/eepsea/papers/bromley.htm

Bruce, J. W. and L. Fortmann (1991) 'Property and Forestry', *Journal of Business Administration* 20(1–2): 471–98.

Bruce, J. W., L. Fortmann and C. Nhira (1993) 'Tenure in Transition, Tenures in Conflict: Examples from the Zimbabwe Social Forest', *Rural Sociology* 58(4): 626–42.

Cohen, M. (2/7/98) 'Tackling a Bitter Legacy', *Far Eastern Economic Review.*

Colfer, C. J. P (1987) 'Change and Indigenous Agroforestry in East Kalimantan', in L. Fortmann and J. W. Bruce (eds) *Whose Trees?* Boulder, CO: Westview Press.

Colfer, C. (with R. G. Dudley) (1993) *Shifting Cultivators of Indonesia: Marauders or Managers of the Forest?* Community Forestry Case Study Series 6. Rome: FAO.

Dauvergne, P. (1994) 'The Politics of Deforestation in Indonesia', *Pacific Affairs* 66(4): 497–518.

Dauvergne, P. (1997a) 'Globalisation and Deforestation in the Asia-Pacific'. Department of International Relations Working Paper 1997/7. Canberra: Australian National University.

Dauvergne, P. (1997b) *Shadows in the Forest. Japan and the Politics of Timber in Southeast Asia.* Cambridge, MA, and London: MIT Press.

Dauvergne, P. (1997c) 'Weak States and the Environment in Indonesia and the Solomon Islands'. Department of International Relations Working Paper 1997/10. Canberra: Australian National University.

Dauvergne, P. (1998) 'Environmental Insecurity, Forest Management and State Responses in Southeast Asia'. Department of International Relations Working Paper No. 1998/2. Canberra: Australian National University.

Departemen Kehutanan (1992) *Sekilas tentang sejarah dan perkembangan Tata Guna Hutan Kesepakatan: makalah Menteri Kehutanan Republic Indonesia pada Rapat kerja Gubernur Kepala Daerah Tingkat I dan Bupati Kepala Daerah Tingkat II Seluruh Indonesia tahun 1992.* Departemen Kehutanan, Direktorat Jenderal Inventarisasi dan Tata Guna Hutan.

Dephutbun (1988) *Keputusan Kepala Kantor Wilayah Departmen Kehutanan dan Perbeunan Kantor Wilayah Propinsi Daerah Istimewa Aceh Nomor: 445/KPTS?KWL-4/1988.* Departemen Kehutanan dan Perkebunan Kantor Wilayah Propinsi Daerah Istimewa Aceh.

Detektip (20/12/96) 'Aparat Mandul Hutan Gunduk'.

Down to Earth (1999) *Quarterly Newletter of the International Campaign for Ecological Justice in Indonesia* No. 40 (February).

Dove, M. R. (1983) 'Theories of Swidden Agriculture, and the Political Economy of Ignorance', *Agroforestry Systems* 1: 85–99.

Dove, M. R. (1986) 'Peasant versus Government Perception and Use of the Environment: A Case-study of Banjarese Ecology and River Basin Development in South Kalimantan', *Journal of Southeast Asian Studies* XVII(1): 113–36.

Dove, M. R. (1993a) 'A Revisionist View of Tropical Deforestation and Development', *Environmental Conservation* 20(1): 17–24.

Dove, M. R. (1993b) 'Smallholder Rubber and Swidden Agriculture in Borneo: A Sustainable Adaptation to the Ecology and Economy of the Tropical Forest', *Economic Botany* 47(2): 136–47.

Dove, M. R. (1996) 'So Far from Power, So Near to the Forest: A Structural Analysis of Gain and Blame in Tropical Forest Development', in C. Padoch and N. L. Peluso (eds) *Borneo in Transition: People, Forests, Conservation and Development*, pp. 41–58. Kuala Lumpur: Oxford University Press.

Dove, M. R. (1997) 'Dayak Anger Ignored. Inequities in State Development in Kalimantan', *Inside Indonesia* 51: 341–78.

The Ecologist (1995) 'Policy Failure: Protecting against Blame', 25(6): 218.

FKKM (22/9/98) 'New Era for Indonesian Forestry, Forest Resource Management Reformation'. http://forests.org/gopher/indonesia/newindo2.txt

Forrester, G. and R. J. May (eds) (1998) *The Fall of Soeharto.* Bathurst, NSW: Crawford House in association with Regime Change and Regime Maintenance in Asia and the Pacific Project and North Australia Research Unit, Research School of Pacific and Asian Studies, Australian National University.

Fox, J. (1993) 'The Tragedy of Open Access', in H. Brookfield and Y. Byron (eds) *South-East Asia's Environmental Future: The Search for Sustainability*, pp. 302–15. Kuala Lumpur and New York: Oxford University Press.

Fox, J. and K. Atok (1997) 'Forest-dweller Demographics in West Kalimantan, Indonesia', *Environmental Conservation* 24(1): 31–7.

Geertz, C. (1963) *Agricultural Involution: The Process of Ecological Change in Indonesia.* Berkeley, CA: University of California Press, for the Association of Asian Studies.

Gellert, P. K (1998) 'A Brief History and Analysis of Indonesia's Forest Fire Crisis', *Indonesia* 65 (April): 63–85.

Gilley, B (14/1/99) 'Sticker Shock', *Far Eastern Economic Review* 162(2): 20–3.

Gonner, C. (1998) 'Conflicts and Fire Causes in a Sub-District of Kutai, East-Kalimantan, Indonesia'. http://www.iffm.or.id/Fire_Causes.html

Goor, C. P. V. (1982) *Indonesian Forestry Abstracts: Dutch Literature until about 1960*. Wageningen: Pudoc.

Guha, R. (1994) 'Fighting for the Forest: State Forestry and Social Change in Tribal India', in O. Mendelsohn and U. Baxi (eds) *The Rights of Subordinated Peoples*. Delhi: Oxford University Press.

Gunarso, P. and J. Davie (forthcoming) 'Can Indonesian Production Forests play a Nature Conservation Role?', in J. Craig, N. Mithcell and D. Saunders (eds) *Nature Conservation 5: Conservation in Production Environments* (in press). Sydney: Surrey Beatty and Sons.

Haryanto (ed.) (1998) *Kehutanan Indonesia Pasca Soeharto: Reformasi Tanpa Perubahan*. Bogor: Pustaka Latin.

Haverfield, R. (1998) '*Hak Ulayat* and the State: Land Reform in Indonesia', in T. Lindsey (ed.) *Indonesia Law and Society*, pp. 42–73. Leichhardt, NSW: The Federation Press.

Hayes, A. C. (1997) 'Local, National and International Conceptions of Justice: The Case of Swidden Farmers in the Contexts of National and Regional Developments in Southeast Asia'. Resource Management Working Paper 1997/14. Canberra: Australian National University.

Human Rights Watch Asia (1997) 'The Horror in Kalimantan', *Inside Indonesia* 51: 9–12.

IFFM/GTZ (1998) 'Fire in East-Kalimantan in 1998'. http://www.iffm.or.id/FiresinEast2.html

Jakarta Post (10/12/98) 'Forestry Bill does not Support Small Firms'.

Jakarta Post (18/12/98) 'Habibie seeks Review of Forestry Policies'.

Jakarta Post (15/1/99) 'Forestry Concessions to be Auctioned'.

Kano, H. (1996) 'Land and Tax, Property Rights and Agrarian Conflict: A View from Comparative History'. Paper presented to the Tenth INFID Conference on Land and Development, Canberra, Australia (26–8 April).

Khanna, V. (15/12/98) 'Flaws in People's Economy'. http://business-times.asia1.com.sg/1/focus12.html

KLH and UNDP (1997) *Agenda-21 Indonesia: A National Strategy for Sustainable Development*. Jakarta: State Ministry for Environment Republic of Indonesia and United Nations Development Programme.

KLH and UNDP (1998a) *Executive Summary Forest and Land Fires in Indonesia*. Jakarta: State Ministry for Environment Republic of Indonesia and United Nations Development Programme.

KLH and UNDP (1998b) *Kebakaran Hutan dan Lahan di Indonesia. Dampak, Faktor dan Evaluasi*. Jakarta: State Ministry for Environment Republic of Indonesia and United Nations Development Programme.

Klinken, G. V. (1998) 'Taking on the Timber Tycoons', *Inside Indonesia* 53: 25.

Kompas (12/4/96) 'Gangguan Hutan Sangat Serius'.

Kompas (3/4/98) '65 Persen Kebakaran Hutan Kaltim karena HTI'.

van Langenberg, M. (1990) 'The New Order State: Language, Ideology, Hegemony', in A. Budiman (ed.) *State and Civil Society in Indonesia*, pp. 121–50. Clayton, Victoria: Centre of Southeast Asian Studies, Monash University.

Latin (1998) 'Reorientasi Sektor Kehutanan Untuk Menduking Pemberdayaan Ekonomi Rakyat'. http://www.latin.or.id/berita_biotrop.htm

Lev, D. S. (1985) 'Colonial Law and the Genesis of the Indonesian State', *Indonesia* 40: 57–75.

MacIntyre, A. (1990) *Business and Politics in Indonesia*. Sydney: Allen and Unwin.

McCawley, T. (18/12/98) 'A People's Economy', *Asiaweek*: 62–6.

McNeely, J. A. and P. S. Wachtel (1991) *Soul of the Tiger Seaching for Nature's Answers in Southeast Asia*. Singapore: Oxford University Press.

Menteri Kehutanan (1995) 'Keputusan Menteri Kehutanan Nomor: 622/Kpts-II/95 tentang Pedoman Hutan Kemasyarakatan' (Decision of the Minister of Forestry No. 622/Kpts-II/95 concerning Guidelines for Community Forestry). Jakarta: Ministry of Forestry.

Moniaga, S. (1993) 'Toward Community-based Forestry and Recognition of *Adat* Property Rights in the Outer Islands of Indonesia', in J. Fox (ed.) *Legal Frameworks for Forest Management in Asia: Case Studies of Community/State Relations*, pp. 131–50. East-West Center Program on Environment Occasional Paper No. 16. Honolulu, HI: East-West Center.

Mubyarto (1992) *Desa dan perhutanan sosial: kajian sosial-antropologis di Prop. Jambi.* Yogyakarta: Aditya Media.

Muryadi, W., W. Panggabean and M. R. Amady (25/8/97) 'Kabut Mengambang, Tetangga Berang', *Forum Keadilan.*

Muryono, S. (10/1/98) 'Kebakaran Hutan dan Elegi Lingkungan Hidup 1997', *Analisa.*

Myers, N. (1980) 'Role of Forest Farmers in Conversion of Tropical Moist Forests'. http://www.ciesin.org/docs/002-106/002-106a.html

Nababan, A. (1996) 'Pemerintahan Desa & Pengelolaan Sumberdaya Alam: Kasus Hutan Adat Kluet-Menggamat di Aceh Selatan'. Paper presented to the Analyisis Dampak Implementasi Undang-Undang No. 5 Tahun 1979 tentang Pemerintahan Desa terdadap Masyarakat Adat: Upaya Penyysybab Kebijakan Pemerintahan Desa Berbasis Masyarakat Adat, Wisma Lembah Nyiur, Cisarua (17–18 September).

Parlindungan, A. P (1997) 'Hukum Pertanahan Dalam Hubungannya Dengan Pembangunan Perhutanan', in D. V. Barus (ed.) *Hutan Rakyat. Hutan untuk Masa Depan*, pp. 199–220. Jakarta: Penebar Swadaya.

Peluso, N. L. (1990) 'A History of State Forest Management in Java', in M. Poffenberger (ed.) *Keepers of the Forests: Land Management Alternatives in Southeast Asia*, pp. 27–55. Manila: Ateneo de Manila University Press.

Peluso, N. L. (1992) 'The Ironwood Problem: (Mis)Management and Development of an Extractive Rainforest Product', *Conservation Biology* 6(2): 210–19.

Peluso, N. L. (1993) 'Coercing Conservation: The Politics of State Resource Control', in R. Lipschutz and K. Conca (eds) *The State and Social Power in Global Environmental Politics*, pp. 46–70. New York: Columbia University Press.

Peluso, N. L. (1995) 'Whose Woods Are These? Counter-Mapping Forest Territories in Kalimantan, Indonesia', *Antipode* 27(4): 383.

Perbatakusuma, E., A. Elfian and S. Lusli (1997) 'Stabilizing People-Park Interaction Areas through Community-based Sustainable Rural Resources Management: Lessons Learned from WWF-DB, FPNC Leuser Conservation Project in Gunung Leuser National Park, Indonesia'. Paper presented to the Regional Workshop on Participation of Local Communities in Protected Area Management UNESCO-ROTSEA-Kehati Foundation-LIPI 1997, Medan, Indonesia (17–20 March).

Poffenberger, M. (ed.) (1990) *Keepers of the Forest: Land Management Alternatives in Southeast Asia.* Manila: Ateneo de Manila University Press.

Potter, L. (1991) 'Environmental and Social Aspects of Timber Exploitation in Kalimantan, 1967–89', in J. Hardjono (ed.) *Indonesia: Resources, Ecology and Environment*, pp. 177–21. Singapore and New York: Oxford University Press.

Potter, L. (1996) 'Forest Degradation, Deforestation, and Reforestation in Kalimantan: Towards a Sustainable Land Use?', in C. Padoch and N. L. Peluso (eds) *Borneo in Transition: People, Forests, Conservation and Development*, pp. 13–40. Kuala Lumpur: Oxford University Press.

Republika (10/12/98) 'RUU Kehutanan Dinilai Sentralistik'.

Republika (28/12/98) 'Enam Juta Hektare HPH akan segera Dilelang'.

Republika (30/12/98) 'Lima Keppres Bidang Kehutanan dan Perkebunan Dicabut'.

Reuters (25/11/98) 'Indonesia Fires Devastate 3 Million Hectares of Forest'. http://forests.org/gopher/indonesia/3milburn.txt

Robison, R. and A. Rosser (1998) 'Contesting Reform: Indonesia's New Order and the IMF', *World Development* 26(8): 1593–609.

Ross, M. (1996) 'Conditionality and Logging Reform in the Tropics', in Robert O. Keohane and Marc A. Levy (eds) *Institutions for Environmental Aid: Pitfalls and Promise*, pp. 167–97. Cambridge, MA: MIT Press.

RUUPK (1998) 'Rancangan Undang-Undang Nomor ... Tahun 1998 Tentang Tehutanan'. http://www.latin.or.id/ruupk.htm
Schindler, L, (1998) 'The Indonesian Fires and SE Asean Haze 1997/98: Review, Damages, Causes and Necessary Steps'. http//www.iffm.or.id/SIN-Paper.html
Schwarz, A. and J. Fredland (12/3/92) 'Green Fingers: Indonesia's Prajogo proves that Money grows on Trees', *Far Eastern Economic Review*: 36.
SCMP (30/9/97) 'Plantations reject Blame for Blazes', *South China Morning Post.*
Scott, J. (1985) *Weapons of the Weak: Everyday Forms of Peasant Resistance.* New Haven, CT, and London: Yale University Press.
Serambi Indonesia (7/8/95) 'Oknum Aparat Diduga Terlibat Rusak Hutan, Pemerintah Lumpuh'.
Simon, H. (1998) 'Indonesian Government Must Immediately Implement Just and Democratic Forest Resource Management'. http://www.latin.or.id/diskusi_fkkm_jogja_english.htm
SINAR (4/10/97) 'Ini Tamparan Luar Biasa'.
SKEPHI and R. Kiddell-Monroe (1993) 'Indonesia: Land Rights and Development', in M. Colchester and L. Lohmann (eds) *The Struggle for Land and the Fate of the Forests.* Penang, Malaysia: World Rainforest Movement; Sturminster Newton, England: Ecologist; London and Atlantic Highlands, NJ: Zed Books.
Southgate, D. and C. F. Runge (nd) 'The Institutional Origins of Deforestation in Latin America'. http://www.ciesin.org/docs/002-407/002-407.html
Spiertz, J. and M. G. Wiber (1998) 'The Bull in the China Shop. Regulation, Property Rights and Natural Resource Management: An Introduction', in J. Spiertz and M. G. Wiber (eds) *The Role of Law in Natural Resource Management*, pp. 1–16. The Hague: Vuga.
Steering Committee (1998) 'Mengembalikan Kalimantan Timur Kepada Rakyat sebagi Bentuk Pendaulatan Rakyat'. Paper presented to the Lokakarya Usulan Kaltim untuk Reformasi Bidang Kehutanan dan Perkebunan, Samarinda, Kalimantan (28–9 July).
Suara Pembaruan (12/1/98) 'Pengusaha HTI Lebih Suka Ekspor Kayu'.
Suara Pembaruan (10/4/98) 'Dephutbun Kekurangan Petugas Jagawana'.
Suara Pembaruan (23/10/98) 'Perambahan Taman Nasional Mengkhawatirkan'.
Suara Pembaruan (3/11/98) 'Pembangunan HTI Dibenahi Untuk Penyediaan Bahan Baku'.
Suara Pembaruan (14/11/98) 'Hutan Lindung Bisa Diusahakan Sebagai Hutan Kemasyarakatan'.
Suara Pembaruan (26/11/98) 'Penyerahan Pengeloaan Hutan Ke BUMN Berbahaya'.
Suara Pembaruan (27/11/98a) 'Pasokan Kayu Bulat Tahun 1999 Turun 25 Persen'.
Suara Pembaruan (27/11/98b) 'Pencurian Kayu Di Jambi Masih Merajalela'.
Suara Pembaruan (26/12/98) 'Masyarakat Sanggau Menunggu SK HPHKM'.
Suhendar, E. (1994) *Pemetaan pola-pola sengketa tanah di Jawa Barat.* Bandung: Yayasan AKATIGA.
Sunderlin, W. D. (1998) 'Between Danger and Opportunity: Indonesia's Forests in an Era of Economic Crisis and Political Change'. http://www.cgiar.org/cifor
Sunderlin, W. D. and I. A. P. Resosudarmo (1997) 'Rate and Causes of Deforestation in Indonesia: Towards a Resolution of the Ambiguities'. http://www.cgiar.org/cifor/publications/occpaper/occpaper9.html
Sylva Indonesia (1998) 'Petermuan Menhutbun dengan Mahasiswa Fakultas Kehutanan UGM Yogyakarta'. http://www.latin.or.id/info_bagi_kudeta.htm
Timber and Wood Products (28/11/98) 'Indonesia's Forest Industry dogged by Questions'.
Tjondronegoro, S. M. P. (1991) 'The Utilization and Management of Land Resources in Indonesia, 1970–1990', in Joan Hardjono (ed.) *Indonesia: Resources, Ecology, and Environment*, pp. 17–35. Singapore and New York: Oxford University Press.
UUPK (1967) 'Undang-Undang No.5 Tahun 1967 tentang Ketentuan-ketentuan Pokok Kehutanan', in B. Ramulardi (ed.) *Hukum Kehutanan dan Pembangunan Bidang Kehutanan*, pp. 359–78. Jakarta: PT RajaGrafindo Persada.
Vatikiotis, M. and A. Schwarz (4/6/98) 'A Nation Awakes: The Inside Story of Suharto's Last Days in the Palace and an Assessment of his Legacy', *Far Eastern Economic Review*: 21–7.
Waldman, P. (27/10/98) 'The Taste of Death', *Asian Wall Street Journal.*

WWF (29/5/98) 'New Estimates Place Damage from Fires at $4.4 Billion'. http://forests.org/gopher/indonesia/newindof.txt

WWF (1998) 'Haze Damage from 1997 Indonesian Forest Fires Exceeds $1.3 Billion, Study Shows'. http://www.worldwildlife.org/new/fires/dam.htm

6 Balancing Politics, Economics and Conservation: The Case of the Cameroon Forestry Law Reform

François Ekoko

INTRODUCTION

The Tropical Forest Action Plans (TFAP)[1] of the mid-1980s changed perceptions of forests world-wide, and World Bank efforts to include forest policy reform in the process of structural adjustment have created a new context for forest policy in many African countries. As a result, most of these countries have put forest policy reforms on their agenda. Some countries, such as Ghana, have completed these reforms, while others, like Cameroon, are immersed in the process; a third group of countries, such as Equatorial Guinea, has just begun the process. This chapter analyses one component of forest policy reform in Cameroon, the formulation of the 1994 Cameroon Forestry Law. Understanding how policies such as these are designed can help improve strategies to guide future reform efforts.

Cameroon's attempt to replace its 1981 Forestry Law was a test case for other Congo Basin countries, and its neighbours can learn from the difficulties it has faced (SAILD, 1995). Despite major efforts by the government, bilateral and multilateral partners, the reform of Cameroon's forestry legislation has stalled. It is therefore important to ask what other countries in the region might do to keep their forestry reforms from being similarly constrained. The Cameroon experience shows that to design successful forest policies, policy makers must understand and take account of the dynamic socio-political context in which they operate, the distribution of power, and actors' vested interests. Otherwise, forestry legislation, conceived as an instrument for sustainable forest management, is doomed to be ineffective.

---

1. The TFAP was initiated in 1985 by FAO-UNDP as an international co-ordination mechanism, with the aim of tackling the problem of tropical forest depletion and promoting sustainable use for people's social and economic benefit. At that time, there were five areas for action: forest industries, fuelwood and energy, forestry in land use, conservation of ecosystems, and institutions. In 1990, the TFAP became the Tropical Forest Action Programme.

The material for this study comes from a variety of sources, including minutes of meetings and debates at the National Assembly, documents from the Ministry of Environment and Forests (MINEF), World Bank correspondence, and personal interviews with Members of Parliament, a former minister of the environment, civil servants at MINEF and the President's Office, a World Bank staff member, and a national logger. The author sought to collect a wide range of opinions from individuals involved in the discussions over the 1994 Forestry Law. Because of the topic's sensitivity, almost all interviewees preferred informal conversations and many wished to remain anonymous.

The first section of the chapter offers a brief reflection on forestry policy in Cameroon, highlighting major elements of both continuity and change in forestry policy. The second, and major, part of the chapter then concentrates on the process of formulation of the 1994 Forestry Law, focusing on the context, issues, actors, tactics and strategies used, as well as the outcomes. The material is organized chronologically, and divided into three periods — the drafting stage, the debate at the National Assembly, and the implementation phase. Discussion of the first period centres on how the World Bank was able to dominate the process during this stage, while other actors, including the Government of Cameroon, were largely passive. The second period illustrates how political and financial interests became more important when the draft law was debated in the National Assembly, and how the French Government, logging companies, and MPs from forested regions took the initiative. In the third period, the Executive Branch of Government (EBG) regained a certain amount of control over the process, but conflicting pressures from different groups kept the 1994 law from being fully implemented. It became clear that the World Bank's capacity to influence the political process and policy implementation on the ground was much weaker than its ability to influence the initial drafting of new forestry legislation.

GENERAL REFLECTIONS ON CAMEROON'S FORESTRY POLICIES

The 1981 Forestry Law (which was reformed in 1994) had a very difficult beginning. Although the National Assembly passed the law in 1981, the EBG issued the decree of implementation only two years later, in 1983. The idea of reforming the 1981 law surfaced as early as 1985, with the inception of the TFAP, but it was not until the late 1980s — with the growing interest and involvement of the World Bank — that reforms in the forestry sector really began.

The government recognized that the 1981 Forestry Law was outdated and had to be reformed. Its key weaknesses included the nature, area and duration of concessions; the shortcomings of the taxation system (seen as confused and inefficient); the lack of attention paid to people's livelihoods; the absence of a clear framework for logging industries; and the inadequacy of conservation measures. The formulation of the 1994 Forestry Law should therefore be understood in the light of three key elements: the importance of forest activities in Cameroon's economy, the role of forest products in people's livelihoods and incomes, and the broader context of policy formulation. We will look first at what was at stake during the elaboration of the 1994 Forestry Law, before examining important elements of continuity and change in Cameroon's forestry policy.

The 1994 Forestry Law and the Underlying Stakes

Cameroon is endowed with a forested area of 19,598,000 ha (FAO, 1997), 18,000,000 ha of which is located in the humid forest zone. During the last decade, the share of forest-related activities in the economy has significantly increased. In 1995, for example, the export of timber products amounted to US$321 m, representing 19.8 per cent of the total value of Cameroon's exports (Eba'a, 1997). The contribution of timber harvesting to GNP has also risen, from US$488 m in 1990 to US$612 m in 1995, the equivalent of 6.7 per cent of GNP. Similarly, taxes from logging activities more then trebled within four years. Government revenue from these taxes rose from FCFA 9.05 bn in 1992/3, to 15.99 bn in 1993/4, 28.7 bn in 1994/5 and 28.95 bn in 1995/6 (ibid.). Logging activities provide jobs for 33,300 people. Non-timber forest products have also grown in importance as a source of income for local populations and market intermediaries. Ndoye et al. (1997/98) suggest that the value of NTFPs marketed in the first half of 1995 was at least US$1.75 m.

Three key actors (the Government of Cameroon, the World Bank, and the local population) had particular stakes in the reform of the 1981 Forestry Law. The government wanted to use the forestry sector as part of its strategy for economic recovery and to increase its revenues through taxation of the forestry sector. The World Bank had two objectives in mind: to experiment with its new forestry policy guidelines in the Congo basin rainforest, the second largest tropical forest system on earth; and to promote the increase of government revenues that would help the State honour its domestic and international financial obligations.

In recent years, empowerment of the local population has been a new feature in the forestry debate, especially in the southern and eastern provinces of Cameroon. Despite numerous obstacles, people have become outspoken in demanding their share of the forest's wealth. Whereas in the past, the local people's use of the forest was confined to bark collection for traditional medicine, gathering of NTFPs, hunting, and conversion of forest land to agriculture, local populations now look to forest taxation as a potential source of revenue. They have learned that taxes from logging companies can be a significant source of income.

Continuity and Discontinuity in Forestry Policy

Three main features have tended to dominate Cameroon's forestry policies: a narrow view of forests as economic resources, a close link between politics and forests, and the strong influence of external actors in policy design. All of these features also apply to the 1994 Forestry Law.

The view of forests as mainly an economic resource has prevailed among policy makers and foreign donors for years. These groups have always emphasized the forestry sector's contribution to economic development (World Bank, 1978). The current government's focus on forestry products as a source of income thus reflects the traditional view of Cameroon policy makers, enhanced by the country's poor economic and financial situation (Horta, 1991). Politics and forests have been closely linked in Cameroon since Independence. In some cases, forest resources have been used to gain allies or reward friends for their support. This is one reason why timber extraction and exports in the English-speaking regions of Cameroon tend to increase at the time of major political events (Sharpe and Burnham, 1996). In other cases, forests have been cleared to achieve objectives such as political stability. An example of this was the burning and clearing of forests in the Nkam division and in the Eastern Province in the 1960s in order to push out rebels and to open up roads for military vehicles and troops (Ebaka, pers. comm., 1996).

The third element of continuity has been the role of external actors in designing forest policy in Cameroon and in the forestry sector itself. Since Independence, international institutions and bilateral donors have provided the theoretical basis for forest policy, as well as funding various projects. They led Cameroon to create the National Centre for Forest Development (CENADEFOR) and the National Office for Forest Development and Management (ONAREF) at a time when forestry sector development was fashionable. They were also largely responsible for Cameroon's 1981 Forestry Law being orientated towards large-scale timber production, rather than conservation.

Over the last two decades, the World Bank has changed its emphasis with regard to forestry policy several times, and each time the Cameroon Government has followed. Prior to 1989, the World Bank advised Cameroon to focus on commercial timber production for national economic development (World Bank, 1978, 1989). Then it advocated social forestry, and then sustainable forest management (World Bank, 1990, 1991). The World Bank's current position on forest conservation in Cameroon is both shifting and ambivalent. On the one hand, the Bank issues environmental guidelines, imposes green conditionality and pushes for legislation protecting forests; on the other hand it is advising Cameroon to increase its revenue by encouraging logging activities, and funding road construction projects that promote deforestation (World Bank, 1989).

External actors have also dominated within the private sphere of Cameroon's forestry sector, with large foreign logging companies playing the dominant role in timber production since the late nineteenth century (Mvogo, 1988).

Despite these elements of continuity, the socio-economic and political context in which they occur has changed. Growing poverty, structural changes in the economy, new agendas of bilateral donors and international organizations, environmentalist awareness campaigns, and Cameroon's process of democratization have led to changes both in the size of the playing field and the nature of the game. Whereas governments and their external allies used to decide the forests' fate with a single, well-articulated voice, a multitude of actors now intervene.

THE FORMULATION OF THE 1994 FORESTRY LAW AND ITS AFTERMATH

We will now analyse the formulation of the 1994 Forestry Law and the socio-economic, cultural, and political context that influenced it. Three main features that emerge from this analysis are: the importance of self-interest and political factors; the ever-changing balance of power; and the uncertainties and precariousness of actors' gains. In Cameroon, there are six main groups concerned with forests in different capacities: the executive, the legislature, multilateral institutions, bilateral donors, logging companies, and ordinary people (local communities, farmers, and forest dwellers). Each played a different role during the three main stages of the 1994 Forestry Law.

Of these six groups, the first two are entitled by the constitution to design policy and to draft, pass and implement laws and policies. When laws are passed by the National Assembly (NA), the Executive Branch of the Government (EBG) issues the decrees and administrative edicts necessary for their implementation. Various ministries and local government representatives will oversee the application of these laws. In the case of forestry laws, the central and local services of the Ministry of the Environment and Forest, the Divisional and Sub-Divisional Officers are responsible for implementation. These organizational arrangements for the implementation of regulations and policies partly explain the tendency of interest and other groups to primarily lobby the EBG.

The Forestry Law that emerged from this process reflects the shifting balance of power at the time it was drafted, debated, and approved. During the drafting stage, the World Bank largely determined the law's content. The NA debates brought in other powerful actors, including French and Cameroonian logging companies, several political parties, and the French Government, and forced the EBG to mediate between them. Since the law's passage, unresolvable conflicts and a weak public sector have led to a virtual stalemate, leaving most parties frustrated. We will now look at each phase in turn.

The Drafting Stage: World Bank Domination

This first period of the Forestry Law's development was characterized by bipolar negotiations (see Aaron, 1973) between the World Bank and the EBG, with occasional participation by logging companies and the Canadian International Development Agency (CIDA), and by the marginalization of local communities and local non-governmental organizations (NGOs).

Two factors determined the balance of power at this stage: the unequal distribution of power between the World Bank and the EBG, and the historical context in which the negotiations took place. The World Bank enjoys a strong bargaining position *vis-à-vis* many developing countries (George and Sabelli, 1994). In addition, the recent greening of its policies and willingness to promote environmentalist views have contributed to a rapprochement between the Bank and some of its fiercest earlier critics, the NGOs. When the Forestry Bill was drafted, the international NGOs backed the Bank on several key issues, including forest reserves, community forestry, and wildlife protection. Logging companies discreetly supported the Bank on crucial issues such as the duration and area of concessions and restrictions on log exports. The main bilateral donors at the time (France, Canada, and the USA) also seemed to share the Bank's basic philosophy regarding the bill, although they differed on specific issues, such as taxation (Le Prohn, pers. comm., 1997).[2]

The World Bank's focus was on the type and the distribution of taxes. It suggested an increase of surface area and stumpage taxes, and the modification of the basis for taxation: the value of the log (the basis for taxation) should be index-linked to its FOB price rather than set administratively (World Bank, 1997). Using some conclusions from the reports prepared by a Canadian-based consulting firm, the Cameroon Government defended the idea of a gradual increase of surface area taxes (confidential conversation, 1997). At this stage, however, no major players seriously opposed the World Bank's views.

Through its development agency, CIDA, Canada funded studies on forest taxation and zoning, while a Canadian institution-building project provided technical support and advice to the EBG during the drafting stage. Elements of this support may be questioned, however; the private Canadian consulting firm involved in the project may have had a conflict of interests, as its own future business opportunities depended on the content of the new Forestry Law. Some people have even portrayed this group as an agent of foreign interests (confidential conversation, 1997).

The second determining factor in the balance of power at the drafting stage was the financial and political vulnerability of Cameroon between 1989 and 1994, a period of serious economic and political difficulties for the

---

2. These issues were not directly linked to the Forestry Law; but were concerned with the vital tools accompanying the law.

Cameroon Government. Table 1 highlights the major financial difficulties of the time: declining fiscal and export revenues and capital flight created a major liquidity problem, and the government was unable to meet its obligations either domestically (salaries and internal debt payments) or internationally (foreign debts) (*Africa Confidential*, 1994).

At the same time, the regime was also facing a crisis of legitimacy. In 1990, the local population and foreign governments began pressing for greater democracy and a more open political system. Between 1990 and 1992, the country was virtually paralysed by a popular campaign known as 'Operation Ghost Towns', and the regime's very survival was at stake. In 1992, President Biya's victory in the presidential elections was seriously questioned both inside and outside Cameroon, and for a short period the USA and several European countries (excluding France) tried to isolate the Cameroon Government. Under such precarious circumstances, the regime was careful to avoid any moves that might further isolate it in international circles; it was forced to seek funding to reactivate the economy as a means of regaining its legitimacy, and secure its alliance with France, its only powerful international ally.

The Government's financial and political worries made perfect ammunition for thc World Bank. The Bank's 'stick' during negotiations over the Forestry Law was frequent correspondence and missions from Washington that threatened to suspend Bank co-operation if Cameroon failed to undertake the necessary economic reforms, meet its financial obligations to the Bank, and comply with green conditionality. The 'carrot' it offered was the prospect that making forestry sector reform part of the structural adjustment process would allow Cameroon to receive World Bank funding for forestry projects and increase government revenues from forest-related activities.

The tone of the letters which the World Bank sent the Cameroon Government during this period reflects its superior position. In a letter dated 29 May 1990, the Bank expressed its concern about the Government

Table 1. Government Financial Operations, 1988 to 1993

	1988/89	1989/90	1990/91	1991/92	1992/93
Total revenue	**562.8**	**478.0**	**506.0**	**502.0**	**480.0**
Fiscal revenue	532.5	452.9	458.0	341.0	317.0
Non-fiscal revenue	30.3	25.1	48.0	161.0	163.0
Total expenditure	**730.9**	**760.9**	**774.0**	**694.0**	**663.0**
Current expenditure of which:	558.9	569.0	592.0	583.0	562.0
wages	290.0	287.0	300.0	296.0	267.0
interest payments	68.0	139.0	130.0	153.0	167.0
capital payments/net loans	172.0	184.0	133.0	84.0	85.0
Basic deficit	**–168.1**	**–282.9**	**–268.0**	**–192.0**	**–183.0**

Source: The Economist Intelligence Unit (1995).

appointing a new director and deputy-director to the National Forestry Office (ONADEF) without first consulting with the Bank; in the Bank's own words 'We take the liberty to remind you that the memorandum recommended ... that the appointment of the Director of ONADEF be suspended' (World Bank, 1991).[3] A few months later, on 27 November 1990, the Bank sent a letter to the Minister of Agriculture reminding him that the new Forestry Law should be based on the recent report of a World Bank consultant. In a letter to the Minister (No. 1473/2) on 13 November 1991, the Bank strongly urged the Government of Cameroon to incorporate all of its comments into the new Forestry Law, concluding: 'The World Bank reminds the Government that to this effect, it must also be consulted as soon as the draft decrees of application are prepared' (World Bank, 1991: 3).[4] Subsequent letters expressed World Bank concerns or dissatisfaction regarding certain of the Law's provisions.

The Bank's fingerprints are also visible on key provisions in the draft Forestry Law. For example, the rationale for the provisions relating to the allocation, duration, and area of concessions closely reflects the Bank's treatment of these issues in the 1992 *World Development Report* (World Bank, 1992). Community forestry, sustainable timber management, and the expansion of the area under forest reserves and parks were also included in the Law at the Bank's request, but after consultation with the Worldwide Fund for Nature (WWF).

Off the record, some journalists, civil servants, and assembly representatives claim that the draft of the 1994 Forestry Law as submitted to the National Assembly was imposed upon the Cameroon Government by the World Bank. Other sources, however, suggest that the content of the draft resulted from negotiations which led to a relative consensus before it was submitted (Le Prohn, pers. comm., 1997). The Bank itself describes its role as being simply an adviser to the Government (Ndomb, pers. comm., 1996).

Background to the Drafting Process

Behind the drafting process of the 1994 Forestry Law lay the Tropical Forest Action Plan (TFAP) for Cameroon, itself inspired by the FAO Tropical Forest Action Plans of 1985. The TFAP for Cameroon was initiated by FAO/UNDP in 1987 and the main report presented in 1988, prior to the revision of the 1981 Forestry Law (FAO/UNDP, 1988). The World Bank's role in drafting the Law can be divided into two phases: an

3. Author's translation: the original French version reads: 'Nous nous permettons de rappeler que l'aide mémoire recommandait que ... la nomination du Directeur de l'ONADEF soit mise en suspens'.
4. The original French reads: 'La Banque Mondiale rappelle à ce propos qu'elle doit aussi être consultée dès que les projets de ces textes d'application auront été préparés'.

initial phase in which the main themes of Cameroon's TFAP were incorporated into the country's structural adjustment package and the adoption of a new forestry law was made part of the conditionality for the structural adjustment loan; and a second phase in which the law was actually drafted.

Between 1989 and 1993, Washington-based World Bank officials and consultants held several important meetings with officials of the Forestry Department (DF) in Yaounde to discuss a project on 'Forests and the Environment'. At first, discussions focused on project activities that might be funded by the World Bank and other donors (MINAGRI, 1989a). Five topics prioritized by the TFAP (forest industries, fuelwood and energy, forestry in land use, ecosystem conservation, and forest institutions) were discussed as possible areas for funding. The missions also discussed the general context of forest policy in Cameroon and possible constraints on the implementation of project activities.

Gradually, the Bank's concerns shifted from project activities related to the TFAP to Cameroon's forest policy in general, and the need for a new forestry law. While the Bank's 'Forest and Environment' team negotiated project components for the first four activities mentioned above, a second team, based in Washington, worked on institutional issues, with particular emphasis on forestry legislation. Eventually, the World Bank concentrated its project-related activities in two areas: ecosystem conservation and institutions. As a result, during the Bank's third mission, the task manager suggested that the project title be changed from 'Forest and Environment' to 'Forests, Parks and Game Reserves'. By the fifth mission, in 1993, the World Bank had laid out all its cards and clearly stated its recommendations regarding the new forestry bill and forest policy.

The World Bank's role during this period was positive in at least two important respects: it ensured that the theme of community forestry was incorporated in the new law and it championed environmental concerns such as the need for sustainable management and protected areas. The Bank helped win the support of logging companies for the idea of a new forestry law by reassuring them that their concerns about issues such as forest management requirements and taxation would be taken into consideration.

For its part, during the drafting stage, the EBG failed to co-ordinate its various departments, to promote participation of actors other than logging companies, to consult its own political allies, or to mediate between potentially conflicting interest groups. One possible explanation for these failures is the EBG's limited capacity; another is its apathy. The EBG was not only financially and politically fragile, its internal organization was also weak. It lacked clear objectives and strategies in its negotiations with the World Bank.[5] It failed to prepare a single document to be used as the basis

5. It might be that the EBG had already decided to submit to the Bank's desires and follow its lead.

for discussing a new forest policy. Even the forest policy documents that were produced in 1992 and 1993 were written at the suggestion of the World Bank. Civil servants from the Forestry Department never met with the President's advisers to discuss the forestry bill. Nor did the President's advisers consult with their party, the Cameroonian Peoples Democratic Movement (CPDM), to assess the bill's political implications, despite its highly controversial and sensitive nature (confidential conversation, 1996). Moreover, the bill was never discussed by the ministerial council before being submitted to the National Assembly.

The few attempts to co-ordinate and exchange information between ministries involved in negotiating structural adjustment programmes and the reform of the forestry sector largely failed. Different sectoral bureaucracies within the EBG (agriculture, forestry, economy and finance, and transport) negotiated separately with the World Bank, and because of their lack of common direction and objectives, they did not fully appreciate or understand what was at stake. In some cases, specific departments within each ministry monopolized the discussions. Some members of parliament claim that the Minister of Environment had difficulty defending the forestry bill in parliament during the general discussion because he was not very familiar with it. It is unclear, however, if this was the case. The Minister at the time insists that his collaborators were heavily involved in the discussions about the forestry bill during this period (confidential conversation, 1996).

Despite its weaknesses, the EBG's Forestry Department did resist certain World Bank suggestions, and refused to consider all the recommendations of the World Bank's 1993 consultants' report. DF staff challenged some of the World Bank delegation's ideas about how large an area was needed for protected area buffer zones, how many international experts should be hired, the role of local forest industries, and plans to provide training for Cameroon nationals in those industries (MINAGRI, 1990).[6]

This resistance was often overcome by the World Bank by repeating its threat to suspend its co-operation or financial assistance (SAILD, 1995).[7] The then newly-appointed Minister of the Environment, who represented the Mouvement pour la Democratie et le Rassemblement (MDR) party — a junior partner in the coalition government — could not afford to risk upsetting the highest political authorities, nor could the President's Office tolerate resistance to World Bank demands on 'minor issues', which might lead to further financial and economic troubles and possible political suicide.

---

6. For example, DF staff wanted fewer foreign consultants, who are very expensive and not always more knowledgeable than local professionals.
7. The Bank threatened to suspend the signing of its third agreement with Cameroon if the latter did not pass a new Forestry Law.

Sleeping Giants and Missing Players

The large logging companies, mainly foreign businesses, were not especially active at the drafting stage. Although they are powerful players in the sector and have formidable lobbying machines, they chose to remain largely outside the initial debates. Very few attended the meetings convened by the EBG during the early discussions of a new forestry law (MINAGRI, 1989c). The highest attendance was reached in July 1989, when directors of the four biggest logging companies (SFID, SFIS, GRUCAM, EFC) attended a meeting in Douala convened by the World Bank (MINAGRI, 1989b). At a later stage they were invited to some meetings at the Forestry Department.

Logging companies in Cameroon were ambivalent about the idea of a new forestry law. They disliked the prospect of having to manage forests, since they lacked the necessary know-how and felt forest regeneration was not their responsibility (MINAGRI, 1989b). They thought three years was too short for a forest concession, but forty years (the initial suggestion of the World Bank and Forestry Department) was too long, given inevitable encroachment by farmers. Some opposed log export bans which would force them to process all their timber in Cameroon. They mistrusted the Cameroon forestry administration and were fearful of a new taxation system that could increase their costs. They perceived farmers and smaller loggers as threats to their business. They were also divided among themselves, by interests (over such issues as the log export ban, the monopoly of *vente de coupe* for nationals, or the attribution of concession by auction), nationality (foreign versus domestic companies) and size (Ghadir, pers. comm., 1996).

If the logging companies were sleeping giants during the drafting stage, forest dwellers and NGOs were almost completely absent from the discussions. Local communities were not consulted, although a number of community representatives were informed of the imminent changes during meetings organized for that purpose. Some communities occasionally sent petitions and letters to government representatives, NGOs, and French officials to complain about some logging companies' activities (*La Voix du Paysan*, 1994).

In a sense, the World Bank behaved as a *de facto* spokesperson for local communities by insisting on the importance of community forestry and popular participation in forest activities. Commenting on an earlier government draft, the World Bank suggested: 'Community forests must be encouraged by the law. Communities managing community forests should have the same rights as the local councils that manage communal forest' (World Bank, 1991: 7).[8] One Bank staff member in particular pressed to ensure that

8. The original French version reads: 'La constitution de forêts communautaires doit être encouragée par la loi. Les communautés gestionnaires des forêts communautaires doivent jouir des mêmes droits que les collectivités locales gestionnaires des forêts communales'.

the livelihoods of forest people were protected, following the logic of the World Bank operational directive on indigenous people (Dyson, 1992).

NGOs were also largely absent during the debate over the 1994 Forestry Law. There are a number of possible explanations for this, including the fear that openly opposing the EBG could have them branded as opposition sympathizers (SAILD, 1995); a belief that the World Bank was already promoting many of their environmentalist views; and their own organizational weaknesses. The boom of local NGOs and NGO activities is a recent phenomenon, especially in the area of environment and development, and many local environmental NGOs have limited technical capacity (Atanga-Ada and Tumejong, 1994).

The French authorities, too, maintained a low profile at this stage.[9] It is possible that, given the sensitivity of the issue, they were adopting a wait-and-see approach; France has, in the past, been somewhat lukewarm over the role of the World Bank in Francophone Africa. It is also likely that the French Government was still developing its position, since French logging companies had conflicting interests: some companies could accommodate the log ban export, for example, while others feared it. Finally, French policy makers might have anticipated that difficulties could arise in the NA and did not want to be seen as interfering or defending the interests of the French logging companies. In that case, their silence could be seen as tactical, waiting for the right moment to intervene.

Four conclusions can thus be drawn regarding the drafting stage of the 1994 Forestry Law: (1) World Bank views dominated; (2) the EBG failed to co-ordinate the policy formulation process or assume a leadership role in policy debates; (3) political considerations, and specifically the regime's political survival, dominated the EBG's concerns during the period; and (4) several key actors played only marginal roles in drafting the law.

As will be shown in the following sections, limitations during the drafting stage came back to haunt the World Bank and the EBG when the law was debated in parliament, and have since restricted its implementation.

The Debate in Parliament: Party Politics, Financial Interests and External Intervention

The debates in the National Assembly on controversial provisions of the draft Forestry Law revealed how strongly forestry policies were entangled with wider politics and with individual material interests. The speeches and public rhetoric were often nothing more than a veil to protect the vested

---

9. At this point, France did not have a clear policy that covered its overall interests in the forestry sector. It was only in 1997, three years after the bill was passed, that France ordered a preliminary study as a first step to designing a forestry policy for Cameroon.

interests of specific groups. This period was characterized by the declining influence of the World Bank, the emergence of new players into the policy formulation process, conflicts of interest between the parties, and a constantly shifting balance of power.

The Forestry Law draft submitted to the National Assembly revolved around five themes: balancing production and conservation, local community participation, increased revenue from forests for economic development, greater investment by logging companies, and sustainable forest management. Thus it took into account the three main components of sustainable forest development: economic development, social justice, and ecological considerations. Nevertheless the draft had significant weaknesses and several very controversial provisions. This led public opinion (alerted by the media) to oppose the draft submitted by the EBG, and to find significant support from members of parliament.

Because the EBG feared opposition to its draft law and possible amendments by parliament which could upset the World Bank and bilateral donors, it worked hard to have the law passed. Its basic strategy was to pressure MPs from the governing coalition to follow party orders on the issue, and to leave as little time as possible for debates in the National Assembly by delaying discussion of the law until the last moment. The Government submitted the bill to the National Assembly just four days before the closing of its ordinary November session, delayed the discussion at the closing meeting of the extraordinary session which followed, and frequently suspended the debates while the sessions were being held.

Two of the principal debates can be used to illustrate the major features in this period of the law's formulation. These debates focused on allocation of timber rights and restrictions on log exports. The first highlights the role of partisan politics, public opinion, and MPs' individual material interests in the formulation process, while the second shows how logging companies were able to use the Cameroonian Government's political dependence on French support to their own advantage.

The Debate over Timber Sales

Section 45, paragraph 3, of the initial draft of the forestry bill (no. 544/PJL/AN) submitted to the National Assembly in November 1993 by the EBG states that: 'Standing timber shall be sold by public auction to the highest bidder, for a non-renewable maximum period of one year' (RC, Bill No. 544/PJL/AN/93: 16). Within the National Assembly, however, MPs of all parties, including those belonging to the ruling government coalition, rejected the use of auctions to sell standing timber (Ridandi, pers. comm., 1996) and agreed to amend the provision as follows: 'Standing timber shall be allocated by the minister in charge of forests on the recommendation of a competent commission, for a non-renewable maximum period of one year' (RC, Law No. 94/01; S. 47: 3).[10]

The MPs who opposed the EBG on this issue argued that Cameroonians would be unable to compete with foreign logging companies in an auction system, so that an auction system would lead to the forest being sold out to foreigners (Nzongang, 1993). The World Bank and some Forestry Department officials, on the other hand, claimed that the auction system would increase state revenue and stop corruption in the granting of forest concessions.

On the surface, the issue was one of nationalism versus economic efficiency and transparency. On closer inspection, however, and despite their patriotic rhetoric, the MPs' actions appeared to be motivated by an interest in partisan political gains and financial benefits. MPs of different political affiliations and regional origins sought to capitalize on the Forestry Law's unpopularity among the Cameroonian public to boost their parties' fortunes in the municipal elections scheduled for June 1994. The Forestry Law was very controversial and many people opposed it because they felt that it legalized the plundering of Cameroon's resources. Rural and urban people alike, irrespective of their ethnicity or political affiliation, feared losing the forest-land of their ancestors, and these cultural beliefs and ancestral ties to land were strong enough to outweigh their loyalty to any political party. MPs from forest regions were particularly active during the discussions and, despite being members of rival political parties, showed a great deal of solidarity. These representatives explained the logging situation in their regions to many of their peers, consulted frequently with one another, and presented the greatest number of amendments (confidential conversation, 1996).

In contrast to the forestry bureaucracy, which scarcely consulted the public while the Forestry Law was being drafted, the MPs expressed the public's fears and concerns during the debates, and thus performed their role as popular representatives. Prior to the debate, many among the rural youth had viewed assembly representatives as traitors for allowing the natural resources of their region to be exploited without any financial compensation. The Forestry Law debate allowed MPs to regain the support of these groups and be perceived as representing their interests.

The MPs also opposed an auction system because of their own personal business interests. Some MPs are involved in logging either directly or indirectly (through relatives) as shareholders in logging companies or owners of logging licences leased by medium-sized foreign logging companies. The possibility of allocating timber through an auction system threatened their interests. The MPs were also interested in protecting their own political prerogatives. The auction system would have reduced their ability to use

10. In the initial draft of the bill, Section 45 deals with timber sales; in the final version of the Law, Section 47 deals with this issue.

personal and political contacts and influence to obtain concessions and other authorizations for their friends, partners, or relatives. It would thus have eliminated an important source of patronage and diminished their power and influence.

Whether or not the parliament's modification of Section 45 of the draft was a blow to the EBG is not completely clear. Some people suggest the EBG itself supported the change, but could not publicly say so because of World Bank pressure. However, no hard evidence has been presented to support that claim.

The National Assembly also managed to frustrate the World Bank (as well as logging companies and the French Government) by passing an amendment calling for the creation of a new Timber Authority. These entities opposed such an authority on the grounds that it was unnecessary and would lead to more bureaucracy and misuse and mismanagement of funds in the forestry sector.

The Debate over Log Export Restrictions

A second example of the entanglement between politics, the distribution of forest wealth, and foreign interests is provided by the issue of the amount of unprocessed logs which could be exported. The draft of the forestry bill submitted to parliament required that no more than 30 per cent of logs be exported, implying at least 70 per cent would be processed domestically. However, during debates in the National Assembly's production and constitutional committees, and a special brain-storming session which was held to discuss the bill, a majority of MPs from both the ruling coalition and the opposition rejected the idea of exporting any logs at all (confidential conversation, 1996). They advocated processing all logs domestically before exporting them.

The MPs argued that processing timber domestically would create new jobs in remote rural areas and revive local economies; increase forest sector profits and tax revenues from industries and workers; and help modernize the sector and attract investment into forest industries. Nationalist MPs also felt that banning log exports was the best way to stop the smuggling of logs through neighbouring countries and the exploitation of Cameroon's forests by foreigners.

Given the mood within the Assembly, the defeat of the EBG on this issue seemed inevitable. This led a few MPs, mostly from the CPDM, the president's party, to propose a compromise in which 80, rather than 70, per cent of logs had to be processed domestically. It also stimulated some big foreign logging companies, whose business and financial interests were at stake, to lobby the EBG, send letters to MPs explaining their position on log exports, and intervene through a new intermediary — the French Government.

It was at this point that a number of French politicians entered the debate, to protect their interests and those of the French logging companies. The French Minister of Co-operation and Overseas Development, who happened to be touring Cameroon at the time, cut short his planned visits to a number of provinces in order to meet with the President of Cameroon and the Speaker of the National Assembly.[11] The President's Office issued a communiqué about these discussions stating that the principal topic was economic co-operation. There are indications, however, that one of the main purposes of the visit was to lobby against log export restrictions (confidential conversation, 1996).

Among the issues discussed during the talks were France's financial rescue package, which allowed Cameroon to pay its arrears to the World Bank, and a French proposal for a debt for nature swap (confidential conversation, 1996). Pearce (1994) suggests that France also proposed to cancel part of Cameroon's debt if the government agreed to allow French companies to maintain exclusive control over logging in some regions and to export unprocessed logs.

Through his presence, the French minister may have sought to reassure French companies operating in Cameroon that the French government would defend their interests and explain their position to the EBG over the issues of log export and concession area.[12] The French minister's visit and discussions with President Biya and the Speaker of the National Assembly probably made the EBG more determined to reject the log export ban proposed by the MPs and to have the forestry bill approved as originally drafted. The discussions took place in a context in which the private financial interests of major French decision makers and the political concerns of the Cameroon regime were tightly entangled. For example, one key French adviser on African affairs was the son of President François Mitterrand, who was also a shareholder of SFID, one of the main logging companies which could suffer if concession areas were smaller and logging exports were restricted (Pearce, 1994). Given the Mitterrands' potential ability to destabilize the regime, the EBG found itself politically compelled to take decisive action to prevent the National Assembly from prohibiting the export of unprocessed logs.

According to Cameroon's constitution, an extraordinary session cannot be extended beyond its scheduled finish. At the same time, under World Bank pressure to meet deadlines established for the law's approval, the EBG ruled out delaying discussions of the bill until the following ordinary session

11. For a minister of another country to visit the National Assembly to meet with the Speaker of the House during a debate over such a controversial issue is highly unusual.
12. The real question, however, is whether the French minister was putting forward an official line, or whether he was representing a position held by influential logging companies whose shareholders were well-known in the decision-making circles of Paris. According to some sources there was no official French forest policy for Cameroon: see footnote 9.

in June 1994. This led to a situation with all the ingredients of high drama: external pressures on the EBG, divisions between the executive branch and its MPs, divergent views among parties within the ruling coalition, public opposition to the bill fuelled by strong nationalistic sentiments, and decisions which could not be postponed. At this point, the only party with an official position regarding the law, or strict guidelines for its MPs on how to vote, was the opposition National Union for Democracy and Progress party (NUDP).

Faced with this hard reality, the EBG abandoned its *laissez-faire* approach and made a major push against a complete log export ban, arguing that log exports were necessary for logging companies to obtain the funds required to develop capital-intensive processing industries. The EBG asked its chief whip in Parliament to hold a meeting of all MPs from the President's party, the CPDM, at which MPs were advised to observe strict party discipline and vote against banning log exports (Noah, pers. comm., 1996). Members of the other parties in the governing coalition, the UPC and MDR, were also asked to back the CPDM during the vote.

These steps enabled the coalition to win the vote, even though some CPDM MPs from forest regions abstained or voted with the opposition. In the approved version of the Forestry Law, the wording on this issue (Section 71, paragraph 1) states that: '70% of logs shall be processed by local industry'. Immediately after this provision was passed, however, the EBG made a concession to the opposition by setting a time limit on log exports, saying that no logs should be exported after 1999.[13]

Implications of the Debate

The Forestry Law debate in the National Assembly underlined the importance of the role played by different actors' political and financial interests, and the significance of cultural factors and nationalistic sentiments in shaping public opinion.

During this stage, the World Bank became a second rank player, unable to intervene forcefully and decisively, while the Cameroon Parliament and French politicians took on major roles, along with the EBG. The World Bank also ceased to be the only actor representing popular interests. Where previously it had been the Bank that insisted on including community forestry and a role for NGOs and other grassroots associations in the law, now it was the MPs who voiced popular concerns about losing control over

13. The log ban mentioned in the 1994 Forestry Law officially came into effect in early 1999. However, there are exceptions that allow logging companies to continue exporting logs under some conditions.

ancestral forests. This apparent convergence of support for community forestry did not lead to an alliance between the World Bank, the National Assembly, and local communities, however, since these groups had different underlying agendas and each group was internally divided.

The EBG's true role during this stage is somewhat obscure. Co-ordination between the EBG and the political parties backing the governing coalition was weak, except during the final moments of the crucial vote on log export restrictions. It remains unclear, however, whether this was mainly due to the EBG's limited institutional capacity or whether the EBG consciously used the National Assembly's theoretical autonomy to implicitly support certain amendments which it could not politically afford to endorse explicitly, in a sense conspiring against its 'own' bill. While the EBG showed determination regarding certain issues, on others it seemed to lack the will to oppose the parliament's amendments, and did not systematically require party discipline from its MPs and the coalition partners.

The Implementation Stage: The Law is Crippled

Since the passage of the 1994 Forestry Law, the EBG has failed to implement some important provisions or has implemented them in a contradictory fashion. For example, after the National Assembly modified the provision for an auction system to allocate timber resources, the World Bank threatened to suspend its co-operation unless the provision was changed again. In response, the EBG reintroduced the auction system through a decree of implementation, thus meeting the Bank's demands, but continued to allocate timber resources administratively when it served certain interests.

During the drafting stage, the World Bank was implicitly backed by the French Government on a number of provisions. This partnership broke down during the NA debates and the implementation phase, as each actor sought to defend its own interests. While the World Bank had worked to increase government revenues by higher taxes on forest resources and to increase the public sector's capacity to regulate unsustainable logging practice, an implicit coalition between logging companies, French politicians, and some EBG officials has sought to keep taxes low and avoid regulations on forest management that increase logging costs.

The Illusion of Winning and Losing

An obvious question to ask of the 1994 Forestry Law is, who benefited and who lost out? At first glance, the French Government did well in the process, since it was able to influence the EBG and the new Forestry Law does not visibly harm large foreign logging companies (many of which are French). The law continues to permit log exports and the government gives logging companies tax holidays. Some government officials and politicians also seem to have won, since the new law protects their financial interests.

The World Bank and local communities might be considered neither winners nor losers. Some provisions favour them, while others do not. The provisions on community forestry are potentially positive for these groups, but may have little effect, depending on the accompanying regulations and how they are implemented. To an extent *vente de coupe* and community forests compete over the same space. This remains a source of worry since national loggers oppose community forests. There is also (in theory) an improvement in the distribution of tax revenues that should benefit local communities.

The Cameroon State, forest dwellers, and the environment were apparent losers. The state will collect less revenue from logging activities; forest dwellers have received no guarantees that their rights will be protected; increased logging will probably accelerate forest degradation and loss of biodiversity.

This type of portrayal of winners and losers, however, is somewhat theoretical and based more on the text of the Forestry Law than on its real implementation. On the ground, the situation is more complex and almost everyone is dissatisfied and frustrated. French authorities, logging companies, and the World Bank are all unhappy about the provisions on concession duration, having wanted concessions to last for twenty-five years rather than the fifteen laid down in the Law. Logging companies are worried about their future after the 1997 legislative and presidential elections.[14] They are uncertain whether their licences will be renewed, how policies related to forest management, zoning, and development planning will affect them, what future tax rates will be, and whether log export bans will be enforced. Although the Forestry Law provides a legal framework favourable for logging companies, the Cameroon Government may not have the capacity or will to keep the companies' concession areas from being invaded by rural people looking for agricultural lands, or to stop local villagers from blocking roads to prevent logging, as happened in 1993 in Kagnol and Attiek, in eastern Cameroon.

The World Bank is upset by the provisions relating to timber sales and the new Timber Authority which is concerned with the level, type and actual distribution of the taxes (World Bank, 1997); it is also frustrated by the EBG's selective implementation of the new law. The MPs are angry about the EBG's reversal on the issue of the auction system in the decree of

---

14. The results of the 1997 legislative and presidential elections gave greater room for manoeuvre to the government, as the ruling party (CPDM) gained a comfortable majority in the National Assembly. So far, however, little has changed as far as government practices are concerned.

implementation, its failure to implement other important sections of the law, and certain omissions in the final published version of the 1994 Forestry Law by the Secretary General of the National Assembly. The EBG feels uncomfortable with what it sees as the 1994 Forestry Law's strong 'conservationist' flavour, the limited financial support from international donors that pushed for more conservationist policies, and the problems surrounding the use of auctions to grant concessions.

NGOs worry that the EBG is proceeding too slowly in setting up the institutions and mechanisms required to implement the 1994 Forestry Law. They also have doubts about whether the 1994 Forestry Law is compatible with Cameroon's new constitution, which emphasizes decentralization, and about possible changes in forestry policies after the 1997 legislative and presidential elections. Local communities feel cheated and complain that the law protects logging companies at their expense. Although they could potentially benefit from a 1995 decree on community forestry, it has proved difficult to establish the procedures and institutions required to implement that decree. In certain regions, these communities are rapidly gaining power and self-confidence. On 27 June 1994, violence broke out between villagers and logging company workers in Bokito (in the Mbam division of the Centre Region) over the exploitation of the Bougnoungoulouk forest (*La Nouvelle Expression*, 1994). This led the EBG to withdraw company licences to log in the area — the first time villagers had achieved such a victory in conflicts with logging companies.

The balance of power on the ground and the willingness of each player to fight for a share of the forest wealth or for its ideas about forests has largely determined how the EBG has (or has not) implemented the 1994 Forestry Law. In this context, the Forestry Law seems no more than a legal framework whose actual impact depends on how the balance of power plays out over time. The actors' eventual gains will depend on their ability to influence the system as it evolves.

Since the balance of power in the forest is constantly changing, it is hard to secure a long-term advantage or to make law correspond with the sociological realities in the field. It was this sociological reality that led the EBG to reverse its positions and support local villagers in the Bokito affair, and that recently led it to assign a large forest concession to two foreign companies using procedures that violate the 1995 regulations which now form part of the Forestry Law (Ferrer, 1996). Victories in the National Assembly or the courts may have little effect or may have unanticipated results in the field. This makes the future impact of the 1994 Forestry Law uncertain.

The World Bank, which had a leading position at the beginning of the process, and then lost it, has been particularly frustrated by this reality. The EBG's direct control over the Law's implementation allowed it to regain power and initiative during the final (and ultimately most important) phase of the Forestry Law, while the World Bank progressively lost control.

CONCLUSION

The concerns leading to the formulation of the 1994 Forestry Law have not disappeared. The need to rationalize the forestry sector, conserve and sustain forest resources, increase the contribution of those resources to development, and improve people's lives, remains intact. However, the process of formulation of the 1994 Forestry Law and the initial steps towards its implementation suggest that those objectives are unlikely to be achieved. The polarization of interests and struggles for power have undermined prospects for a broad alliance in favour of sustainable forest management. Two sets of factors explain this failure: poor policy formulation and the presence of strong and conflicting material and political interests.

Throughout the process, the EBG failed to lead, co-ordinate, and mediate, and no other actor was in a position to do so. This failure jeopardized the design, approval, and implementation of the 1994 Forestry Law. In general, the formulation process was characterized by insufficient co-ordination between government institutions, limited consultation of key actors, short-term political considerations, and decision making based on weak information. The policy formulation process was also undermined by the inability of some officials to adapt to a political context that no longer functions as a one-party system. Despite being highly centralized, conflicting interests within the Cameroon ruling élite have fragmented the state, and the state lacks the capacity to withstand pressure from determined foreign actors.

The World Bank shares responsibility for the Forestry Law's limited implementation. It failed to understand Cameroon's socio-political culture and adopted a top-down approach. Rather than seeking a genuine consensus with the EBG and the National Assembly, it sought to impose its views on them. This made some nationalist MPs resentful of the Bank, which worked to the latter's disadvantage when the MPs were able to exert significant power during the NA debates (Ntae, 1993).[15] The Bank also tried to promote local communities' interests without consulting them, and underestimated both the ability of the EBG to manoeuvre, and the capacity of other players to defend their interests.

The World Bank mistakenly behaved as if Cameroon still had an authoritarian one party system, assuming that the EBG could easily impose its views on the National Assembly and the public.[16] In reality, the social

---

15. In the words of Christopher Ntae, an MP from the NUDP party, currently in the opposition: 'The [forestry] bill was tailored to suit the whims and caprices of our money lenders'.
16. This chapter was written before the 1997 legislative and presidential elections, at a time when multiparty politics was gaining ground. The result of the legislative and presidential elections may well lead Cameroon to a dominant party system that looks like a one party system.

structures underlying African states are complex and democratization has made them more complex still. To successfully achieve reforms in sensitive sectors, one must understand these structures' dynamics and their multiple poles and channels of power.

This chapter has concentrated on the formulation of the Forest Law. Other factors that influenced the quality of the Law and its implementation, but which fall outside the scope of this analysis, include Cameroon's weak law enforcement capacity; the low educational levels of many policy makers; the absence of mechanisms to provide members of parliament with required information (such as seminars or parliamentary assistants for MPs); and a lack of research on the part of decision makers with regard to different constraints and the balance of power in the field, as well as to the opinions of stakeholders.

Given the dominance of financial and partisan political interests in the political process, one wonders whether it is possible to formulate policies that reflect the interests of ordinary people, particularly those who depend on forests. Is an alliance between actors supporting these groups feasible, and who might take part? For instance, could the World Bank, NGOs, rural communities, and forest dwellers form an alliance, and how might it function? Could it overcome opposition from foreign and domestic vested interests? Or could there be a rapprochement between local communities and logging companies? Where might the Cameroon government fit in?

In the future, the prospect of democratic elections, movement towards greater decentralization of power, and the threat of violent actions by local communities will probably lead to a gradual change in the behaviour and attitudes of the government, politicians, and logging companies. When local villagers blocked roads in 1993, the EBG sent troops to repress them. With increased movement towards democracy and decentralization, future governments may be less willing to take this type of action to protect logging companies.

In whatever way the different scenarios eventually play out, the future of Cameroon's forestry policy and its forests will depend both on changes in the overall political context and on the ability of different actors to build effective alliances during policy formulation, approval, and implementation. Lessons from Cameroon will be important in the preparation of institutional changes in the forestry sector of other countries of the Congo basin including Gabon, Equatorial Guinea and Congo.

REFERENCES

Aaron, R. (1973) *Paix et Guerre entre Nations*. Paris: PUF.

Africa Confidential (1994) Vol. 35, No. 8 (April).

Atanga-Ada, A. and Tumejong (1994) *Les Organisations Non Gouvernementales au Cameroun*. Yaoundé.

Dyson, M.(1992) 'Concern for Africa's Forest People: A Touchstone of a Sustainable Development Policy', in K. Cleaver et al. (eds) *Conservation of West and Central African Rainforests,* World Bank Environment Paper No. 1, pp. 212–21. Washington, DC: The World Bank.

Eba'a Atyi, R. (1997) 'Cameroon's Logging Industry: Structure, Economic Importance and Effects of Devaluation'. Yaounde: Centre of International Forestry Research (CIFOR) in collaboration with The Tropenbos Foundation (Cameroon Programme).

Economist Intelligence Unit (1995) *Country Profile: Cameroon.* London: Economist Intelligence Unit.

Ferrer, V. (1996) 'Contribution to an Electronic Seminar of Ekoko: Poverty and Deforestation in the Congo Basin Rainforest'. Tokyo: United Nations University (August).

Food and Agriculture Organization (FAO) (1997) *State of the World's Forest.* Rome: FAO.

Food and Agriculture Organization (FAO)/United Nations Programme for Development (UNDP) (1988) *Plan d'Action Forestier Tropical, Cameroun, Volumes I, II, III,* FO: DP/CMR/001, FO: DP/CMR/002, FO: DP/CMR/003. Rome: FAO/UNDP.

George, S. and F. Sabelli (1994) *Faith and Credit: The World Bank's Secular Empire.* London: Penguin Books.

Horta, K. (1991) 'The Last Big Rush for the Green Gold: The Plundering of Cameroon's Rainforest', *The Ecologist* 21(3): 142–7.

La Nouvelle Expression (1994) No. 171 (July).

La Voix du Paysan (1994) No. 33 (October).

Ministère de l'Agriculture (MINAGRI) (1989a) 'Rapport sur la Mission de la Banque Mondiale du Mois de Juillet 1989, sur le Projet "Forêt-Environnement"'. Yaoundé: MINAGRI (August).

Ministère de l'Agriculture (MINAGRI) (1989b) 'Compte-rendu de la Réunion des Représentants du Gouvernement avec les Exploitants Forestiers à Yaoundé le 15 Juin 1989'. Yaoundé: MINAGRI (30 June).

Ministère de l'Agriculture (MINAGRI) (1989c) 'Compte-rendu de la Réunion des Experts de la Banque Mondiale avec un groupe d'Exploitants Forestiers à Douala, le 5 Juillet 1989'. Yaoundé: MINAGRI (25 July).

Ministère de l'Agriculture (MINAGRI), Direction des Forêts (DF) (1990) 'Compte-rendu de la troisième Réunion sur le Projet "Forêt-Environnement"'. Yaoundé: MINAGRI (February).

Mvogo Tabi, A. (1988) 'La Protection de l'Environnement dans la Coopération des pays d'Afrique Centrale: le Cas de la Flore et la Faune'. Doctoral thesis (3rd cycle). Yaoundé: Institut des Relations Internationales.

Ndoye, O., M. Ruiz Pérez and A. Eyebe (1997/98) 'The Markets of Non-timber Forest Products in the Humid Forest Zone of Cameroon'. ODI Rural Development Forestry Network Paper 22c. London: Overseas Development Institute.

Ntae, C. (1993) quoted in *Cameroon Tribune* No. 1786, 22 December.

Nzongang, A. (1993) quoted in *Cameroon Tribune* No. 1778, 12 December.

Pearce, F. (1994) 'France Swaps Debt for Rights to Tropical Timber', *The New Scientist* 141(1910): 7.

République du Cameroun (RC) Loi No. 94/01 du 20 Janvier 1994 portant Régime des Forêts, de la Faune et de la Pêche.

RC(AN) (1993) Forestry Bill No 544/PJL/AN (November).

SAILD (1995) 'La Nouvelle loi sur les Forêts, Etude réalisée par le SAILD'. Yaoundé: SAILD.

Sharpe, B. and P. Burnham (1996) 'The Institutional Context of Sustainable Forest Management in a Zone of Heavy Commercial Timber Exploitation: The Case of South East Cameroon'. Unpublished manuscript.

World Bank (1978) 'Forestry Sector Policy Paper'. Washington, DC: The World Bank.

World Bank (1989) 'Cameroun: Rapport sur le secteur de l'Agriculture'. Rapport N 7486-CM. Washington, DC: The World Bank (November).

World Bank (1990) *The World Bank and the Environment: First Annual Report.* Washington, DC: The World Bank.

World Bank (29 May 1990). Letter to Cameroon Government.

World Bank (27 November 1990). Letter to MINAGRI.

World Bank (13 November 1991). Letter No. 1473/2.

World Bank (1991) 'Forest Sector Policy Paper'. Washington, DC: The World Bank.

World Bank (1992) *World Development Report: Environment and Development.* New York: Oxford University Press for the World Bank.

World Bank (1997) 'The Evolution of Cameroon's New Forestry Legal, Regulatory and Taxation System: Report'. Washington, DC: The World Bank.

7 People in Between: Conversion and Conservation of Forest Lands in Thailand

Jin Sato

INTRODUCTION

Academic studies aiming to illustrate the problems of deforestation in Thailand commonly compare figures for the country's forest cover in the 1960s or even earlier, which was reportedly more than 50 per cent, with the present figure, which is around 15 per cent. In trying to understand the decline in forest cover, regression analyses are carried out using different abstract variables such as population density, poverty and literacy. Although this approach, from the aggregate to the particular, may be useful in analysing how predetermined categories contribute to forest loss, it bypasses the definition of these categories (what is poverty? who is poor?), fails to capture the diversity within each conceptual cluster (why, for instance, do groups with identical population densities have different levels of success in managing their forests?), and is blind to the politics operating among the various groups of people to whom these variables apply. Yet, it is in this diversity and through the politics of resource control that we find elements critical for understanding environmental change.

A more effective analysis begins with the study of a specific people residing in a specific location, who are likely to be caught between various interests and power relations representing forces beyond the locale. The analysis of 'ambiguous lands' and the people who inhabit them is particularly revealing for understanding environmental deterioration in Thailand. 'Ambiguous lands' are those which are legally owned by the state but are used and cultivated by local people. They do not fit neatly into the private property regime based on fictions of exclusive use rights and alienability, and consist of the residual lands of state simplification processes on land tenure. State simplification, as described by James Scott, denotes acts by state officials to create a standard grid whereby complex and illegible local systems can be centrally recorded and monitored (Scott, 1998).

The ambiguous property status of such land makes it attractive to a variety of actors, including villagers searching for unoccupied arable lands in the frontiers, government departments on the look-out for new project sites, and conservation agencies sensitive to new areas in need of protection. Ambiguous lands can be considered geographical spaces in which various 'stresses' from social, economic, and political demands are represented. They are the frontiers of state simplification projects, where various initiatives have attempted to solve the problems of deforestation. Analysis of the alleged solutions is a promising window through which to look at how problems are defined in the first place. My primary interest, therefore, is not to identify the driving force of conservation and development as such, but to investigate what is happening in the spaces *between* them. In other words, I am interested in the specific nature of the spaces and peoples that tend to fall between the powerful forces that are commonly found in rural areas of tropical countries.

The attractiveness of ambiguous lands invites us to interrogate the distributional implication of their allocation. Distributional consequences are significant, given that thousands of people either use or actually live on state-owned resources. The economic characteristics of these lands further complicates the equity issue. Despite legal enclosure, many state-owned lands are in reality easy to access, and the actual users and their ecological impacts are difficult to assess. Moreover, local commons, access to which is usually limited to local residents, often contribute in an important way to the larger ecosystem. From the state's point of view, however, the same resources may be considered too scarce and valuable to permit the locals to utilize them, so that certain local commons such as forests become vulnerable to state enclosure. These issues raise questions about how the benefits of commons should be distributed among unequal power holders. Political motivations driving state enclosure and the costs imposed on local people are often obfuscated under the rubric of the seemingly neutral term 'environmental conservation',[1] making an analysis of the distributional dimension even more necessary.

Scholars of common property have tended to emphasize the question of 'who has access to what'. It is important to go further than this, and ask why certain people *end up* using certain types of resources and not others. The basic conditions that determine the utilization of various forest resources in Thailand have largely been shaped by the state's simplification of tenure on forests and lands. To identify these conditions, the following section will examine the history of state simplification in Thailand, showing how the two types of ambiguous lands — state-owned but privately-cultivated lands, and communal lands — were created,[2] and outlining the government activities

---

1. Scholars have tended to pay more attention to the sustainability of regulatory systems and their ecological implications than to equity and distributional aspects within a community and between communities. For an exception, see Jodha (1992).
2. All land in Thailand is either privately owned or owned by the state. All common lands or local commons, therefore, are state owned. In this chapter, I will use the term 'communal lands' to indicate the local commons that are created or guaranteed by the government to be used by the people. The term 'common lands', on the other hand, refers to lands that people have been using customarily without state recognition.

that take place on these lands. This will be followed by an examination of how the Karen, one of the hill peoples living on the ambiguous lands, have struggled to survive between the forces of capitalistic development and forest conservation. Two Karen villages were selected for a detailed study of forest use and dependency — the type of critical data often missing in defining the problem. These data are used to compare how market access and the differences in household wealth within the selected communities influence their connection to forests. The results of the studies suggest that the state's effort to reduce the forest dependency of the Karen, or even to evict them from the forests, has not led to the stated objective of conservation. The final section presents some wider implications with reference to James Scott's thesis on state simplification.

LOCATIONAL SIMPLIFICATION AND THE 'LAND IN BETWEEN'

Pressures from the Conversion of Unoccupied Common Lands

From the Ayutaya period (1350–1767), all lands in Thailand were theoretically considered the property of the king. However, custom and tradition over several centuries gave the so-called freeman the right to take as much land as he and his family could cultivate, but often not exceeding 25 *rai* (1 *rai* = 0.16 ha) (Wales, 1934: 121). Agricultural land accounted for only 2 per cent of the land in Thailand around 1850, and throughout the country there was an abundance of unused land, which needed only to be cleared and cultivated (Ingram, 1971: 12). Underpopulation was seen as the main obstacle to national development. The Harvard sociologist, Carle Zimmerman concluded from his pioneering economic surveys of rural areas in Thailand that 'there is *still great under-population in Siam* and that health work is needed to promote more population because these *greater numbers are necessary to an improved agricultural technique in many of the areas*' (Zimmerman, 1931: 225, emphasis in original). With a sparse population in relation to abundant land, there was no need for either farmers or the government to demarcate property.

Slavery was widespread until the beginning of the twentieth century, and the ownership of labour rather than land guaranteed the power of feudal chiefs. The relationship between patrons and freemen was personal, not territorial (Ingram, 1971: 13). A dramatic incorporation into the global market economy began after the 1855 Bowring Treaty with the British, which opened up the global rice market for Thailand. Rice was exported primarily to western Europe and then to China (Thompson, 1941). Lands previously considered 'wastelands' gradually gained market value and became targets for occupancy and competition. Teak extraction by European companies also accelerated in the north: with the destruction of European oak forests, the Europeans began to use teak for wooden sailing vessels, and later for the decks of iron steamships (Falkus, 1989: 133).

The population grew together with the commercialization of agriculture in lowland areas. The rising value of land began to generate conflicts over its use and ownership in the late nineteenth century, and the government finally launched a massive cadastral survey in preparation for a western-style property system based on individual holdings of land. The first cadastral survey began in 1901 and became the basis for tax collection in cash instead of in kind. Administrative and legislative action was taken by the central government to provide more clearly-defined property rights to agricultural land, primarily to facilitate the expansion of rice cultivation and resolve disputes over newly developed areas (Brown, 1988). Through the rapid adoption of a western-style property system and by functioning as a buffer between the British and the French, Thailand was able to maintain its independence and resist colonialization.

Various surveys aimed to create a direct one-to-one relationship between land and owner (or user). Certain types of land, however, tended to escape this type of state simplification. These were lands that had been used as collective property: roads, grazing lands, coastal areas, etc. It was only in the 1930s that the government created new legislation regarding the use and ownership of these communal and unoccupied lands.[3] The primary objective of this new legislation was not so much to guarantee communal rights for all people but rather to secure public lands for state interests and prevent them from being privatized by farmers and other private interests. These lands legally became the property of the state with the status of 'collective assets of the nation', and villagers were officially required to register for permission to use them.

It is interesting to note that forests were not explicitly included in the list of 'collective assets'. This may reflect the fact that up until the 1930s the state attached no economic value to forest or to other types of collective lands (such as grazing lands) that demanded protection. Furthermore, forestry was the exclusive domain of the forestry department and some foreign logging companies (especially in areas with valuable timber). General forest legislation, apart from that which protected valuable species such as teak, remained vague, as did the definition of 'forest' itself. The 1941 Forest Act defined forest as 'all land that does not belong to any individual based on the land law'. Extent of tree cover was not a condition that defined forests in a legal sense. This definition of forest and state property allowed for contradictions and ambiguities between those who legally owned the land and those who actually used it. These 'ambiguous' areas became the source of many conflicts. The government attempted numerous forms of

---

3. The Civil and Commercial Code of 1932 and the Waste Land Act of 1935 were the first to recognize the status of lands other than private tenure. The world economic depression in the late 1920s put many farmers into severe debt, and accelerated land concentration as well as the need to regulate common lands, which had not been an issue for centuries. This was particularly important in the densely populated central region (Kitahara, 1973).

land settlement projects to bring order to these areas, but were mostly unsuccessful. The goal of these projects was to stabilize the population and suppress anti-government movements, as well as to deflect the people from continuing to encroach on state property.

Pressures from Conservation Policies on Common Lands

Aside from serving as open-access reserves in case of land shortages, certain forests have served as local commons to provide an essential source of fuel, food, medicine, and materials for building houses, boats, baskets, and so on (Falkus, 1990: 69). As with agricultural lands, forests were abundant in most parts of the country; in certain locations where a scarcity of forest was perceived, local people developed a communal mechanism to regulate the use of forests to avoid over-exploitation or privatization. In the northern region, such communal forests are to be found in watershed areas; in the northeast there are sacred 'ancestor forests', connected to religious beliefs that have prevented them from being cultivated or appropriated privately (Ganjanapan, 1998).

At the state level, the large-scale planned conservation of trees began in the late nineteenth century, motivated by the expanding European teak industry in the northern regions. Until that time, the logging of teak in the north had been conducted at the whim of feudal chiefs in Laos and largely operated by Burmese and Chinese merchants. As the teak industry became profitable, questions of lease and ownership, royalty collection, and governmental authority became prominent issues.[4] Western timber companies demanded that the central government impose strict restrictions on timber extraction in order to protect the long-term prosperity of their industry, precipitating an unusual unity of interest between the Siamese government and the western logging companies (Brown, 1988: 117).

As part of the westernizing policies pursued by King Chulalongkorn, the government invited H. Slade, an English forester from India, to serve as the first director-general of the Royal Forest Department (RFD), established in 1896. His task was to regulate the over-exploitation of teak in the north. Unlike other exportable resources such as tin, which needed more foreign investment, the teak business was a target for regulation and constraints. At that time, forest cover was estimated to exceed 70 per cent of the country (Feeny, 1989). The total staff of the RFD charged with the care of those forests was less than thirty in 1908 and still only one hundred by 1934.[5] Increasingly strict regulations imposed by the RFD to protect

---

4. The trade treaties signed during the mid-nineteenth century with western countries had imposed a financial burden on the Thai government. The increase of trade was too gradual to be relied on to produce sufficient tax funds. This situation forced the government to concentrate on improving the efficiency of its tax collection (Vella, 1955).
5. In 1995, there were approximately 18,000 staff members at RFD (RFD, 1996).

British operations favoured large foreign investors, who had the resources to sustain the increasingly expensive logging enterprise, to the exclusion of small domestic companies. Western companies dominated the teak industry until the 1950s.

The RFD gradually introduced legislation to classify forests. These laws reveal government perceptions of important (and unimportant) issues during each period. With the Forest Protection Act of 1913, the RFD extended its protection to cover non-teak trees for the first time. In 1938, it introduced the first territorial conservation policies to secure forest reserves for future logging operations. Territorial zoning was strengthened by the first Five-year National Social and Economic Development Plan in 1961, in which the government designated 50 per cent of the country as state-owned 'permanent' forest. Since then, the government has continued to classify land and forest as a way of determining who will have access to what.

Under the state territorialization of forest lands, no legislation was prepared to protect the communal rights of local people. All common lands were legally state property; with the lack of legal recourse and the ambiguity of property relations, local commons (such as forests, grazing areas, etc.) were increasingly incorporated into national parks and wildlife sanctuaries. Since the late 1980s, in reaction to attempts to simplify local resource systems, there have been counter-movements led by NGOs and universities calling for a new bill to guarantee the communal control of forests by farmers. The enactment of the bill is still pending, and the battle between villagers and the state is still intense in many parts of the country.

Battles over the 'Land in Between'

There are two dominant forces behind the simplification of forest lands: converting them and conserving them. Both forces are guided primarily by the government, although international conditions also have a strong influence on them. Global market demands prompted the expansion of rice cultivation, and later cash crop production, into the forest frontier. The creation of protected areas and the scientific regulation of forests are responses to international environmental movements, particularly in view of the increasing availability of donor funds for the environmental sector. These two dominant forces tend to have particular impacts on certain groups of people. For example, ethnic minorities and landless farmers have not only been sidelined in the process of economic development, but have also been burdened with the costs of conservation initiatives in the very locations where they sought to escape from poverty.[6]

---

6. Migration may, for example, be due to population growth accompanied by a reluctance to subdivide land; debt foreclosure; loss of tenancy rights; insufficient capital to purchase land on the part of young families; or forced displacement caused by infrastructural schemes and plantation projects (Hirsch, 1990: 36).

Forests have served not only as an economic alternative but also as a political haven for disaffected groups fleeing political repression, particularly when communist movements were active in the mid-1970s (Hirsch, 1990).

The shift in property relations can be illustrated by picturing exclusive state lands as one dimension, and private farm lands as another. The ambiguous area, or the land in between the two, is owned by the state but utilized by local villagers. When land was abundant and population scarce, the public lands used exclusively by the state were limited to areas for mining, military bases, and teak harvesting. Private lands with exclusive property rights were also limited because the majority of the land was, in principle, open-access, and guaranteed by usufruct rights based on customary laws. A large portion of land, therefore, belonged to the state yet was *de facto* privately accessible (ambiguous). As the economy became more market-orientated, however, private assets gained in importance and the legal structures that protected private rights were extended, resulting in the expansion of private farm land. This expansion was strongly supported by development policies and foreign aid. The large-scale introduction of commercial crops including sugar cane, coffee, tapioca, maize, and eucalyptus into private lands played a major role in soil erosion. The state, meanwhile, strengthened its exclusive control over public lands, the area where forest land was most prevalent.

My focus is on what is happening to the land in between. Typical of these ambiguous areas are the fringes of national parks and other types of protected areas now often referred to as 'buffer zones'. Buffer zones, therefore, are another product of state simplification. In recent literature on protected area management, fringe areas of core conservation zones are seen as important for providing an additional ecological layer, and also as places where the basic needs of the local population can be met (Sayer, 1991).

Buffer zones attract much attention not only because of their relation to the core protection area, but also because they happen to fall under two contradictory national policies in Thailand: land reform and forest conservation. Competition between the Agricultural Land Reform Office (ALRO) and the RFD in the buffer zone has become acute: on the one hand, government policy since 1975 has been to redistribute land to landless farmers; on the other hand, a parallel policy aims to expand protected areas to cover up to 25 per cent of the total land area. Based on statistics from 1996, the protected area is still less than 15 per cent (RFD, 1996). The problem is that there is no land to provide to the landless other than the forest reserves which were previously granted to logging concessions. Within the forest reserves, areas no longer suitable for plantation and areas largely encroached on by farmers are supposed to be handed over to the ALRO. The ALRO will then decide how to distribute this land to the landless poor. From the RFD's point of view, however, forest reserves in the buffer zones should be the first to be included in the expanding protected area. Unfortunately, there are no clear scientific criteria to determine which lands should be given to the farmers through the ALRO, and which should remain under the RFD's control. For the ALRO, deforestation is not an issue; in a sense, it is a necessary part of its solution to resolve the landless problem.

PEOPLE IN BETWEEN

The Karen under Simplification

There are nine official categories of hill people in Thailand (Bhruksasri, 1989), including the Karen population, estimated to number about 600,000 (Hayami, 1996). In the eighteenth and nineteenth centuries, the Pwo Karen in western Thailand were considered allies by the Thai king. They served the king as scouts in anticipation of the Burmese invasions from the west, as well as providing valuable forest products such as ivory, cinnamon, and cotton (Gravers, 1994). The British also valued the Karen's knowledge of timber and their expertise in mobilizing elephants for their harvesting in the mountains and jungles of the frontiers. King Chulalongkorn's administrative reform in the early twentieth century, however, marginalized the Karen's previous position in the Thai polity (Jorgensen, 1996). During the 1960s, a communist insurgency extended to the Thai border, and the economically marginalized hill people found themselves on centre stage once again. In the 1970s, many of the Karen joined communist groups which made them politically dangerous 'tribes' in the view of Thai society.

The Huai Kha Khaeng (HKK) wildlife sanctuary extends over the Uthai Thani and Tak provinces of western Thailand. It is a central component of the kingdom's integrated protected area, which covers 14,000 km^2, including five wildlife sanctuaries, eight national parks, and three would-be national parks.[7] Cubitt and Stewart-Cox note that 'for a country that is widely known and often criticized for its deforestation, this is astonishing. It would take a grueling hike for about a month to get from north to south of [HKK]' (Cubitt and Stewart-Cox, 1995: 124).[8] The HKK forest was designated a wildlife sanctuary under Thai law in 1972 and later, in 1992, became the first natural UNESCO world heritage site in Thailand.

---

7. For purposes of comparison, Yellowstone National Park is about 5,500 km^2.
8. HKK and the adjacent Tung Yai Naresuan world heritage sites consist of a complex mosaic of evergreen and deciduous forests. They lie at the meeting point of four biogeographic zones and derive elements of their flora from the west and north (Himalayan) and south (Sundaic) regions. At least 120 mammals, of which 5 are endemic, 401 bird species, 41 species of amphibians, and 1207 species of fresh water fish have been recorded (Nakasathian and Stewart-Cox, 1990).

Only 22.1 per cent of the total population in the buffer zone surrounding the sanctuary today was born in the area. They are all Karen (CUSRI, 1992). Originating in Burma, a portion of the Karen population resided inside the sanctuary along the Kwa Yai River until they were evicted from the area in the late 1970s. Government intervention with hill people intensified in the late 1960s when communist propaganda reached marginal areas on the hills. Various government schemes, such as cash crop promotion, health care delivery, and education services, were introduced to co-opt the hill people and keep them from turning to the communist party.

In terms of the general discourse on the Karen's connection with forests, they, along with other hill people, have long been identified as 'forest eaters' (Gravers, 1994). As early as 1923, a commentator of Karen economy noted that: 'In some places the same clearing is cultivated three years in succession, but in others a new clearing is made every year. In the past, large tracts of forest have been destroyed in this way, and even now the Forest Department has taken the matter up, a good deal of destruction takes place in valuable forests' (Andersen, 1923: 55).

More recently, the Thailand Development Research Institute (TDRI) argued that 'for the Royal Thai Government, the hill tribes pose a series of profound political, social and ecological problems. Much highland deforestation ... can be laid directly at their door' (TDRI, 1987: 80). A people once blamed for their backwardness and opposition to modernization and development are now being blamed from an environmental standpoint. The proposed strategy which flows from this position is, naturally, to move people away from the precious forests.

> As described by Mr. Chatchawan Phitsamkham, the superintendent of the Huai Kha Khaeng Wildlife Sanctuary, people have been attracted to the forest and its resources like ants to a lump of sugar. The ICAD (Integrated Conservation and Development) activities we propose are intended to turn people *away* from the protected areas, to attract them to 'new and sweeter lumps of sugar' outside the forests. (World Bank, 1993: 11, emphasis in original)

Having been evicted from the HKK and now squeezed into the 'buffer zone' of an internationally recognized protected area, the Karen are not only denied their traditional farming techniques, but also modern chemical agriculture for fear they will damage the forest. This is a 'late developer's trap' — you cannot go back but neither can you move forward in the same way that 'modern' farmers have done. Michael Dove summarizes the dilemma as follows: 'the challenge is to achieve the benefits achieved by past paths while *not* following them' (Dove, 1994: 1069, emphasis in original).

Due to a lack of documentation, it is difficult to describe how the Karen used to interact with forests before their eviction from HKK. However, with the aid of reports by three Danish anthropologists (Ewers, 1994; Gravers, 1994; Jorgensen, 1976, 1996) who conducted their fieldwork in the Karen community in the early 1970s, and my own interviews with elderly people in the villages, a general picture emerges. Rotational shifting cultivation of upland rice was the central economic activity of the people. Their production portfolio consisted of yam, taro, cotton, tobacco, chillies, and various vegetables in the same plot of approximately 5 *rai* (1 *rai* = 0.16 ha) per household. Banana trees and sugarcane often surrounded swidden agriculture plots (Gravers, 1994). Metal products (such as knives), guns, salt, and certain cloth had to be purchased from the local market in the distant town in exchange for their agricultural products. Although little is known of the extent to which the Karen impacted on the forests in the past, their long-term continuous residence in the same area demonstrates, at least, a successful implementation of the swidden system. The rotation cycle of a fallow period was between ten and fourteen years. As a villager once said to me with some frustration, 'if our farming system was such a bad thing, the Huai Kha Khaeng forest would never have become a world heritage site; it would have disappeared by now'.

Despite the blame assigned to hill people and local villagers living close to rich forests, there are almost no data on the nature of their forest use. This is surprising because, where real problems are perceived, states usually gather data that will enable them to address those problems. A project of state legibility is often directed towards obstacles so that they can be effectively targeted and manipulated, as Scott demonstrates with the Nazi's racial map of Jews (Scott, 1998: 78). Official statements frequently refer to the destructive activities of local villagers, but the persistent absence of empirical data on villagers' actual forest use suggests that the state may not perceive villagers as serious threats at all, but uses them to deflect public attention away from corruption, large-scale infrastructure development, and illegal logging by public officials themselves.[9]

The following section presents an analysis of the kind of data that have been completely missing from the debate over whether to allow the Karen to live in the buffer zone, which are also important in assessing the legitimacy of the state's project in dealing with local peoples. The analysis focuses almost solely on the material connections between the Karen and the forests because this, over and above cultural attachments and non-material linkages that people have to their forests, seems to be the central concern for policy makers.

Changes in the Karen Economy

I studied the Karen communities adjacent to the HKK forest complex from March 1996 to March 1997. There are more than twenty villages within a 5 km radius of the HKK boundary. Four of these villages are the Karen

---

9. For example, the RFD began to count the number of people living inside the conservation forests only in the 1990s.

communities with the longest histories in the area. They are also the most dependent on forest resources. From the four villages, I selected two as intensive study sites for the following reasons: (1) the populations of both villages are about the same and reasonably large, reflecting various types of livelihood strategies within each village; (2) although their distance from forested areas is the same, village A has inferior access to roads and markets, enabling me to measure the influence of the market economy in relation to forest use; (3) both villages have similar histories of migration and initial property holdings, both having been evicted from the HKK in the 1970s.

The 1970s were a critical turning point for the Karen economy. A Thai logging company built the first road that penetrated into the villages. The road had a significant impact in pushing the subsistence economy into the market economy. Most notably, middlemen could now reach the formerly inaccessible villagers, which encouraged the Karen to plant cash crops. Mono-cropping of maize increased the number of weeds, and ploughing was needed to prepare the land. Repeated use of the same land made the soil harder and more difficult for animals to plough, so that farmers had to rent tractors from those who owned them, mainly Thai moneylenders in the lowlands. The introduction of tractors reduced the opportunity to harvest different crops throughout the season. Whereas traditionally there had been a time lag between the planting of rice and the maturing of other crops such as vegetables and potatoes, the need to plough mechanically forced the Karen to give up intercropping.[10] From a rice-based economy, which allowed for cultivation of vegetables, sweet potatoes, sugarcane and some tobacco and cotton within a manageable labour cycle, the shift to mono-cropping of maize and cash crops implied a more intensive labour schedule, with two plots to be worked simultaneously (Jorgensen, 1976; author fieldwork).

Meanwhile, highland development policies initiated by the state from the early 1960s did not have their intended effects. Attempts to introduce paddy rice techniques as a way to push the Karen out of shifting cultivation largely failed, with few farmers adopting them.[11] Coffee plantation projects in the

---

10. The approximate costs of maize production can be calculated. To cultivate 5 *rai*, which is the break-even size to make any profit, farmers would need two rotations of ploughing by tractor. Tractors cost between 750 and 1750 *baht* to rent for the first round and around 500 to 600 *baht* for the second round. Purchase of seeds came to about 900 *baht* for 15 kg, and another tractor rental added 400 *baht* (80 *baht* per *rai*). If the farmer were cultivating more than 15 *rai*, he would need additional cash to hire extra labour, a practice that has become more common than relying on traditional labour exchanges. The reciprocal nature of labour exchange becomes unsustainable when certain people gain access to mechanical equipment and the land distribution becomes unequal within a community. Hiring labour then becomes more convenient, because it precludes the obligation to return labour in exchange.
11. There were two main reasons why farmers did not adopt this high-yielding technique: (1) the difficulty in obtaining flat land to retain water; and (2) the difficulty of mixing a variety of crops to satisfy their subsistence needs.

Table 1. Basic Data on the Two Research Villages (1995)

	Khongsao Village (A)	Ban Mai 2 Village (B)
Population	224	230
Average annual income[1]	18,270 *baht*	33,026 *baht*
Self-sufficiency of rice	64%	39%
Frequency of labour exchange[2]	3.7 days per year	3.5 days per year
Ave. size of land under cultivation	6.3 *rai*	10.4 *rai*
Access to the market	difficult access by car	accessible by car
Village access to electricity	no	yes

Notes:
1. The high average income figure results from the inclusion of high salaried people (school-teachers, clinic doctors); without these non-farming people, the average income of village A would be around 10,000 *baht* and that of the village B around 20,000 *baht*.
2. 'labour exchange' represents the average number of labour days exchanged among villagers during the planting season of July and August.
(1 *baht* = 3 ¢; 1 *rai* = 0.16 ha)

1980s also failed because of inadequate market connections. At the same time, forcing people to move out of the core area may have had an unintentional negative impact on the forests. The RFD considers forest fire as one of the major threats to the HKK today, with the sanctuary being damaged by fire every year. Ironically, because there are no people resident in the area, there is nobody to detect and fight forest fires until the flames are large enough to attract the attention of the forest fire unit in the park.[12]

The responses of the Karen to ever-tightening government regulation of forests (such as the abolition of shifting cultivation) and land (the imposition of a private property system) divide the Karen roughly into two groups: those who have intensified their cash crop production to escape from poverty, often at the expense of their subsistence production; and those who could not afford to follow this strategy and were forced to depend on other means, such as working on other farmers' residual land, and borrowing money from local moneylenders. Many of them are unable to work in the cities to earn cash because they lack citizenship, personal connections, the language skills or the confidence.

The ALRO is now responsible for implementing the integrated conservation and development project (ICDP) in the buffer zone area. Their latest approach to both conserving the HKK and responding to the needs of the farmers, particularly those falling under the second category above, is to try to shift people's economic orientation away from the forests by promoting income-generating activities in and outside the buffer zone. This new approach,

12. In 1994 alone, 68,271 *rai* (10,923 ha) reportedly burned down inside the sanctuary (Noikorn, 1998). While I was conducting my fieldwork, a villager informed me that fire fighters themselves often set fires in the forests because their daily wages double when they are actively fighting fires.

although recognizing the basic right of people to live in the area, still assumes that local people are the main abusers of forests. Poverty, together with a purported lack of knowledge about how to farm appropriately, are believed to be driving people to unsustainable resource use (PEM Consult, 1996). The next section will examine the validity of these assumptions.

Karen Forest Use and Dependency

Before discussing the kinds of interventions that might be conducive to conserving forests, it is important to understand how people interact with them on a daily basis. Unfortunately, the exact nature of forest dependency by local residents has seldom been investigated, and no site-specific information exists on the subject. Previous attempts to measure forest dependency in other areas of Thailand have tended to suffer from one or more of the following shortcomings: (1) the research span is often limited, and year-long variability is not taken into account; (2) when potential seasonal biases are taken into account, the size of the samples is often too small; (3) where the samples are large enough, the selection of households often ignores the economic stratification within villages, which might strongly affect levels of forest dependency; (4) data collection often relies on the memories of villagers or the recording of daily forest resource consumption by the villagers themselves, but such data are not reliable when one is dealing with the use of 'illegal' resources.

In an attempt to avoid these shortcomings, my own measurement of forest dependency of local villagers is based on different wealth ranks within villages; this makes comparisons possible, and questions the common demonization of poverty *per se* as the central cause of excessive resource exploitation. I developed two indicators to capture forest dependency. *Income dependency* is the percentage and amount of income people obtain from selling forest products. It is important to note that we measure the *relative* proportion of income derived from forests:[13] when one villager derives all his income from forests whereas another derives only 50 per cent, we can say that the former is more dependent on forests even though the two may have an identical income in absolute terms.[14] *Livelihood dependency* is

13. Measuring income was more challenging than expected. Villagers in general do not know exactly how much they earned last year. The total income from maize, which was the central source of income for the majority of villagers, was relatively easy to estimate because they sell this crop once a year. In contrast, bamboo shoots are sold on a day-to-day basis in smaller units. Based on accounts of villagers, I cross-checked information with the middlemen who had some statistics of purchases to come up with the best estimate.
14. Unlike an ordinary economic survey that often relies on income data (often unavailable), or the amount of land holding, wealth ranking allows us to observe how villagers themselves view wealth and poverty in their community. It also minimizes the snap-shot effect of relying on a single quantifiable measure.

measured by the variety and intensity of forest product consumption in terms of food contained in each meal. Direct use of forest resources requires labour investments and must, therefore, reflect the importance attached to the activity by villagers. I did not measure the amount of firewood or charcoal people consumed in the area because there seemed to be little variation among the households.

For income dependency, information was collected from all households through informal interviews and cross-checking. The choice of indicator for measuring livelihood dependency was difficult, but I eventually decided to observe the consumption of meat, for three reasons. Firstly, meat consumption reflects both the household's connections to the market and to the forest; meat was likely to be income elastic and could be expected to reveal differences in consumption between rich and poor. Secondly, people prefer to eat meat if they have a choice; it is an important source of protein, although they do not get to eat it often. The obstacles to meat-eating include the labour required to hunt, forest guards, and/or the cash required to buy meat at the local market. Thus, meat consumption will likely reflect the general economic well-being of a household. Thirdly, from the viewpoint of the forest guards, animals are a forest resource to be protected. Forest animals are thus contested resources and consumption of them reflects the desperation of villagers under adverse circumstances.

Using a wealth-ranking exercise, twenty sample households were selected from each of the two villages (forty in total). With the help of two assistants from each village, a number of randomly selected meals was observed to test the frequency of forest meat consumption in each of the following seasons: thirty meals between July 20 and August 20 (the busiest time of the year for planting seeds and preparing the soil), another thirty meals in October (a rainy month with no major work, economically very difficult), and an additional twenty meals from February to March (during the dry season and after the harvest, when people are generally better off). These sampling periods reflected various seasonal conditions that may affect villagers' access to the forest. The Karen in this area normally take two meals per day with little difference in the content of each meal. This provided a total of eighty samples for each household. Village assistants were permitted to choose which meal to observe each day.[15]

From the ranking exercise and discussions with villagers, I was able to extract criteria for wealth (and poverty) that people implicitly use to evaluate each other. Some of the common criteria were size and neatness

---

15. Forest animals that are hunted and eaten are mostly 'illegal' in a strict sense. To increase the reliability of the data, I waited about a month before selecting villagers to help with data collection. Based on my discussions with them, I developed a matrix for documenting observations: different symbols were used for meat that came from the forest and that from the market to obtain a rough idea of the relative proportion of food originating from each source.

of house, amount of land, number of children (fewer is better in this case), debt, income, family labour power, and self-sufficiency in rice. These elements were combined in complex ways to come up with the total ranking of households. From these indicators, I learned that the amount of land under cultivation is strongly correlated with income dependency on forests.[16] My hypothesis is that because the busiest time for planting and weeding overlaps with the bamboo harvest season, those with larger land holdings cannot afford to allocate labour for obtaining forest resources; differences in dependency are not necessarily due to wealth. The results of the data collection based on this method are presented in Tables 2 and 3.

Some observations can be made from these data. In terms of income dependency based on the amount of land, less wealthy families tend to depend more on forest products (i.e., bamboo shoots). This is not surprising given the fact that their private workable land is limited. Forest products are important not only in terms of supplementing income but also in equalizing the flow of income throughout a year (bamboo shoots generate income during the months of July and August when no other income

Table 2. Income Dependency Based on Amount of Land under Cultivation

	Village A (limited access to the market)		**Village B (better access to the market)**	
Area under cultivation (*rai*)	Annual income from forest (avg. *baht*)	% share of income from forest	Annual income from forest (avg. *baht*)	% share of income from forest
More than 20	3500 (n = 1)	14%	1512 (n = 12)	4
15 to 19	1780 (n = 10)	11%	2833 (n = 6)	8
10 to 14	2000 (n = 3)	22%	2916 (n = 6)	17
less than 9	1664 (n = 22)	34%	4552 (n = 19)	35

Notes:
1. The price of bamboo shoots in village A is lower than that of village B by 20% because of bad road access.
2. Area under cultivation is different from area under occupation, because many farmers do not have enough capital to fully cultivate their land. The figure includes only cash crops.
3. Those who have a regular salary (e.g. schoolteachers) are excluded from the sample. It was interesting to find that even a relatively large family did not split up its labour to maximize its income from the two sources (i.e. forest and agriculture). It was natural for them to work together in the same place as a family.

16. Initially, I analysed income dependency based on wealth categories. All households were dependent on selling bamboo shoots at a similar level. I needed to look further into the composition of 'wealth' and decided to use the size of land as a criterion to differentiate villagers.

Table 3. Livelihood Dependency: Proportion of Meat derived from Forest, as Percentage of Total Meat Consumption, by Wealth Rank (one-year weighted average)

Wealth Rank	Village A	Village B
1 (sufficient)	39%	33%
2	60%	33%
3 (average)	39%	44%
4	56%	37%
5 (poor)	52%	46%

Note: Figures are calculated based on the weighted average of frequency in each season. For wealth rank 2, the sample was n = 1, so it is likely to be more biased than others.

sources exist). In terms of livelihood dependency, no clear disparities were found between the higher and the lower-ranked. However, access to the market and good roads seems to differentiate the level of dependence between the two villages. Furthermore, the year-long study showed that, regardless of wealth and market access, most of the villagers' meals are composed not of meat but of rice, chilli, and some vegetables. Meat consumption depends on the season, on the villagers' willingness to go into the forest, and on mere luck. Much of the villagers' protein intake comes from fish and small animals such as wild chicken that are not the central target for conservation. There is thus no justification for characterizing the Karen as poachers of wild animals. At the same time, however, the frequency of meat intake, although lower than expected, should not give the impression that forest resources are unimportant to Karen lives. For many, especially the poor, the forest remains their only source of livelihood.

The dependency on forests observed among the majority of the population in the research site, set against the strict policing of the forests by the RFD, indicates that the Karen have limited opportunities to generate income and secure food. Hunting activities are often carried out not in the forest but in the Karen's fields during harvesting, when wild animals such as pigs come out to feed on the crops.[17] There is almost no selling of animals inside or outside the village. Timber use is also very limited — most houses are constructed with bamboo. Encroaching farmland into the conservation areas is too risky and too easily detected by the guards. On the other hand, some Karen do take advantage of the ambiguous demarcation of the sanctuary and look for forest products around the border. In many

17. In this sense, at least, forests are not only helping the local people, but the local people are also helping the forests. I thank Professor Michael Dove for encouraging me to explore this insight, although it is not possible to do so in full in this chapter.

cases, they know where to look for the products while escaping the eyes of the forest guards.

Shifting cultivation has almost disappeared in this area, mostly because, as a result of state conservation policies, there is no place to shift to. Furthermore, shifting cultivation is ill-suited to the prevailing land registration and economic system where rewards are given to those who cultivate the same land every year and generate cash. In this sense, the nature of the Karen's forest dependency has changed dramatically from farming in their forests to collecting resources from the state forest. The separation of farmland and forest land is becoming increasingly sharp. Ironically, Karen living on the fringes of the HKK are now the target of development projects aimed at reducing the impacts of their *non-traditional* agriculture, which they only recently adopted after being forced to abandon their traditional shifting cultivation.

The living conditions of marginal hill people on the edge of biologically rich forests illustrates a concentration of stress from various forces. The twin facts that hill peoples are concentrated in the northern and western forested areas and that they traditionally practice shifting cultivation have encouraged most government officials to see them as the principal cause of deforestation (Rigg and Stott, 1998: 108). However, one can also argue that if the forested areas survived for so long with hill people living in and around them, the Karen cannot be *the* cause of environmental destruction. The future connection of the Karen with the forests has to be examined in the larger context of increasing privatization of lands and intensifying calls for preservation of biodiversity. The increasing scarcity of resources and space for livelihood were not naturally generated: they were socially produced by the establishment of the sanctuary and the regulations of land use.

SOME WIDER IMPLICATIONS: SIMPLIFICATION AND DIVERSITY

In developing some of the theoretical implications of this case study, James Scott's 'state simplification' thesis serves as a useful point of departure. Scott argues that the state often fails to improve the living condition of its own people by over-simplifying the 'local':

> The more I examined these efforts at sedentarization, the more I came to see them as state's attempt to make a society legible, to arrange the population in ways that simplified the classic state functions of taxation, conscription, and prevention of rebellion ... But it is harder to grasp why so many well intended schemes to improve the human condition have gone so tragically awry. (Scott, 1998: 2–4)

Whether states typically intend to improve the quality of life for their populations is open to question, but before deploring the repeated failures of state projects, it is certainly useful to focus on how the state actually *prepares* simplification. The investigation of state preparation may allow observers to minimize the negative consequences derived from wrong interventions. For this preparation, Scott coins the terms 'legibility'. Legibility, according to Scott, is a condition of manipulation and invention of units that are visible to the state (Scott, 1998: 183). In this study, I have dealt primarily with two subjects of state simplification that are related to forests: land and the treatment of hill people. Three general lessons concerning state simplification can be drawn from this case study.

First, the state does not always simplify or impose legibility but often takes advantage of existing illegibility. States choose when and where to impose simplification to make a system more legible and controllable. Simplification is only a means to achieve the competing goals of the government. Confiscation of communally-utilized land, swidden fallow, and their subsequent transformation into protected areas, occurs because of farmers' lack of established tenure and political power, not because of the illegibility of local systems. The ecological consequences of this are clear. If the forests such as those of the Karen are taken away and the pioneering settlers are rewarded with lawful land certificates, what incentive is there for local people to manage forest lands sustainably? Despite its purported objective to conserve forests, state intervention often provides a negative incentive for local people to do so themselves.

Second, state simplification affects different people differently. We have seen in Thai history that forest conservation in the form of state imposed restrictions to forest access often entailed denying access to local people while granting access to societal interests such as energy, mining, and plantation industries — interests which have had more damaging impacts on the survival of forests. In the past, restrictions on access to forests worked to the advantage of European companies, the only groups capable of paying the costs for expensive logging operations. Today, protected areas without local residents often become the best candidates for dam construction sites. Large-scale illegal logging inside the national parks by high-ranking officials is a frequent occurrence.[18] International organizations are pouring funds into the government to fill this incentive gap without questioning the ways in which the state attempts to conserve forests. The consequences of restricting access to forests are by no means equal.

Third, in the residual lands produced by the state simplification process, competition occurs not only between villagers and the state, but also among the government departments themselves. In this sense, one department's attempt to make a system more legible may be creating illegibility for other departments within the government. As we have seen, there is inherent

18. The most recent incident was reported in the Salween national park in Mae Hong Son province adjacent to the Burmese border. Thousands of logs were secretly brought to Burma where they were stamped and smuggled back to Thailand as if they had been legally imported. This operation allegedly involved high-ranking officials from the military, the forestry department, and the police department (Kaopatumtip, 1998).

conflict between the RFD and the Agricultural Land Reform Office over control of the buffer zone, even though they belong to the same ministry. Functional specialization and increasing divisions in the government bureaucracy create diverse groups of interests even within the RFD. Some officials, though limited in number, are sympathetic to the plight of villagers and support progressive legislation on community forestry that grants people lawful access to forests.[19]

Especially since the ban on logging, forestry projects have emphasized the preservation of resources and not the ways in which resources are used by different people. Hill people have benefited little from the 'development' of the highlands, despite the valuable resources in these areas. In addition, hill people have been blamed for the deforestation that has destroyed about half of the nation's forests, even though they make up only 1 per cent of the population. The level of attention and intervention devoted to certain locations and people are quite independent of their measurable characteristics (soil, climate, ecology; population, poverty, distance from forests). It is the political economy that makes a particular combination of resources and people more prominent than others. I contend that this particular combination of marginalized people residing next to a rich forest has not occurred by accident but has been structurally produced.

CONCLUSION

State simplification often complicates more than it simplifies. I have attempted to show in this study that, despite the state's consistent effort to divide lands clearly into public and private spaces, irreducible and perhaps growing ambiguity remains not only at a national level but also within villages and within state agencies. This ambiguity is due partly to the vaguely worded legislation and policies that allow for various and often conflicting interpretations. It is also the result of the failure of the state to incorporate the ambiguity and establish communal property as part of the system. In areas where forests are still critical to local economies, state provision of these as communal forests may be the first move towards conservation; but, for conservation to be effective, it must go hand-in-hand with the suppression of illegal logging by élites and the building of capacity for the locals to organize themselves. This is not because élites necessarily do more damage than farmers, but because failure to build this capacity will weaken the credibility and the moral foundation underlying all legislation on forest

---

19. Diversity applies to NGOs as well. Environmental NGOs are often characterized as emphasizing either villagers' rights (the light green) or wildlife (the dark green). However, in opposing mining in wildlife reserves, for example, both types of NGOs and local villagers can all agree (Achakulwisut et al., 1998). Attention to issues rather than organizations will provide more room for inter-organizational alliances.

access. The role of local people in forest conservation (and destruction) should be understood not only from the angle of how they interact with forests directly; it should also be assessed in terms of the function they perform as a buffer between the larger political/economic forces and forests.

The global discourse on environmental issues, particularly since the 1980s, has significantly affected how upland minorities are categorized. These shifts in categorization, however, have almost nothing to do with how the Karen actually interact with forests or how they define themselves in a wider context; they serve solely as political tools to justify new forms of intervention. Moreover, international concern over scarce resources provides the state with an additional reason to intervene, and to utilize whichever discourse serves its interests (Peluso, 1993). Resource politics also affects the position of the people living close to the resource in question. The Karen have traditionally resided in between the lowland Thais and the upper highlanders such as the Hmong. They have placed themselves in between Burma and Thailand, not only geographically but also politically. Now they also find themselves situated in between conservation and development.

In terms of the direction of future research, the categorization of NGOs, the state, and local people as distinct entities is no longer useful for the analysis of forest control. Each of these is often composed of conflicting subgroups and factions. The heterogeneity of the state, on certain occasions, provides room for manoeuvre from the viewpoint of local villagers. By exposing diverse departmental interests, one might find ways in which farmers could take advantage of potential allies within the state apparatus to counter other, more constraining influences of the state (such as the farmers' general preference for land reform rather than forest protection). How to match up components with similar interests and create alliances is a key question.

The human capacity to develop institutional mechanisms to manage a resource that is locally perceived as abundant yet globally defined as scarce is being challenged. History shows that once the scarcity of a certain resource is detected by powerful agents, it is doomed to be overexploited (Dove, 1993). Incentives to conserve a scarce resource, therefore, go hand in hand with incentives to exploit it all the more, to the exclusion of local people. Can we initiate an equitable, long-term management strategy for a resource which has not been perceived as scarce? Scholars have the freedom to shift between the various viewpoints that define 'scarcity' — but they also have an obligation to contextualize these points of view and address the socio-political implications of privileging one definition over another.

Acknowledgements

The research upon which this essay is based was made possible by the Regional Community Forestry Training Center (RECOFTC) in Bangkok, to which I was affiliated as a visiting scholar from August 1995 to March 1997. I would like to thank Dr Somsak and Dr Vitoon for their assistance. I am also grateful to Dr Wirachai of the Agricultural Land Reform Office (ALRO) and Acharn Somporn for their generous assistance in village selection and contacts with local people. An earlier version of this paper was presented at the Agrarian Studies Colloquium at Yale University on 2 April 1999. I am grateful to all the participants of the colloquium, and to Steve Striffler, Eliza Darling, Seth Cook, James McCann, Pierr Minn, Kay Mansfield and an anonymous referee for helpful comments.

REFERENCES

Achakulwisut, A., V. Chinvalorn, and K. Inchukul (1998) 'Call to End Mining in Wildlife Reserves: Government Urged to Revoke Concessions', *Bangkok Post* 30 May.

Andersen, J. P. (1923) 'Some Notes about the Karens in Siam', *Journal of Siam Society* 17: 51–8.

Bhruksasri, W. (1989) 'Government Policy: Highland Ethnic Minorities', in J. MacKinnon and B. Vienne (eds) *Hill Tribes Today: Problems in Change*. Bangkok: White Lotus-Orstom.

Brown, I. (1988) *The Élite and the Economy in Siam: c.1890–1920*. Singapore: Oxford University Press.

Cubitt, G. and B. Stewart-Cox (1995) *Wild Thailand*. Bangkok: Asia Books.

CUSRI (Chulalongkorn University Social Research Institute) (1992) 'Social and Economic Studies of the Communities in the Buffer Zone of Huai Kha Khaeng Wildlife Sanctuary'. Social Research Institute, Chulalongkorn University (in Thai).

Dove, M. (1993) 'A Revisionist View of Tropical Deforestation and Development', *Environmental Conservation* 20(1): 17–24.

Dove, M. (1994) 'North–South Relations, Global Warming, and the Global System', *Chemosphere* 29(5): 1063–77.

Ewers, K. (1994) 'Politics of Biodiversity Conservation in Thailand: Global and Local Discourse'. Paper presented at the Workshop on Environmental Movement in Asia, Leiden (27–9 October).

Falkus, M. (1989) 'Early British Business in Thailand', in R. P. T. Davenport-Hines and G. Jones (eds) *British Business in Asia since 1860*, pp. 117–56. Cambridge: Cambridge University Press.

Falkus, M. (1990) 'Economic History and Environment in Southeast Asia', *Asian Studies Review* 14: 65–79.

Feeny, D. (1989) 'Agricultural Expansion and Forest Depletion in Thailand, 1900–1975', in J. Richards and R. Tucker (eds) *World Deforestation in the Twentieth Century*, pp. 112–43. Durham, NC: Duke University Press.

Ganjanapan, A. (1998) 'The Politics of Conservation and the Complexity of Local Control of Forests in the Northern Thai Highlands', *Mountain Research and Development* 18(1): 71–82.

Gravers, M. (1994) 'The Pwo Karen Ethnic Minority in the Thai Nation: Destructive "Hill Tribe" or Utopian Conservationists?', in *Asian Minorities: Three Papers on Minorities in Thailand and China*, pp. 21–46. Copenhagen Discussion Papers No. 23. Copenhagen: The Center for East and Southeast Asian Studies, University of Copenhagen.

Hayami, Y. (1996) 'Karen Tradition According to Christ or Buddha: The Implications of Multiple Reinterpretations for a Minority Group in Thailand', *Journal of Southeast Asian Studies* 27(2): 334–49.

Hirsch, P. (1990) *Development Dilemmas in Rural Thailand*. Singapore: Oxford University Press.

Ingram, J. (1971) *Economic Change in Thailand 1850–1970*. Stanford, CA: Stanford University Press.

Jodha, N.S. (1992) 'Common Property Resources: A Missing Dimension of Development Strategies'. World Bank Discussion Paper 169. Washington, DC: The World Bank.

Jorgensen, A. B. (1976) 'Swidden Cultivation among Pwo Karens in Western Thailand', in S. Egerool and P. Sørensen (eds) *Lampang Reports*, pp. 275–87. Copenhagen.

Jorgensen, A. B. (1996) 'Elephants or People: The Debate on the Huai Kha Khaeng and Thung Yai Naresuan World Heritage Site'. Paper presented at the 48th Annual Meeting of the Association for Asian Studies, Honolulu, Hawaii (11–14 April).

Kaopatumtip, S. (1998) 'Breaking the Vicious Cycle', *Bangkok Post* 19 April.

Kitahara, A. (1973) *Land Law of Modern Thailand: A Preliminary Discussion on the Land Law System before the Second World War*. Tokyo: Institute of Developing Economies (in Japanese).

Nakasathien, S. and B. Stewart-Cox (1990) *Nomination of the Thung Yai-Huai Kha Khaeng Wildlife Sanctuary to be a UNESCO World Heritage Site*. Bangkok: Wildlife Conservation Division, Royal Forest Department.

Noikorn, U. (1998) 'Forest Fires Rage at Sanctuary: Insufficient Staff Seen as Main Problem', *Bangkok Post* 11 March.

Peluso, N. (1993) 'Coercing Conservation: The Politics of State Resource Control', in R. Lipschutz and K. Conca (eds) *The State and Social Power in Global Environmental Politics*, pp. 46–70. New York: Columbia University Press.

PEM Consult (1996) 'Project Document: Huai Kha Khaeng Complex — Integrated Conservation and Development Project'. Copenhagen: DANCED/Ministry of Environment and Energy.

RFD (Royal Forest Department) (1996) *Forestry Statistics of Thailand*. Bangkok: Royal Forest Department.

Rigg, J. and P. Stott (1998) 'Forest Tales: Politics, Policy Making, and the Environment in Thailand', in U. Desai (ed.) *Ecological Policy and Politics in Developing Countries: Economic Growth, Democracy, and Environment*, pp. 87–120. New York: State University of New York Press.

Sayer, J. (1991) *Rainforest Buffer Zones: Guidelines for Protected Area Managers*. Gland, Switzerland: IUCN.

Scott, J. (1998) *Seeing like a State: How Certain Schemes to Improve the Human Condition Have Failed*. New Haven, CT: Yale University Press.

TDRI (Thailand Development Research Institute) (1987) *Thailand: Natural Resources Profile*. Bangkok: TDRI.

Thompson, V. (1941) *Thailand: The New Siam*. New York: Macmillan.

Vella, W. (1955) *The Impact of the West on Government in Thailand*. Berkeley, CA: University of California Press.

Wales, H. G. (1934) *Ancient Siamese Government and Administration*. London: Bernard Quaritch Ltd.

World Bank (1993) *Conservation Forest Area Protection, Management, and Development Projects*. Vol. 4. Prepared by MIDAS Agronomics Company, Bangkok, Thailand.

Zimmerman, C. (1931) *Siam Rural Economic Survey*. Bangkok: Bangkok Times Press.

8 Resettlement, Opium and Labour Dependence: Akha–Tai Relations in Northern Laos

Paul T. Cohen

INTRODUCTION

The Golden Triangle region of mainland Southeast Asia (comprising parts of northeastern Burma, northern Thailand and northern Laos) presently produces about 40 per cent of the world's illicit opium; in the early 1970s, it accounted for as much as 70 per cent. Most of this opium is grown by minority ethnic groups (Hmong, Yao, Lisu, Akha, Lahu, Wa), whose livelihoods are dependent on a highland swidden economy. As the principal cash crop of the highlands of this region opium has been the crucial economic nexus in inter-ethnic relations. Until the mid-1980s, when opium production began to decline steeply due to eradication and crop-substitution programmes, opium commerce in northern Thailand linked lowland traders and highland communities in complex credit transactions with debts usually settled in kind, in opium, maize or rice (Cohen, 1984; Miles, 1973). Opium is a labour-intensive crop and it was common for some highland communities to augment household labour with labour from neighbouring ethnic groups in order to maximize opium production. The Yao, for example, used to purchase non-Yao children (especially Tai and Khmu) to increase the size of the household workforce (Miles, 1973). More commonly, other opium-growing highlanders, such as the Hmong and Lisu, would hire in labour when necessary. These wage labourers were usually of ethnic groups that did not grow opium (Karen, Lua, Northern Thai) and were, more often than not, opium addicts (Cohen, 1984; Cooper, 1984; Dessaint and Dessaint, 1975; Durrenberger, 1974; Geddes, 1976; Lee, 1981).

Opium is also the main cash crop for the Akha of Muang Sing district in northwestern Laos. Yet, these Akha do not grow large quantities of opium (indeed, less than the consumption needs of addicts) even though the government has not yet adopted a punitive policy towards opium cultivation. They do not hire the labour of other ethnic groups and rarely of neighbouring Akha villages. Rather, many Akha communities — those who have settled on the lower slopes of the highlands — have come to provide a dependent labour force for Tai villages in the lowlands in a way that contributes significantly to the expansion of wet-rice production and the lowland surplus rice economy. The aim of this chapter is to explain this dependency; it focuses on the factors that have encouraged the movement of Akha down to the lower slopes (government forest preservation and resettlement policies, and low productivity of swidden agriculture); the economic and social costs of resettlement; the lack of government development assistance; and the role of opium production, exchange and consumption.

MUANG SING

Muang Sing is a district (*muang*) in the northern province (*khwaeng*) of Luang Nam Tha in Lao PDR. The district covers an area of 1,650 km^2 and has a population density of 14.2 persons per km^2. Muang Sing town, the administrative and economic hub of the district, lies at about 680 metres above sea level, in the centre of a small but well-watered and fertile valley. Rugged mountains that reach 2000 metres in height and extend to the Nam Tha plain to the east and the Mekong River to the west flank the Muang Sing valley and the smaller Muang Mom valley to the north. Muang Sing town is only 12 km by road to the border with China and the Dai Autonomous Prefecture of Yunnan province. Other roads, often impassable in the wet season, link the town to the provincial centre of Luang Nam Tha and to the small market town of Chiang Kok on the Mekong River, about 60 and 70 km distant respectively.

The total population of Muang Sing is approximately 23,500 and is ethnically quite diverse: Akha (sixty-eight villages), Tai Lue (twenty-six villages), Tai Neua (five villages), Tai Dam (one village), Yao (five villages) and Hmong (three villages). Officially, the Tai groups are classified as *Lao lum* ('lowland Lao') and the Akha, Hmong and Yao as *Lao sung* ('highland Lao'). All but three of the Tai villages are situated in the Muang Sing valley and all but four of the non-Tai tribal villages are located in the highlands and foothills.

The Akha speak a Tibeto-Burman language. According to Akha legend, they began to slowly migrate from their ancestral homeland in Tibet more than 2000 years ago into southern Szechuan and Yunnan in China and more recently into Burma, northern Thailand and northern Laos (Grunfeld, 1982: 22–23). Estimates of the total Akha population in these areas vary greatly, from 300,000 to 640,000 (Alting von Geusau, 1988: 217; Grunfeld, 1982: 23; Kammerer, 1989: 269). In Laos the Akha number approximately 60,000, almost all from the northern-most provinces of Phongsaly and Luang Nam Tha. Chazee (1995: 154) reports that while some Akha entered Laos from 1850 onwards, the majority came from Yunnan and Burma after 1900. However, oral histories of Akha villages in Muang Sing indicate that some villages were established there in the early nineteenth century by migrants from Burma, and elders of two Akha villages claim that their ancestors arrived in the Muang Sing area between 700 and 800 years ago (Gebert, 1995: 55–89).

Tai Lue settlement of Muang Sing has two sources. In about 1885, Chao Fa Sirinor, the ruler of the small Lue principality of Chiang Khaeng on the Mekong River, relocated his capital to Muang Sing and brought with him about 1000 settlers (mainly Lue). Subsequently, the majority of Lue who settled the Muang Sing valley migrated from the southern principalities of Sip Song Panna: Muang La, Muang Phong, Muang Yuan, Muang Hun and Muang Mang.[1]

Like their Lue neighbours, the Tai Neua speak a Tai dialect and are also Theravada Buddhist. However, they have a different Tai script and some of their customs distinguish them from the Lue. Today the highest concentration of Tai Neua is in the Dehong Dai-Jingpo Autonomous Prefecture of western Yunnan (bordering Burma). The first Tai Neua to inhabit Muang Sing were subjects of the ruler of Chiang Khaeng and were sent ahead of the Lue as an advanced guard of settlers, probably in the late 1870s.

The Yao are repatriated refugees (from the failed CIA-backed Yao rebellion in Luang Nam Tha in the 1960s); prior to this period there had been a number of long-established Yao villages in Muang Sing in the vicinity of Chiang Khaeng. The Tai Dam (Black Tai) and Hmong are relative newcomers, all settling in Muang Sing since the early 1990s. The Tai Dam migrated from Phongsaly province, the Hmong from Xieng Khouang and Houaphan provinces.

Village Research Sites

This chapter is based on anthropological research carried out in 1995 and 1997 among the Tai Lue of the Muang Sing valley; it concentrates specifically on the multiple and complex economic relations between the Lue village of Ban Tin That and the Akha villages of Ban Sopi Mai and Ban Yang Luang, both located in the foothills and within a distance of about 3 km from Ban Tin That (see Figure 1).[2] Ban Tin That lies 4 km from Muang Sing town on the main road to Luang Nam Tha. The village was established

1. The Tai kingdom of Sip Song Panna (once known as Muang Lue) was established in the twelfth century AD. Sip Song Panna, with its capital at Chiang Rung, was able to maintain a high degree of autonomy by paying tribute to both Chinese emperors and Burmese kings. It was only permanently incorporated into the Chinese state in the late nineteenth century as a consequence of boundary treaties between China, France and Britain. It is now the Xishuangbanna Dai Autonomous Prefecture within Yunnan province.
2. In 1995, I only had time to interview the headmen of Ban Sopi Mai and Ban Yangluang. However, in late 1997 I returned to study Akha–Tai relations more from the Akha perspective and completed a survey of all households in these two Akha villages.

Figure 1. District of Muang Sing.

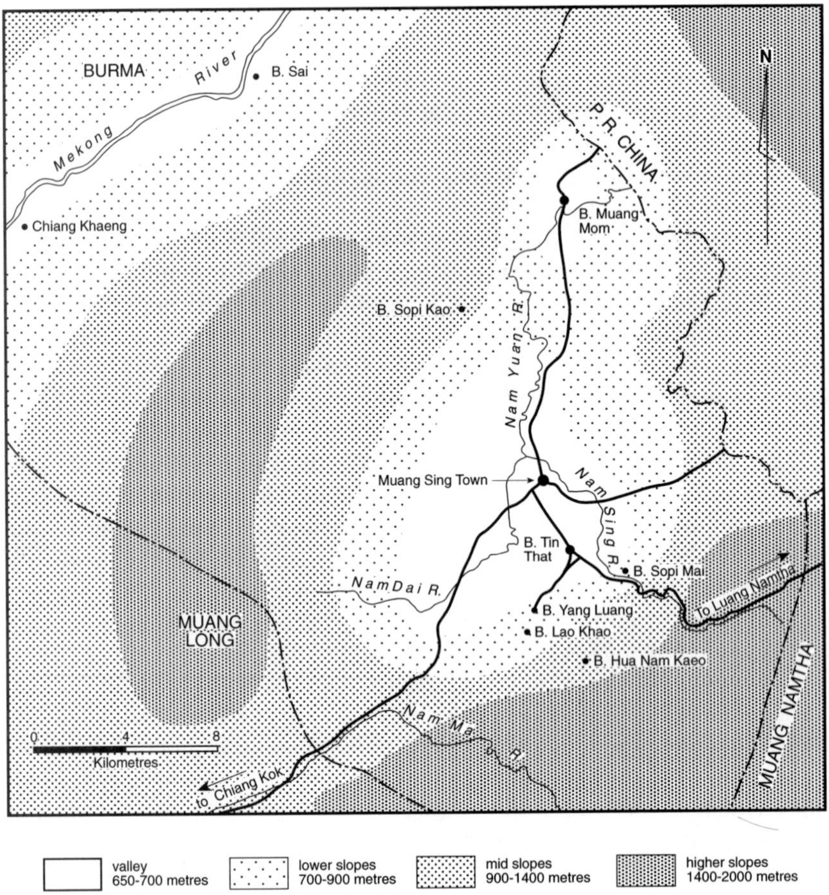

some seventy-five years ago by Lue from a village in Muang Sing that had earlier been settled by migrants from Muang Hai and Muang Hun in Sip Song Panna. More recent immigrants fled from Sip Song Panna to Ban Tin That in response to the collectivization campaign in China in 1958. In late 1996, the village comprised seventy-five households.

Ban Sopi Mai (thirty-nine households in late 1997), was established in 1994. Some villagers had split off from the parent village (Ban Sopi Kao) to move to another highland location and then moved to the present site following an epidemic. However, although Ban Sopi Mai has only been recently settled, members of the original village had cultivated wet rice and opium here for almost twenty years. Ban Yang Luang was established in early 1990, following the burning down of the original highland village of Hua Nam Kaeo, about half an hour's walk away. Before this relocation,

Akha of the parent village had been working wet-rice fields at the edge of the Muang Sing valley for about thirty years. In late 1997 Ban Yang Luang comprised sixty-four households.

'Traditional' and 'Lower Slope' Akha Villages

In 1995 the Lao-German Cooperation Project[3] (LGCP) carried out a socio-economic baseline survey of thirty-eight villages in Muang Sing, of which thirty-three were Akha. The report distinguishes between 'traditional' and 'lower slope' Akha villages. The 'traditional' villages are up to two days walk from the lowlands and only have occasional contact with the market centre of Muang Sing. These villages are dependent on swidden cultivation of dry rice and opium, with only limited use of cash and 'almost no hiring out of labour'. The 'lower slope' villages have a much greater reliance on wet-rice cultivation, have more frequent contact with Muang Sing town and its market and also with lowland Tai villages, and are much more enmeshed in a cash economy, including wage labour (Gebert, 1995: 6, 11). The Akha villages in my study, Ban Sopi Mai and Ban Yang Luang, fit the 'lower slope' appellation, both in terms of their location and economic attributes.

However, the usefulness of this dichotomy and contrast can be questioned. Michael Epprecht surveyed nineteen Akha villages in Muang Sing in 1996, with six selected for more intensive study (Epprecht, 1998: 24). All but two of these villages are located at elevations between 900 and 1400 metres in the remote highland sub-districts of Chiang Khaeng and Ban Sai, far from Muang Sing town. His research shows that these so-called 'traditional' Akha villages are economically very marginal. Population pressure on land has led to short fallow cycles and low dry-rice yields of less than 700 kg per ha (ibid.: 43) — somewhere between a half and a quarter of the yields of wet rice in the lowlands. Only a small percentage of households produce enough rice to meet subsistence needs, and opium is often used to buy rice. However, opium production is also low: an average of 547 gm per household, which is less than half the amount required (1.2 kg) for a household with one addict (ibid.: 67, 87). The combination of low levels of dry-rice production and low opium production forces a high percentage of households to rely on trade, sale of livestock and wage labour, with 36 per cent of households hiring out their labour (ibid.: 43, 87, 124). The purely topographical label of 'mid slope' may better describe these highland villages, avoiding the pristine, static connotations of the term 'traditional'.[4]

---

3. A joint project between the German aid organization, GTZ (Deutsche Gesellschaft für Technische Zusammenarbeit) and the Ministry of Public Health, Lao PDR.
4. According to Alting von Geusau, the Akha prefer a mid-slope location to site their villages and this preference is expressed in Akha mythology and literature (1988: 225).

This brings us to the issue of why large numbers of Akha have chosen to establish new settlements at the edge of the lowlands. According to the Lao-German Cooperation Project report, they 'have moved down to the lower slope areas at the request of the government authorities, and were also promised assistance if they were ready to cultivate flat land and abandon shifting cultivation practices' (Gebert, 1995: 11). In the cases of the Akha villages of Ban Yang Luang and Ban Sopi Mai similar requests and promises were made by government officials, according to the headmen of these villages (see also Chazee, 1995: 156).

Lao PDR government policy on shifting cultivation was enunciated at the Sixth Party Congress in 1996: 'Shifting cultivation (also known as slash-and-burn agriculture) is a problem the Government wants to address. Peoples whose livelihoods depend on shifting cultivation must be settled in areas where they can be allocated land to earn a living' (Gebert, 1996: 12). The government is committed to ending shifting cultivation by the year 2000, the rationale being that shifting cultivation is a major cause of deforestation and erosion and that forestry is a major source of state revenue, contributing about 40 per cent of export earnings (Fisher, 1996: 42; Kaneungnit et al., 1996: 267). More recently, the policy of resettlement has been reaffirmed in the government's 'focal site' strategy, as part of its Rural Development Programme for the period 1998 to 2002. The State Planning Committee defines the focal sites strategy as 'the bringing together of development efforts in an integrated and focused manner with a clearly defined geographical area, aiming at eradication of poverty and at promoting sustainable development' (*Watershed*, 1998: 51). The resettlement component of focal sites emphasizes the relocation of highland people to permanent settlements in the lowlands as a means of eliminating shifting cultivation. By 1997 the Government had already established sixty-two focal sites (including 320,000 people) and planned to establish eighty-seven sites by 1999. Of these, 50 per cent involve resettlement (ibid.: 51).

The policy is not very realistic, given the high percentage of mountainous land in Laos (especially in the north). However, as far as resettlement is concerned, there is often a significant divergence between national policy and local implementation in the context of the highly decentralized nature of political power in socialist Laos. For example, provincial and district officials in Xieng Khouang province appear to have pursued a very vigorous and determined policy of resettlement of the highland population of Nong Haet (mainly Hmong) to lowland areas. By contrast, local officials in Muang Sing had, by 1996, abandoned the policy of resettling Akha (Gebert, 1996: 1, 16).

The change in policy in Muang Sing seems to have been due less to official concern with the shortage of potential wet-rice land in the Muang Sing valley than to the desire to avoid further inter-ethnic conflict between the long-settled Tai and immigrant highlanders. In 1992 the Hmong deputy governor of Luang Nam Tha province announced on national radio that there was an abundance of flat land in Muang Sing which could be cultivated with wet rice. This encouraged an influx of hopeful Hmong migrants from Xieng Khouang and Houaphan provinces. In the absence of clear-cut legislation, there was considerable disputation over land and water between the Hmong newcomers and Tai and Akha villagers.

It is clear that, even before the policy of resettlement in Muang Sing was abandoned, government pressure and promises of assistance alone could not explain the decision of many Akha to move down to the lower slopes. The low productivity of swidden agriculture (rice and opium) in the highlands must have also been an enticement to move to the lowlands to develop wet-rice fields. This was especially true for the young: Epprecht found a large dent in the population pyramid in the 15–24 age group, which he explains in terms of the 'worsening prospects for the future of their subsistence agriculture due to an observed diminishing soil fertility in the area and increased scarcity of general resources' (Epprecht, 1998: 65–6).[5] This is confirmed by my research in Ban Yang Luang. The establishment of the village in 1990 was preceded by dissension between older and younger men of the Akha village of Hua Nam Kaeo. The senior men wanted to rebuild the highland village after it was destroyed by fire, while the younger men argued for a new village on the lower slopes where they believed their economic prospects would be better.

Lower Slopes and Higher Hopes

Whether Akha resettlement to lower slopes has been a response to government pressure and promises or to low agricultural productivity, or to a combination of these factors, the question remains: to what extent have their expectations of a more secure livelihood been realized? The answer requires some analysis of the economic and social characteristics of Akha villages of the lower slopes.

---

5. In theory it would be possible for the Akha to respond to declining dry-rice yields by substantially increasing opium production and using opium profits to purchase rice. Opium is an ideal crop in situations of land shortage, as it can be grown continuously on the same plot for many years without a significant decline in yields. This is precisely how the Hmong responded to resource scarcity in the western highlands of northern Thailand in the 1960s and 1970s (Cohen, 1984; Cooper, 1984; Geddes, 1976; Lee, 1981). This option may be closed to the Akha due to an incapacity to mobilize sufficient labour, either by hiring workers or by increasing household size. The eleven 'traditional' Akha villages included in the LGCP survey averaged only 5.4 members per household (Gebert, 1995: 7), whereas studies of Hmong villages in northern Thailand report household sizes of 8.0 (Geddes, 1976: 110), 8.4 (Lee, 1981: 71), 7.6 (Cooper, 1984: 3, 7) and 7.7 (Tapp, 1989: 9). An alternative explanation is that the Akha of Muang Sing and elsewhere simply give priority to rice over opium (Epprecht, 1998: 71; Feingold, 1970: 330).

Wet-Rice and Dry-Rice Cultivation

Economic profiles of the twenty-two Akha lower-slope villages included in the LGCP baseline survey reveal that the development of wet rice in the lowlands has been a problem, with less than half of all households owning wet-rice fields in 1995.[6] Consequently, these villages still rely heavily on dry-rice cultivation. By contrast, Ban Sopi Mai and Ban Yang Luang are much better off with regard to wet-rice cultivation; 82 per cent of households own wet-rice fields. This can probably be explained by the fact that wet-rice cultivation has been long established in both villages — more than thirty years and twenty years respectively, compared to an average of only 6.7 years for those in the baseline survey. Another factor could be the concentration of the LGCP Akha villages at the edge of the much smaller Muang Mom valley where less flat land is available.

The wet-rice fields of Ban Sopi Mai (30.4 ha) are almost all irrigated by the Nam Sing River. Some 74 per cent of the Ban Yang Luang fields (64.92 ha) are rainfed, with only a small area (17 ha) irrigated by the Nam Kaeo stream. The low volume of water from this stream has limited irrigation and caused considerable tension over the years between the Akha of Ban Yang Luang and the Lue of Ban Tin That (Chazee, 1995: 156). At 2.44 tonnes per ha, wet-rice yields for the two Akha villages in 1997 were only marginally less than the yields for the Lue village of Ban Tin That (2.56 tonnes per ha). However, the Lue have significantly larger landholdings — an average of 1.63 ha for Ban Tin That,[7] compared to 1.12 ha for the two Akha villages. The relative success of wet-rice cultivation in Ban Sopi Mai and Ban Yang Luang has reduced dependence on unproductive dry rice. Only 43 households out of 103 cultivated dry rice in 1997, producing only 20 tonnes at an average of 220 kg per household (465.1 kg per producing household).

Opium Production and Consumption

Epprecht reports that 93.3 per cent of the 433 households he surveyed (almost all mid-slope villages) grew some opium in 1996. The average production per household of the 414 producing households was 547 gm (Epprecht, 1998: 75). For the 103 households of Ban Sopi Mai and Ban Yang Luang the average opium yield in 1997 was 443.25 gm per household for the sixty-nine households that grew opium. Thus the difference in yields between mid-slope and lower-slope opium-growing households is not that

---

6. Specifically, 44.38 per cent of 543 households from seventeen villages for which information was available (Gebert, 1995: 68–89).
7. These data and most other economic data for Ban Tin That are based on a one-in-three household survey of the village (i.e. 25 out of 75 households).

great. The major difference is the decline in lower-slope villages in the percentage of households that grow opium. The LGCP report notes that only ten of the twenty-two lower-slope villages surveyed cultivated opium (Gebert, 1995: 67–89), and Epprecht (1998: 75) claims that 'many households of most of the villages settled in the valleys do not engage in poppy cultivation any more'. Ban Sopi Mai is consistent with this trend: only fifteen (38.4 per cent) out of the thirty-nine households reported a harvest of opium poppy in early 1997. However, Ban Yang Luang seems exceptional, with a high percentage of opium-growing households (84.3 per cent, or fifty-four of the sixty-four households). It may be, too, that government surveillance of the more visible and accessible areas at the periphery of the lowlands has prompted under-reporting and disguised the amount of opium cultivated by lower-slope villages.[8] Whatever the case, there can be little doubt that there is considerable pressure on Akha villages of the lower slopes to continue to cultivate opium. This pressure emanates from the persistence of high levels of addiction and opium consumption.

Rates of opium addiction among the Akha of Muang Sing are very high: about 9.3 per cent for the total Akha population of the district, compared to only 2.8 per cent for the Hmong and 3.5 per cent for the Lue (Gebert, 1996: 15). In a survey of 155 households from six mid-slope Akha villages, Epprecht found that 9.8 per cent of the total population and 21 per cent of the adult population (twenty years and above) were addicts (1998: 84). This compares with the LGCP survey finding of 14.2 per cent of the adult population for the eleven mid-slope ('traditional') villages surveyed. The rate for the lower-slope villages is only marginally higher at 14.4 per cent (Gebert, 1995: 47).

At the beginning of 1998, there were twenty-one opium addicts in Ban Sopi Mai, giving a rate of addiction for the thirty-nine households of 12.2 per cent of the total population and 27.6 per cent of the adult population. There were fifty-one addicts in the sixty-four households of Ban Yang Luang, with a rate of 15.3 per cent of the total population and 36.7 per cent for the adult population.[9] More than two-thirds (69.4 per cent) of the addicts of both villages are male. The rate of addiction in Ban Yang Luang is exceptionally high and is equal to the highest rate of addiction reported by Epprecht for the mid-slope villages he surveyed. There has also been a rapid increase in Ban Yang Luang in the number of addicts — from thirty-four in 1994 to forty-one in 1996 and fifty-one in 1998, representing 10.3 per cent, 12.5 per cent and 15.3 per cent of the total population respectively.[10] Overall, we

8. When I interviewed him in 1996, the headman of Ban Sopi Mai claimed that the village had stopped growing opium nearby three years earlier as a result of government pressure.
9. I follow the Akha in defining an 'addict' (in Lao '*khon tit fin*') simply as a person who consumes opium on a regular, daily basis.
10. The 1994 figure is based on Chazee (1995: 166); the other figures are from my own surveys in November 1996 and January 1998.

can conclude that levels of addiction are much higher among the Akha than among other ethnic groups in Muang Sing and that the resettlement on the lower slopes has certainly not led to a diminution in the incidence of addiction.

The Increase in Disease

According to Epprecht, opium addiction among the Akha of the mid slopes can only be explained fully in terms of the interaction of a complex of individual, social and economic factors. However, he does single out one underlying cause to be 'a general hopelessness and sense of malaise' arising from economic insecurity and poor health (Epprecht, 1998: 99ff). Although many Akha who settled on the lower slopes have improved their economic prospects through access to wet-rice land — especially the villages of Ban Sopi Mai and Ban Yang Luang discussed above — these gains have been largely negated by increased human and livestock mortality that has accompanied resettlement. This is evident from the comparison in Table 1 between mid-slope and lower-slope Akha villages, based on data from the LGCP survey.

The child mortality rate for the lower-slope Akha villages is extremely high, comparable to the highest rates in the world of about 300 (UNICEF, 1994: Table 1), and is in stark contrast to a rate of only 70 for the Muang Sing valley, inhabited predominantly by Tai (Gebert, 1995: 32). The high human mortality of lower-slope Akha villages is due to deterioration in water supply and sanitation and increased incidence of malaria at lower altitudes. According to the headman of Ban Yuang Luang, eighty people

Table 1. Human and Livestock Mortality in Mid-Slope and Lower-Slope Villages

	Mid Slope	**Lower Slope**
Human epidemics	6	24
Epidemic deaths	199	749
Child mortality[1]	133	326
Livestock epidemics	19	39

[1] Deaths per 1,000 live births for children under five
Source: Gebert (1995: 29, Table 7; 32, Table 8). [11]

11. As noted earlier, the LGCP survey included 33 Akha villages, 22 mid-slope ('traditional') and 11 lower-slope. Livestock epidemics are those since resettlement or during the past ten years.

died, mainly from malaria, within two years of the establishment of the new village in 1990. In his mind the trauma of these years and the subsequent physical and psychological toll of continuing illness and death are major factors in the high and rapidly rising incidence of opium addiction in his village.[12]

Livestock losses, mainly from haemorrhagic septicaemia, have also been very high in both villages. In Ban Sopi Mai fifty water buffalo died in 1995; forty-nine died in Ban Yuang Luang the following year. The death of a single buffalo constitutes an enormous economic loss to individual households — in monetary terms almost as much as the average household's annual production of rice. Livestock epidemics have also precluded the Akha from exploiting the thriving cross-border livestock trade with Thailand. These epidemics could be prevented by vaccination. Local officers of the Ministry of Agriculture have vaccinated extensively throughout Tai villages in the lowlands; in 1997 all but two of the water buffalo and oxen of Ban Tin That were vaccinated (with only two deaths). Yet no attempt has been made to vaccinate livestock in the two nearby, accessible Akha villages of Ban Sopi Mai and Ban Yang Luang.

Other forms of government assistance to the Akha (such as credit, agricultural extension and health services) have also been very limited. The few agricultural innovations that have occurred in lower-slope Akha villages have resulted from informal contact between Akha (new rice strains from China) and between Akha and Tai farmers and traders (knowledge-sharing of wet-rice cultivation techniques, and new crops such as water melons). Inadequate government assistance is due in part to the scarcity of financial and material resources,[13] although traditional Tai attitudes of superiority and disdain towards the Akha have also encouraged indifference to the plight of the Akha.

Dominance of the Lowland Tai

Leo Alting von Geusau has described the Akha as a 'perennial minority', originally in relation to small but stronger valley kingdoms and later to 'predatory modern states' (1988: 215–16). The highland Akha of Muang Sing have historically occupied a similar position of political and economic subordination to the lowland Tai, especially the Tai Lue.

12. The headman reported the death of ten people in 1997 in Ban Yang Luang, all from malaria.
13. GTZ has contact with some 60 Akha villages and is well resourced. It has given some assistance to lower-slope Akha villages (for example, building of schools) but its policy has been to concentrate on the more remote Akha villages on the assumption that lower-slope Akha villages are in a better position to receive government assistance.

A French report at the turn of the century commented on 'the "natural proclivity" of the Lue to surrender themselves to pomp and pageantry as long as it gave them the "illusion of being a great people" or at least being at the top of an ethnic hierarchy which placed the montagnards at the bottom' (see Gunn, 1989: 62). In the nineteenth century, under Lue feudal rule, the Akha were classified along with other non-Tai groups as *Kha* (literally 'slaves'). A special Lue official was assigned to supervise Akha corvée labour of fifteen days per year, later replaced by taxes in silver, animal meat and opium (Nguyen Duy Thieu, 1993: 11–12).

Today the Akha of Muang Sing in theory have political equality as 'highland Lao' (*Lao sung*) in the multi-ethnic state of Lao PDR. In practice, their political power is limited to that of village headmen at the lowest rung of the local administrative hierarchy, dominated by Tai Lue and Tai Neua officials. The lowland Tai Lue and Tai Neua have also maintained their economic dominance through a prosperous surplus rice economy. Population pressure on land is minimal, with a lowland population density of only 51 per km^2. This low density probably owes much to the political turmoil of the 1960s caused by the Pathet Lao 'liberation' (*pot poi*) of Muang Sing and the CIA-backed Yao rebellion, conflicts which led to the mass exodus of Tai lowlanders from Muang Sing to the Royalist strongholds of Bokeo province. This loss of population has only been partly compensated by the influx of Lue from Sip Song Panna as a result of the collectivization campaign in China.

The potential area for wet-rice cultivation in the lowlands of Muang Sing has been estimated at 6500 ha, but less than half of this potential (2975 ha) has been realized. A favourable land situation has encouraged an ongoing process of expansion of wet-rice fields, especially by villages at the rim of the Muang Sing valley.[14] These conditions have served to limit landlessness, inequalities of landholdings, and tenancy. A high proportion of Tai Lue and Tai Neua villagers in Muang Sing produce rice surplus for sale, even without the benefits of 'green revolution' technology. For example, nineteen of the twenty-five households I surveyed in Ban Tin That (76 per cent) sold rice from the 1995 harvest, at an average of 1.78 tonnes sold per household. Villagers also earn cash income from the sale of crops (such as peanuts, chilli peppers, garlic, and sugar cane), from livestock, and from weaving, as is the case for Tai from other villages. However, by far the largest source of cash income is the sale of rice. Furthermore, the market for the sale of rice to buyers from China has expanded significantly since the Lao PDR/China border agreements of 1992, which stimulated cross-border trade.

14. This is evident from my surveys of the Lue villages of Ban Tin That at the eastern edge of the valley and of Ban Na and Ban Nam Dai at the northern and southern ends of the valley respectively. A visit to the new Hmong village of Don Mai at the western edge of the valley revealed recent extensive development of wet-rice fields in the vicinity.

CONTEMPORARY TAI–AKHA ECONOMIC RELATIONS

Today, lowland Tai villages and lower-slope Akha villages are interlinked by multi-stranded economic relations based on cash transactions, barter and multiple currencies. For example, the Tai regularly exchange woven cloth, pigs, chickens, cakes, and noodles for Akha cotton (a low yield but disease-resistant variety). Tai Lue of Ban Tin That sell woven cloth (white and black) to Yao villagers for cash, but not to neighbouring Akha, who do their own weaving. However, these Lue are still dependent for cotton on nearby Akha (including those from Ban Sopi Mai and Ban Yang Luang). The cotton is bought with cash in exchange for prepared food (such as cakes) or clothes (e.g. sarongs) and, most commonly, in exchange for chickens. The Akha do raise chickens but demand invariably exceeds supply, as the Akha make sacrifices to ancestors up to twelve times per year (Hansson, 1992: 186).

Wage labour is the dominant form of economic exchange between lowland Tai and lower-slope Akha. Epprecht found that 39 per cent of households surveyed in the 'traditional' (mid-slope) Akha villages hired out labour. However, with the exception of one village, 'the bulk of all wage labour in the target area happens within the Akha community itself', with 31 per cent being internal and only 8 per cent external (Epprecht, 1998: 124, 125). Lower-slope Akha villages are very dependent on wage labour outside the village. This is emphasized in the LGCP report (Gebert, 1995: 4, 5, 11) but not quantified. For the lower-slope Akha villages of Ban Sopi Mai and Ban Yang Luang in my own study, the hiring-out of labour within the village comprised only 19.3 per cent of all wage labour (1665 person days) in 1997, compared to 80.6 per cent for external wage labour. Almost all (96.2 per cent) of the external wage labour was for neighbouring Tai villages.[15]

Wage labour for the Tai includes various tasks in the rice cultivation cycle: transplanting, which consists of the uprooting (*lork ka*) and planting (*dam na*) of seedlings, reaping (*kiao khao*), threshing (*ti khao*), and the carrying of rice (*baek khao*) from the fields to village granaries. Other types of hired labour are: cutting firewood (*tat feun*), the digging of ditches (*khut hong*) around rice fields as a barrier against the intrusion of livestock, land clearing (*buk boek*) for new rice fields, and the making of house-posts (*peng sao heuan*). The amount of labour expended for each type of work is indicated in Table 2.

Akha labour hired by the Tai for the transplanting, reaping and threshing of rice is most commonly paid in cash, though sometimes in rice or in opium. Occasionally payment is made in tobacco, chickens and even in

15. Ban Yang Phiang, Ban Nam Kaeo Luang, Ban Nam Kaeo Noi, Ban Kum, Ban Patoy and Ban Tin That, with a heavy concentration on the last two villages.

Table 2. Ban Sopi Mai and Ban Yang Luang: Akha External Wage Labour, 1997 (in person days)

	Male	Female	Total
Uprooting	31	4	35
Planting	57	312	369
Reaping	163	322	485
Threshing	6	3	9
Carrying	16	8	24
Firewood cutting	0	0	0
Ditch digging	125	45	170
Land clearing (wet-rice fields)	194	0	194
House-post carpentry	18	0	18
Other	22	16	38
Total	632	710	1342

sickles. In Ban Tin That, in 1997, the rates of payment per day were 1500 kip (about 80 US cents) in cash, or 10 kg of unmilled rice, or 2 *saleung*[16] of opium. A midday meal is provided whatever the mode of payment. These rates may vary in relation to the degree of competition for labour within a village at particular times of the agricultural cycle — higher rates may be offered during the harvest when competition for labour peaks. Tai villages closer to China, where Akha can get higher wages, and Tai villages near Akha villages which receive assistance from the Lao-German Cooperation Project, were paying 2000 kip per day.[17]

Addict households (that is, households with at least one addict) provided 83.5 per cent of external labour (in terms of person days of labour) of Ban Sopi Mai and Ban Yang Luang in 1997, although they comprised only 56.3 per cent of all households. For addict households, male external wage labour exceeded female overall (577 to 541 person days). More than half the male labour (55 per cent) was for digging ditches and the clearing of land to develop new wet-rice fields. This is onerous, back-breaking work that Tai farmers prefer to avoid and is undertaken almost exclusively by male Akha addicts. Payment is usually higher than for other types of hired labour and is made in cash or opium and on a daily or piecework basis.

The Economic Advantages of Akha Labour for the Tai

Akha villages of the lower slopes provide Tai lowlanders of Muang Sing with a pool of dependent, cheap and predominantly addict labour. Akha

16. One *saleung* of opium = 2.72 gm.
17. In Ban Tin That, in 1997, the rate for Akha labour for the carrying of rice from the fields was between 100 and 250 kip per bag.

labour also has more specific economic advantages. Firstly, Tai villages, particularly those at the edge of the valley, are in the process of gradually clearing new land to extend wet-rice cultivation, in response to the expanding market for rice. Since this clearance is being carried out almost exclusively by the labour of Akha opium addicts, Akha addicts play a crucial role in the growth of the Tai surplus rice economy. Secondly, while the Tai depend almost entirely on the Akha for the heavy tasks of clearing land for new rice fields and for the digging of ditches, Akha are employed for other work, such as the transplanting and harvesting of rice, only to supplement the Tai's dominant use of household and co-operative exchange labour of family, kin and neighbours. This supplementary labour has the advantage of being temporary and flexible, and can be used by Tai villagers according to the labour demands of a particular year,[18] a particular time of year (such as harvest-time, when speed is necessary to avoid crop destruction by the grazing of neighbours' water buffalo), and to cover variations from year to year in the size of the household labour force.

Tai villagers attempt to hire Akha labour as cheaply as possible, entering into various forms of barter and cash transactions in which the Akha are significantly disadvantaged. Akha addicts often exchange rice for opium during the rice harvest (November–December), when opium is in short supply.[19] After the opium tapping in January and February, when rice stocks are low and depleted by earlier sales, Akha addicts exchange opium for milled rice.[20] On occasion, the Akha also sell rice and opium to the Tai for cash. The Akha lose out in these transactions because they are compelled to sell when the prices of opium are at their lowest and buy when prices are at their highest (often with more than 100 per cent difference between lowest and highest prices). These types of transactions were reported not only by the Tai Lue of Ban Tin That but by other Lue villages as well.

Tai villagers also acquire opium from the Akha in exchange for woven cloth, fish, cigarettes, cakes and noodles. Tai women exchange some of these goods for opium in the Akha opium fields during the tapping. By these means and those mentioned above, Tai Lue villagers of Ban Tin That were able to accumulate between about 60 and 120 *saleung* of opium per year. Little of this is traded out of the village; almost all is used to hire

---

18. My twenty-five household survey of Ban Tin That indicates that there was a significant decline between 1996 and 1997 in the use of Akha labour (from the villages of Ban Sopi Mai, Ban Yang Luang and Ban Lao Khao) for planting (*dam na*) and reaping, from a total of 248 person days in 1996 to 148 in 1997. The reason given by Lue farmers for this decline was that rainfall was both late and low in 1997 and they were forced to restrict the area of rice planted.
19. In 1996 and 1997 exchange rates varied between 1 and 3 *saleung* of opium for 10 kg of unmilled rice, depending on the level of opium prices and price variations throughout the year. 1996 prices ranged from approximately 500 kip to 1200 kip and 1997 prices from 300 kip to 700 kip.
20. In 1996 the exchange rate in Ban Tin That was 1 *saleung* of opium for 1.7 kg of milled rice.

Akha labour, especially that of opium addicts, for ditch digging and land clearing.

The Economic Consequences of Opium Addiction

The LGCP report argues that addict households of lower-slope Akha villages are less productive than non-addict households because the dependence on wage labour for the Tai of addict households limits their chances to improve their own fields. This leads to further rice shortages and more dependence on wage labour, thereby creating a 'vicious cycle' from which they cannot escape (Gebert, 1995: 44). Sanit Wongprasert, in his study of a Lahu village in northern Thailand (with an addict population of 16.5 per cent of the total population), also claims that opium addiction reduces household agricultural productivity. He reports that 'non-addicts cultivate larger fields of all crops including opium, commit more labour to agricultural tasks and get higher yields' (Wongprasert, 1989: 164).

However, my data for the Akha villages of Ban Sopi Mai and Ban Yang Luang demonstrate that addict households are no less productive than non-addict households. On the contrary, they own slightly more wet-rice land on average than non-addict households and they produce marginally higher rice and opium yields. Addict households are also better off with respect to livestock ownership, no doubt due to the fact that addict households tend to be older, larger and have more workers. These findings are summarized in Table 3.

There is a very limited market for rice land and I know of no cases from Ban Sopi Mai or Ban Yang Luang of Akha addicts who have sold their land to support their habit. Water buffalo are more vulnerable to sale by addicts but this does not seem to have significantly depleted the buffalo stocks of addict households (see Table 3). Furthermore, addiction does not appear to

Table 3. Agricultural Resources and Productivity: Comparison of Addict and Non-Addict Households of Ban Sopi Mai and Ban Yang Luang

	Addict households	**Non-addict households**
Total no. households:	58	45
Households owning wet-rice land (%)	86.2	77.7
Av. area wet-rice land owned (ha)	1.17	1.04
Wet-rice yields (1997/tonnes per ha)	2.52	2.33
Av. opium yield per household (1997/gm)	460.36	426.25
No. persons per household	5.83	4.05
No. workers per household	2.85	1.96
Av. age head of household	39.3	32.05
No. water buffalo per household	1.18	0.91
No. pigs per household	1.8	1.31

greatly impair work capacity. There are only five relatively young addicts (below the age of forty) who claimed their addiction made them too weak to work.[21] Four out of the five rented out their land to kin or to nearby Tai villagers. Opium smoking has the potential to disrupt agricultural work and limit household productivity, given that addicts prefer to have three smoking sessions a day. However, the evidence from my conversations with addicts is that most adjust their smoking routines to work demands, by reducing the time of a session, eliminating lunch-time smoking, and eating opium instead of smoking it.

It is possible that the productivity of addict households is maintained by the exploitation of the labour of non-addict members, particularly of females who, as noted above, constitute a much smaller proportion of the addict population. According to the LGCP report: 'A woman is also just as likely as a man to hire out her labour, and if she has the misfortune of being married to an opium addict, the woman may hire out her labour just to earn opium for her addicted husband' (Gebert, 1995: 21). This is true, but only for the small number of male addicts who are incapable of work. As noted above, for addict households male external wage labour exceeds female overall, due to the preponderant use of male addict labour for ditch digging and land clearing. The use of female hired labour from addict households for transplanting and reaping of rice far exceeds that of males (481 to 219 person days). However, it has been noted that Akha women generally 'are expected to work longer and more' in the fields (Epprecht, 1998: 64) and that, for rice cultivation in particular, 'women's contribution is usually greater' (Kammerer, 1988: 39). I would hesitate to conclude, at least on the basis of my data, that addict households exploit female labour more than non-addict households.

Although addict households in Ban Sopi Mai and Ban Yang Luang are no less productive than non-addict households they do, nevertheless, suffer from significantly higher rice deficits. For both villages, 51.7 per cent of addict households had rice deficits in 1997 (with an average of five months per household), but only 30.9 per cent of non-addict households experienced deficits (with an average of 4.2 months). It is worth noting that in Ban Yang Luang half the addict households that had rice deficits (eight out of sixteen) produced a surplus of rice but ended up short of rice as a result of exchanging rice for opium. The main negative economic consequence of opium addiction for addict households may therefore relate not to agricultural productivity but to the economic costs of opium consumption. The costs to a household of supporting an addict are considerable. A single addict consumes about 1.5 kg of opium per year, which is approximately equivalent to the amount of rice needed to feed an average household. Given

21. Chazee reports that out of the 34 opium addicts in Ban Yang Luang in 1994 only 2 were incapable of working (1995: 166).

limited opium production (on average less than a third of consumption needs), addict households experience great difficulty in satisfying both the household requirements for subsistence rice and the imperative opium needs of addicts.

The cost of opium addiction has also been raised by the increased price of opium itself. During the 1980s and early 1990s, opium prices were very low; in 1990, Akha in Muang Sing were paid 4 *saleung* of opium for a day's work. Between 1992 and 1996 prices rose steeply, so that in 1994 a day's labour fetched only 1 *saleung* (Escoffier-Faveau, 1994: 19). In 1997 payment had increased to 2 *saleung*, as a result of a slight fall in the price of opium, but this was still barely enough to cover the daily smoking needs of a single addict.

Low opium production, high costs of opium and minimal alternative sources of income (such as the sale of livestock) place great pressure on addict households to engage in wage labour, especially labour for neighbouring Tai villagers. However, income from external wage labour is limited both by the labour demands of Tai employers and by the labour requirements of the Akha's own wet-rice fields, particularly during the harvest period.[22] Moreover, as noted above, many Akha households are further disadvantaged by having to sell rice to buy opium for addict members and then later having to sell their modest supplies of opium to meet rice shortages, selling cheap and buying dear in both transactions.

CONCLUSION

An unproductive swidden economy and enticement by government officials (in response to the current forest preservation policy) have combined to encourage many Akha of mid-slope villages of the highlands of Muang Sing to resettle at the periphery of the lowlands. However, the high hopes of a more secure livelihood from this movement downhill have not been fulfilled. Some lower-slope Akha villages, such as Ban Sopi Mai and Ban Yang Luang, have quite successfully adopted wet-rice cultivation in the lowlands, but this gain has been largely negated by the costs, economic

22. There is no conflict between the work schedules of dry and irrigated wet rice as dry rice is planted about two months earlier (in May) and harvested up to a month earlier (October). Conflict in labour demands between Ban Tin That farmers (who grow mainly irrigated wet rice) and Ban Yang Luang farmers (who grow mainly rainfed wet rice) is minimized by the fact that the Akha of Ban Yang Luang have to wait for good rains and usually plant and harvest about one month later than their Lue neighbours. Nevertheless, there is some overlap in early December in the harvesting work of the two villages. There is a much greater overlap in the rice cultivation cycles of Ban Tin That and Ban Sopi Mai, as both villages depend on irrigated wet rice. However, many Akha of Ban Sopi Mai grow an early-maturing variety of non-glutinous rice from China that allows them to harvest early and free up some household labour for external wage labour.

and social, of high human and livestock mortality, lack of government assistance, and persistently high rates of opium addiction. In a situation of limited alternative sources of income, opium addiction has forced lower-slope Akha villages into wage labour dependence on lowland Tai villages and entanglement in disadvantageous commodity transactions that re-inforce this dependence.

Ironically, the movement of Akha downhill, with promises and expectations of 'development', has arguably been of much greater benefit to the lowland Tai. Lower-slope Akha villages provide the lowland Tai Lue and Tai Neua with a pool of cheap and temporary labour that they can utilize according to changing agricultural needs. Most importantly, Akha labour, especially that of addict households, has been crucial in enabling lowland Tai to extend wet-rice cultivation and to tap the expanding commercial market for rice.

The negative experience of the Akha of Muang Sing with regards to resettlement is by no means exceptional. A 1997 UNDP report, *Basic Needs for Resettled Communities in the Lao PDR*, includes a survey of sixty-seven displaced villages from six provinces. All these communities identified serious problems with resettlement: devastating epidemics (particularly from malaria), loss of assets, rapid debt accumulation, reduced rice supplies, intensified competition for land, and lack of government resources to provide needed assistance. Such findings have led to criticism of resettlement strategies by UNDP and other international development agencies, and their reluctance to fund resettlement projects (*Watershed*, 1998: 53). Furthermore, Lao PDR government policy itself appears to be rather inconsistent. Despite government pronouncements about the necessity of resettlement to save the forests, Gebert notes that the Ministry of Agriculture and Forestry 'seems to have a policy more towards improving highland agricultural practices, thereby improving their sustainability in the mountains' (Gebert, 1996: 12). The Department of Livestock and Fisheries (of MAF) appears to have been the driving force in this alternative policy and has strongly advocated the promotion of livestock production and fish culture within the highlands as a means of stabilizing and reducing shifting cultivation (see Chapman et al., 1998).

Acknowledgements

The field research on which this chapter is based was part of a joint university project, 'Trans-Mekong Corridors: Economic, Social and Environmental Change in the Mekong Region', funded by a grant from the Australian Research Council. I would like to acknowledge the support of the ARC and also that of the Institute of Research on Culture and Society, Ministry of Information and Culture (in particular, the Director, Houmphanh Rattanawong), and the Department of Livestock and Fisheries, Ministry of Agriculture and Forestry (in particular, the Director General, Singkham Phonvisay, and Dr Bunthong Bouahom, Director of the Livestock Division). I would like to thank Khamphaeng Thipmountaly, Nan Chaiseng, Phat Lamcan and

Bunleuam Noracak, from the above ministries, for their research assistance in the field. In Muang Sing I received various forms of assistance from Gunther Kohl and Wulf Raubold of the Lao–German Cooperation Project for which I am grateful. Acknowledgement is also due to Wulf Raubold, Michael Epprecht and the referees for their helpful comments on a draft of this chapter.

REFERENCES

Alting von Geusau, L. A. (1988) 'The Interiorizations of a Perennial Minority Group', in J. G. Taylor and A. Turton (eds) *Sociology of 'Developing Societies' in Southeast Asia*, pp. 215–29. Basingstoke: Macmillan.

Chapman, E. C., Bounthong Bouahom and P. K. Hansen (eds) (1998) *Upland Farming Systems in the Lao PDR — Problems and Opportunities for Livestock, Proceedings of an International Workshop held in Vientiane, Laos (18–23 May 1997)*. ICIAR Proceedings no. 87. Canberra: Australian Centre for International Agricultural Research.

Chazee, L. (1995) *Atlas des Ethnies et des Sous-Ethnies du Laos*. Bangkok: Laurent Chazee.

Cohen, P. T. (1984) 'Opium and the Karen: A Study of Indebtedness in Northern Thailand', *Journal of Southeast Asian Studies* 15(1): 150–65.

Cooper, R. (1984) *Resource Scarcity and the Hmong Response*. Singapore: Singapore University Press, National University of Singapore.

Dessaint, W. Y. and A. Y. Dessaint (1975) 'Strategies in Opium Production', *Ethnos* 40(1–4): 153–68.

Durrenberger, P. (1974) 'The Regional Context of the Economy of a Lisu Village in Northern Thailand', *Southeast Asia, An International Quarterly* 3(1): 569–75.

Epprecht, M. (1998) 'Opium Production and Consumption and its Place in the Socio-Economic Setting of the Akha People of North-Western Laos: the Tears of Poppy as a Burden to the Community?'. PhD thesis, Institute of Geography, Faculty of Natural Science, University of Berne.

Escoffier-Fauveau, C. (1994) 'Opium Addiction in Luang Namta and Bokeo Provinces'. Report for Norwegian Church Aid, Vientiane.

Feingold, D. (1970) 'Opium and Politics in Laos', in N. S. Adams and A. W. McCoy (eds) *Laos: War and Revolution*, pp. 322–39. New York: Harper and Row.

Fisher, R. J. (1996) 'Shifting Cultivation in Laos: Is the Government's Policy Realistic?', in B. Stensholt (ed.) *Development Dilemmas in the Mekong Subregion: Workshop Proceedings*, pp. 42–8. Melbourne: Monash Asia Institute, Melbourne.

Gebert, R. (1995) 'Report of the Muang Sing Socio-Economic Baseline Survey for the Lao–German Cooperation Project, Muang Sing Integrated Food Security Programme'. Vientiane: Ministry of Public Health, Lao PDR/Technical Cooperation, Federal Republic of Germany.

Gebert, R. (1996) 'Report of a Mission to Lao PDR on Institutional Strengthening and Integrated Measures in the Field of Opium Supply and Demand Reduction'. Vientiane: GTZ.

Geddes, W. R. (1976) *Migrants of the Mountains*. Oxford: Clarendon Press.

Gunn, G. C. (1989) 'Rebellion in Northern Laos: The Revolts of the Lu and the Chinese Republicans (1914–1916)', *Journal of the Siam Society* 77(1): 61–5.

Grunfeld, F. V. (1982) *Wayfarers of the Forest: The Akha*. Amsterdam: Time-Life Books.

Hansson, I. (1992) 'The Marginalisation of Akha Ancestors', *Pacific Viewpoint* 33(2): 185–92.

Kammerer, C. A. (1988) 'Shifting Gender Asymmetries among the Akha of Northern Thailand', in N. Eberhardt (ed.) *Gender, Power, and the Construction of the Moral Order: Studies from the Thai Periphery*, pp. 33–51. Madison, WI: Center for Southeast Asian Studies, University of Wisconsin.

Kammerer, C. A. (1989) 'Territorial Imperatives: Akha Ethnic Identity and Thailand's National Integration', in J. McKinnon and B. Vienne (eds) *Hill Tribes Today: Problems of Change*, pp. 259–301. Bangkok: White Lotus; Paris: Orstom.

Kaneungnit Tubtim, Khamla Phanvilay and P. Hirsch (1996) 'Decentralisation, Watersheds and Ethnicity in Laos', in R. Howitt, J. Connell and P. Hirsch (eds) *Resources, Nations and Indigenous Peoples: Case Studies from Australia, Melanesia and Southeast Asia*, pp. 265–77. Melbourne: Oxford University Press.

Lee, G. Y. (1981) 'The Effects of Development Measures on the Socio-Economy of the White Hmong'. PhD dissertation, University of Sydney.

Miles, D. (1973) 'Some Demographic Implications of Regional Commerce: The Case of North Thailand's Yao Minority', in R. Ho and E. C. Chapman (eds) *Studies in Contemporary Thailand*, pp. 253–71. Canberra: Research School of Pacific Studies, Australian National University.

Nguyen Duy Thieu (1993) 'Relationships between the Tai-Lua and Other Minorities in the Socio-Political System of Muang Xinh (Northern Laos)'. Paper presented at the 5th International Conference on Thai Studies, SOAS, University of London (5–10 July).

Sanit Wongprasert (1989) 'Opiate of the People? A Case Study of Lahu Opium Addicts', in J. McKinnon and B. Vienne (eds) *Hill Tribes Today: Problems of Change*, pp. 159–72. Bangkok: White Lotus; Paris: Orstom.

Tapp, N. (1989) *Sovereignty and Rebellion: The White Hmong of Northern Thailand*. Singapore: Oxford University Press.

UNDP (United Nations Development Programme) (1997) *Basic Needs for Resettled Communities in the Lao PDR*. New York: UNDP.

UNICEF (United Nations Children's Fund) (1994) *The State of the World's Children*. New York: Oxford University Press.

Watershed (1998) 'Is Resettlement Resettlement?', *Watershed, People's Forum on Ecology* 4(1): 51–4.

9 Environmentalists, Rubber Tappers and Empowerment: The Politics and Economics of Extractive Reserves

Katrina Brown and Sérgio Rosendo

INTRODUCTION

One of the best-known examples of grassroots environmental action is the movement of rubber tappers which emerged in Brazil during the 1980s, fighting for the conservation of forests through the establishment of extractive reserves, which can be defined as 'conservation units that guarantee the rights of traditional populations to engage in harvesting forest products such as rubber and fruits' (Anderson, 1992: 67). The creation of extractive reserves has been promoted as 'among the most important strategies for forest conservation' (Hecht, 1989: 53). The designation of extractive reserves has gained support from a diverse array of actors, particularly conservation and environmental organizations who regard it as an opportunity to put into practice an explicit linkage between conservation and development. The rubber tappers' struggle for rights to natural resources in these areas gained world-wide media attention at a time when deforestation, especially in Amazonia, was becoming a major issue for northern environmentalists.

This chapter examines the concept of empowerment within the context of these initiatives. It investigates the alliances formed between environmental NGOs and other agencies and rubber tappers, and looks at how far rubber tappers have been empowered as a result of the intervention of these organizations. The evidence presented in the chapter derives from research in the Western Brazilian State of Rondônia (see Figure 1) and involves a case-study of a project supported by one of the largest international conservation NGOs, the World Wide Fund for Nature (WWF). This project is a partnership between WWF, local rubber tappers' communities represented by the Rondônia Organization of Rubber Tappers (OSR) and its member Associations, and a regional environmental NGO (ECOPORÉ). Research involved examination of key institutions, including the OSR and other organizations located in Porto Velho, the administrative centre of Rondônia, and case studies of three selected extractive reserves (see Figure 1).

The chapter begins by discussing theoretical perspectives on empowerment, and the different dimensions of empowerment outlined in the literature on conservation and development, including the way in which empowerment has been interpreted and implemented in conservation projects. It distinguishes two dimensions of empowerment — political empowerment and economic empowerment — and examines how each of these has been affected by the alliances between rubber tappers and external agencies in the case of extractive reserves in Rondônia. Through this analysis some issues concerned with the effectiveness of extractive reserves in providing long-term livelihood security for rubber tappers and the opportunities for enhanced economic and social welfare are highlighted.

EMPOWERMENT AND CONSERVATION: THEORETICAL PERSPECTIVES

Empowerment can be defined as a process through which 'people, especially poorer people, are enabled to take more control over their own lives, and secure a better livelihood with ownership of productive assets as one key element' (Chambers, 1993: 11). This implies finding means to facilitate and assist the efforts of resource-poor groups to meet their needs, either through their own organizations or through pressure on the State or other groups to make them act in their interests (Johnson, 1992). Empowerment has become a popular concept in people-orientated conservation, or conservation that attempts to integrate development and environmental protection goals (Pimbert and Pretty, 1994). Ultimately, empowerment should work towards what Chambers (1993: 92) calls a 'sustainable livelihood security'. He defines livelihood as 'a level of wealth and stocks and flows of food and cash which provide for physical and social well-being' (ibid.). A sustainable livelihood includes security against sickness, against early death and against the threat of poverty. It also includes assets or entitlements which can be used to meet contingencies such as sickness and accidents. This implies 'secure command over assets as well as income' (ibid.).

In addition to livelihood security, the concept of empowerment places a strong emphasis on access to the political structures and formal decision making necessary, for example, to enable people to gain control over land and resources. It is also concerned with access to markets and incomes that allow people to satisfy physical and material needs. In other words, effective empowerment should happen at both the political and socio-economic levels. The extent to which these two aspects are connected and interdependent is relatively poorly explored in the literature. There is thus little exploration of the dynamics of empowerment, although it could be argued that there is an assumption that economic empowerment results from political empowerment, and that this assumption underscores interventions such as those in the case-study discussed later in this chapter.

In addition to the two dimensions of empowerment and their achievement, we are concerned here with the ways in which external organizations can enable empowerment. In participatory development contexts, empowerment is sometimes interpreted in the sense that some can act on others to give them power (Nelson and Wright, 1995). It has been argued that disempowered people, due to structural constraints of various kinds, are normally incapable of identifying their own interests and acting upon them. Batliwala (1994: 131), for example, states that 'the demand for change does not usually begin spontaneously from the conditions of subjugation. Rather, empowerment must be externally induced, by forces working with an altered consciousness and an awareness that the existing social order is unjust and unnatural'. Empowerment in this light is seen as something that can be planned in order to bring about a desirable outcome, very much as a service that can be delivered.

However, an approach where one group, be it an environmental organization or a development agency, can bestow power upon another is problematic, since as Rowlands (1995: 104) argues, 'any notion of empowerment being given by one group to another hides an attempt to keep control'. Indeed, in conservation practice external agencies often seek to impose their own agendas on local people, even when they use the language of participation and empowerment in their projects (Pimbert and Pretty, 1994). True empowerment has to come from within. A generic interpretation of empowerment would be a 'process by which people become aware of their own interests and how these relate to those of others, in order both to participate from a position of greater strength in decision making and actually influence such decisions' (Rowlands, 1995: 102). Essentially, then, empowerment is a process that cannot be imposed by outsiders or planned 'from above'. This does not preclude external support which, when appropriate, can enhance and encourage empowerment. External intervention can assist in the empowerment process, but it must be local people who decide in what ways.

Empowerment is increasingly seen as both a means and an end in people-orientated conservation projects. The WWF forest conservation project with the Rondônian rubber tappers has four main objectives: supporting the institutional development of the tappers' grassroots organizations; organizing and mobilizing local communities of rubber tappers within the extractive reserves; promoting the legal establishment and demarcation of the reserves by the state government; and testing and implementing activities that improve the socio-economic well-being of reserve residents without causing environmental degradation (WWF, 1995). The concept of empowerment emerges as a strong theme in these stated objectives. The project is concerned with enabling the rubber tappers to gain access to political decision making (in issues related to the establishment of extractive reserves), and with improving their socio-economic welfare. It also includes goals such as institutional development. The following section will examine the context of Rondônian rubber tappers and their experience with the intervention of organizations such as WWF.

Rondônia in Perspective

The western Brazilian state of Rondônia encompasses an area of 243,000 km^2 (see Figure 1). Forests occupy approximately 75 per cent of the state's territory. The dominant vegetation is upland forest (*terra firme*), although savanna (*cerrado*) covers some 9 per cent of the total, and floodplain (*várzea*) another 9 per cent. The climate is hot and humid with a mean annual temperature of 26°C. The rainy season lasts from November to May with annual rainfall ranging from 1800 to 2200 mm. Only 10 per cent of the area is considered suitable for annual or permanent cultivation (Browder et al., 1996).

In the late nineteenth century Rondônia became an important centre for the production and commercialization of rubber. Although remote, the

Figure 1. Rondônia State and the Location of Extractive Reserves

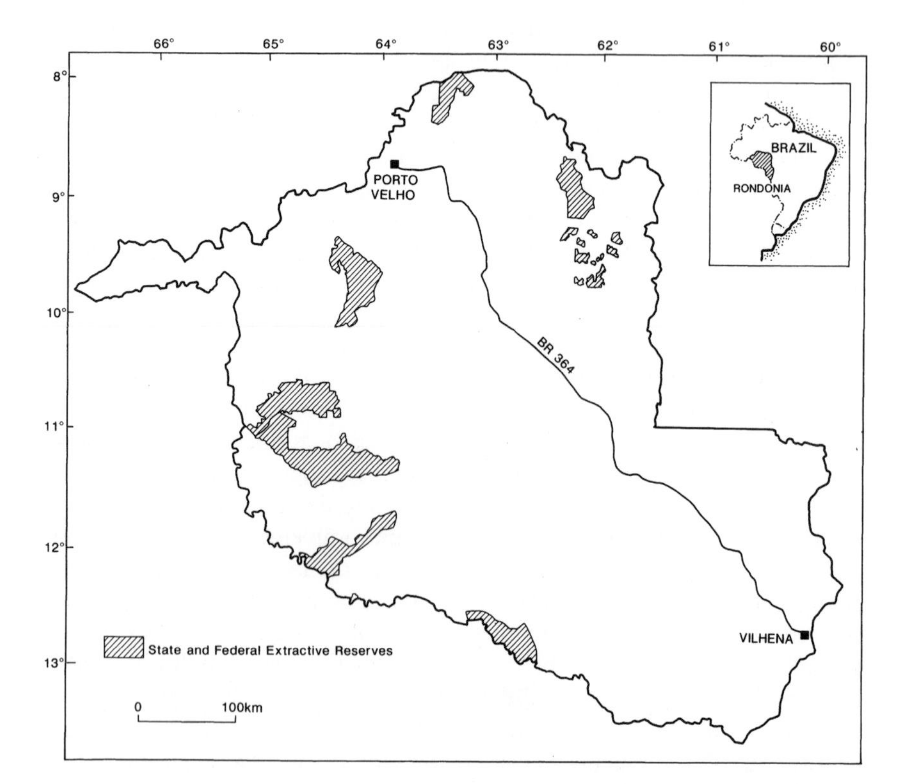

region was then strongly connected to world markets. In fact, the rubber trade was controlled by foreign companies and the entire production was exported. Labour to exploit the region's forests came from Northeast Brazil where large numbers of rural workers were recruited to tap rubber. Around 1915 the Amazonian rubber boom ceased due to competition from Asian plantation rubber and Rondônia's economy collapsed. During the Second World War a renewed demand for Amazonian rubber occurred when Allied countries were cut off from Asian rubber markets. After the War rubber extraction declined but never halted completely. In 1950 the region's economy still depended primarily on the production of rubber for the internal market (SEDAM, 1986).

In the 1960s Brazilian policy makers began plans to 'develop' and 'modernize' Amazonia. Extraction of forest products had no place within the government's policies for the region. In fact, the extractive economy was seen as synonymous with backwardness and a hindrance to socio-economic progress (Homma, 1993). The building blocks of Amazonian development were considered to be agriculture, cattle ranching and mining, and in subsequent years these activities were vigorously promoted (Hecht and Cockburn, 1989). The push to develop Rondônia started in earnest in 1958 with the exploitation of cassiterite deposits. In 1968 a precarious dirt road (the BR-364) linking the region to the rest of the country was completed, making southern markets and labour more accessible. In the years that followed, small farmer settlement programmes were created bringing a significant influx of migrants to the region (Martine, 1990). A variety of subsidies was also established for the creation of cattle ranches, making forest clearing for this purpose a profitable investment (Mahar, 1989).

These policies had significant social and environmental impacts. At the social level Rondônia's population increased enormously from 70,000 in 1960 to 500,000 in 1980 (Benchimol, 1989). In the process, the social profile of Rondônia was completely altered and the traditional and indigenous populations living in the region became a minority.[1] New social groups included colonist farmers, ranchers, loggers and miners. As the variety of social actors in Rondônia increased so did the range of interests in resource use: opposing interests gave rise to conflicts which became particularly severe in the 1980s. The environmental impacts included high rates of deforestation as a result of forest conversion for agriculture and cattle pasture (Cleary, 1991).

---

1. 'Traditional peoples' refers to non-tribal social groups long established in Rondônia. The term as it is used here comprises two main groups: *ribeirinhos* or riverine people living in floodplain areas whose livelihoods depend primarily on a combination of permanent agriculture and fishing; and rubber tappers. Both groups are the result of a rich racial mixture of Europeans, Africans and Indians. These people are often referred to as *coboclos*, defined by Schmink and Wood (1987: 40) as 'racially mixed population that grew with the migration to the region during the rubber boom'. 'Indigenous peoples' or 'Indians' are used to refer to the native or tribal population of Amazonia.

In 1981 the government initiated POLONOROESTE, a project financed by the World Bank. The principal objective of POLONOROESTE was to asphalt the BR-364; it also included various provisions for promoting small farmer settlements, protecting the environment and supporting indigenous communities. The project encountered a series of severe problems, including a major influx of migrants beyond the handling capacity of government agencies; alarming rates of deforestation; high rates of abandonment among migrants settled in colonization projects; and invasions into conservation and indigenous areas. POLONOROESTE also gave rise to intense conflicts resulting from the indiscriminate occupation of land inhabited by forest-dependent communities by more powerful groups such as land speculators, loggers and cattle ranchers.

The chaotic social and environmental situation created under the POLONOROESTE programme gave rise to some of the most vociferous protests against any project ever financed by the World Bank. Protests came from national non-governmental environmental organizations and human rights groups and their international allies, especially North American environmentalists, and soon had international political repercussions. Under political pressure, the Bank suspended disbursements on its POLONOROESTE loans in early 1985 (Rich, 1994).

The involvement of the environmental movement in local struggles over natural resources also had an important impact on the social groups adversely affected by the development 'model' of Amazonia. It created conditions for the establishment of strategic alliances between environmentalists and forest-dependent people.

Extractivism and Extractive Reserves

Rubber tappers live in isolated areas deep within the forest. Their household areas, called *colocações*, are small clearings surrounded by forest and rubber trails, where one or two families live. Although their main cash income derives from the sale of rubber, the tappers, or *seringueiros*, utilize a variety of livelihood strategies to secure their subsistence, including the gathering of a wide range of forest products, shifting agriculture and small-scale animal husbandry. In the course of their long settlement in the forest, *seringueiros* have adopted methods for using resources that, for the greater part, are well adapted to the local environment, and are comparable to those of the Indians. They have developed what many view as a 'sustainable' production system, known as extractivism (Allegretti, 1990; Schwartzman, 1992).

However, rubber extraction has historically been associated with severe social and economic exploitation (Hecht and Cockburn, 1989). Rubber tappers are among the poorest and most marginalized sections of civil society in Rondônia and in the Brazilian Amazon in general. This situation results from a combination of factors: in the past, from the unequal relations of production in which extractivism was undertaken, the *aviamento* or debt peonage system;[2] more recently, from the prevalence of unfavourable marketing structures and the falling prices of extractive products in relation to the cost of living (Assies, 1997). Within this context rubber tappers have been a traditionally disempowered social group through economic relations, physical isolation, poor or no access to social services, and neglect from policy makers (Hecht and Cockburn, 1989; Melone, 1993).

During the 1980s the government's development model for the region contributed to the rubber tappers' disempowerment. This development strategy privileged cattle ranching, agriculture, and logging over other forms of land use with less ecologically adverse impacts (Hecht and Cockburn; 1989; Homma, 1993). Schmink and Wood (1987) describe the conflicts inherent in such policy, where the goals of expanded production and short-term accumulation are fundamentally at odds with environmental conservation. Despite the poverty associated with extraction of forest products, even in the 1990s this activity still provides a livelihood, however meagre, for approximately 5000 rubber tappers in Rondônia.

When information about the sustainability of extractivism and the struggles of the rubber tappers to protect the forests was disseminated in the West, environmental groups enthusiastically joined forces with this politically and economically disempowered group of Amazonians (Melone, 1993). The result of this alliance was the proposal for an innovative model for sustainable land use in Amazonia — extractive reserves. These would be 'public lands designated for the specific purpose of sustainable use of forest products such as rubber, brazil nut, and palm heart by the resident population' (Allegretti, 1990: 253). Modelled on the idea of indigenous reserves, or areas where the rights of native peoples to their traditional lands are guaranteed, extractive reserves rapidly became regarded as holding the promise of 'reconciling economic development and environmental conservation' (Anderson, 1992: 67).

However, extractivism also has its sceptics. Southgate et al. (1996: 16) observe that:

> living standards among the rubber tappers of Bolivia and Brazil are miserable, comparing poorly with the meagre socio-economic norms of rural Amazon. By contrast, profits generated through non-timber extraction tend to lodge at the top of the marketing chain. The Manaus Opera House is lasting testimony to the wealth accumulated by exporters during the Amazon rubber boom.

---

2. Hecht and Cockburn (1989) provide a detailed account of the *aviamento* system. Hugh-Jones (1992: 69) notes the wide range of different relationships which exist between traders and describes a continuum between barter and debt-peonage which depends on the asymmetries of power between different parties.

Observations such as these imply that although extractivism may be profitable for some people and in some instances, markets for extractive products need to be regulated (Assies, 1997). Furthermore, most studies of extraction look only at non-timber forest products and fail to consider the potential of sustainably harvested timber, as well as farming, hunting and fishing. Extraction may be an important element in a diversified livelihood strategy: Assies (1997), for example, describes rubber and Brazil nuts as key components of the 'agro-extractivist cycle', along with agricultural activities. However, there may be resistance on the part of conservation organizations to these activities, which are traditionally seen as conflicting with conservation goals.

Support for the establishment of extractive reserves has grown and now includes a diverse array of NGOs, researchers, financial institutions and policy makers (Anderson, 1992). The World Bank has also endorsed the concept, especially following national and international protests against its involvement in the socially and environmentally disastrous POLONOROESTE programme. The intense public pressure of US environmental groups supporting the rubber tappers was crucial for this shift in World Bank policy (Rich, 1994). The Bank, in turn, provided the political leverage through which pressure could be applied upon the Brazilian government for the establishment of extractive reserves. In effect, future World Bank loans to Brazil would be conditional upon the creation of a number of extractive reserves. In the late 1980s the World Bank announced a US$167 m loan (from a total budget of US$228.9 m) to the Rondônia Natural Resources Management Project (PLANAFLORO), aimed at repairing the damage caused by POLONOROESTE.

The main objective of PLANAFLORO is to promote a new approach to 'sustainable development' in the state of Rondônia through a series of initiatives for the protection and management of natural resources, such as socio-economic and ecological zoning, promotion of agroforestry systems, recovery of degraded lands, environmental protection, sustainable forest management, environmental education, support to indigenous communities, and creation and management of extractive reserves and other conservation areas (Rondônia, 1994).

The extractive reserve movement in Rondônia has already achieved the creation of nineteen extractive reserves covering an area of approximately 885 million hectares and is currently pressing the state government for the creation of five more reserves totalling over one million hectares of tropical forest. While environmental conservation is an important function of these areas they should also be able to provide for the economic survival of the 434 families who inhabit them. PLANAFLORO has put in place some enabling conditions for the creation of extractive reserves in Rondônia. However, as will be further discussed below, these are neither a sufficient condition for their establishment nor an assurance of their effective implementation. WWF support to rubber tappers in Rondônia evolved within this context.

Recognizing the validity of the rubber tappers' aspirations, the WWF project has developed activities to enhance their incomes either directly or in association with PLANAFLORO related actions. These initiatives have the potential to empower the rubber tappers in economic terms and thus contribute to improving their 'sustainable livelihood security'. Implementation is channelled through locally-established rubber tappers associations. Extractive reserves are thus seen as a means of reaching multiple objectives for conserving forest and providing a livelihood for extractivist communities. They have become a focus for external agencies, and a symbol of the struggle of marginalized people to maintain their way of life against the powerful forces of loggers and ranchers. However, they may not be viable, in economic or political terms, without outside support. The following sections of this chapter examine how alliances with external agencies have supported the establishment of extractive reserves, and how these have benefited rubber tappers. The political empowerment of rubber tappers through alliances to establish extractive reserves is first examined, before some of the impacts of intervention on the livelihoods and welfare of the rubber tappers and the economic dimensions of empowerment is discussed.

POLITICAL EMPOWERMENT: ACTORS, 'PROJECTS' AND PUBLIC POLICY ADVOCACY

Social Mobilization and Organization

In Rondônia there is a long history of resistance to deforestation involving *seringueiros*, especially during the implementation of POLONOROESTE. However, in the past resistance was isolated and often resulted in unsuccessful actions. It was only in 1989 that efforts to organize the rubber tappers were given a strong impetus by the National Council of Rubber Tappers (CNS), following government plans to establish a number of reserves in the state.

In February 1989 the CNS, with support from national and international organizations,[3] promoted an important meeting of rubber tappers in the municipality of Guajará Mirim where a number of workshops and discussions concerning the problems faced by the extractivist population of the area took place. In particular, the concept of extractive reserves, their applicability to the region, and the necessary steps for their creation were extensively debated, emphasizing the need for community organization as a means to legitimize proposals for their establishment.

---

3. These included rural workers' unions and environmental organizations such as the Institute of Amazonian Studies which played a key role in supporting the rubber tappers movement in Acre (see Melone, 1993).

Similar meetings were subsequently arranged in other parts of Rondônia. As a result, several local leaders emerged who were prepared to take over the social mobilization and organization work initiated by the CNS, and to establish horizontal linkages between extractive communities, giving the rubber tappers' movement in Rondônia a context specific direction. The OSR is the result of this organizational work.

Great emphasis has been put on continuing the process of mobilization and organization of extractivists, particularly by promoting the creation of local rubber tappers' Associations. The importance attached to social organization arises partly from the legal requirements for the creation of extractive reserves. Government rules stipulate that only representative organizations of the local inhabitants of a given forest area can formally request that area to be declared an extractive reserve. Moreover, because extractive reserves remain the property of the state and only the rights of use are transferred to rubber tappers through a community-use title, a representative organization acting on behalf of the reserve inhabitants is needed to claim that land-use title from the government.

An important part of the process of organization of rubber tappers is the 'base community work' or *trabalho de base* which happens at base community meetings (*reuniões de base*). The creation of a local Association, for example, is normally preceded by a number of *reuniões de base* in which facilitators explain the purpose of and procedures for the creation of an extractive reserve and motivate the local inhabitants to organize themselves into an Association in order to legitimize their claims. The role of the OSR as the state-level representative of rubber tappers and that of the local Associations as its subdivisions is also explained. When the extractive areas are threatened by land grabbers, loggers or cattle ranchers, rubber tappers are encouraged to resist and organize confrontations to defend them. Community base work is thus both a consciousness-raising and an organizational activity.

Once the Association is created, base community meetings continue to be held periodically in the extractive reserves. They fulfil three important functions. Firstly, they are the participatory channels through which local communities can express their needs and priorities as well as their views on matters that affect them, thus generating important information that feeds back into the Associations and OSR. Secondly, they are vehicles for the dissemination of information regarding the present and future plans of the OSR and Associations and the activities and negotiations in which they are engaged. Thirdly, they are also arenas for discussion and decision making on important issues of the community's general concern.

Coalition Forming and Empowerment

Silva (1994: 703) has argued that Latin American societies are heavily penetrated by and dependent on external actors, and that this means that a source of international support for domestic actors is often critical for the relative power of coalitions. His comparative study of pro-grassroots development coalitions in Brazil, Mexico and Peru highlights the factors which determine the make-up of coalitions as well as their effectiveness. In addition to alliances with powerful international actors, a key enabling factor in the dynamics of coalitions was found to be the initial position of key government actors and policy makers, and the institutional frameworks within which coalitions operate. In the case of Rondônia this framework was provided in the contract provisions of PLANAFLORO which anticipate the creation and legal establishment of a number of extractive reserves in the state of Rondônia, as well as resources to promote their social and economic viability (Rondônia, 1994). In theory, therefore, when WWF initiated support for the rubber tappers in 1991 there were already some enabling conditions in place for the establishment of extractive reserves. However, during negotiations for the formulation of PLANAFLORO, several NGOs in Rondônia, with the support of national and international organizations, raised doubts about the capacity of the project to achieve its proposed objectives. Their main concern was the lack of participation of the project's intended beneficiaries — who include not only rubber tappers but also small farmers and indigenous peoples — in the formulation of the initial PLANAFLORO proposal. This was considered unacceptable in a project which, according to the World Bank, should constitute a 'model' of popular participation for subsequent Bank ventures (Millikan, 1995).

Some of these NGOs had already been working with communities of rubber tappers, small farmers and Indians and were aware of the social and economic reality of their situation. They envisaged that without effective participation, meaning a mechanism through which local people could express their needs and priorities, there was the danger that the full social and environmental potential of PLANAFLORO would not be realized, and, more seriously, that project beneficiaries would become the losers in the process. As a result of popular mobilization in Rondônia a number of representatives from NGOs and rural peoples' organizations, including the rubber tappers, were invited to a meeting with the state government and representatives of the World Bank in June 1991. The outcome of this meeting was the signing of a 'protocol of understanding' which established forms of participation by civil society in the planning, monitoring and evaluation of PLANAFLORO (Millikan, 1995).

However, true participation requires that peoples' views are effectively taken into account (Nelson and Wright, 1995). Many popular organizations do not have the voice to make their claims heard by policy makers. In other words, although the World Bank and the government of Rondônia opened the channels for the participation of disadvantaged rural groups in PLANAFLORO this was no guarantee that their needs would be met.

Negotiations for PLANAFLORO take place within a highly technocratic environment. Generally, participation in these negotiations requires involvement in bureaucracies and in the elaboration of proposals, discussions, meetings and so forth. Many rural groups possess neither the skills or the information to take part in these institutional practices where technical or legal language is the norm, effectively excluding those who do not master such discourses. In short, for participation of rural civil society to have an impact in the planning of PLANAFLORO, rural groups are required to speak the language of policy makers. For rubber tappers' leaders, for example, this is not easy; apart from other disempowering circumstances, they also have little formal education, which can limit the extent of their participation.

NGOs often have the knowledge and expertise that grassroots groups need to negotiate, lobby and pressure government institutions for benefits or services. In some instances they also have the means to organize large networks of support constituencies that go beyond national borders and take popular claims as far as multilateral lending institutions. This creates opportunities for strategic alliances between different sectors of civil society. Such is the case of the partnership project, in which environmental NGOs and rubber tappers work together for a common goal — the protection of the forests.

This is not to say, however, that the interest of each partner in forest conservation has the same basis. It is therefore important to understand how environmentalists and rubber tappers pursue their goals. This is where the concept of 'actor strategies' becomes useful (Long, 1992: 36; Long and van der Ploeg, 1994: 79). This refers to the way social groups use their available power resources, or their knowledge and capability, to resolve their particular problems. The rubber tappers' strategy to further their interests, in particular the establishment of extractive reserves, has been to use their image as the defenders of the forest as a tool to raise their profile and bring attention to their struggle. They have successfully enrolled other actors (i.e. WWF and other environmental NGOs) in their 'projects', getting them to accept particular frames of meaning, such as the sustainability of extractivism and its role in conserving the environment.

Environmental NGOs also have their own agenda, including conserving nature and ecological processes, which is translated into particular strategies. They have also discovered the strategic value of supporting the rubber tappers, where the rubber tappers become the means to achieve an end — conservation. Conkin and Graham (1995) highlight the mutual benefits of such alliances and observe how alliances between environmentalists and Amazonian indigenous groups have legitimized the involvement of first-world environmentalists in the affairs of distant nations, as well as advancing environmental causes. Assies (1997: 40–1) claims that alliances between international environmental groups and rubber tappers in Acre transformed local struggles for land rights into a global *cause célèbre,* but warns that such a strategy is in danger of fostering particular forms of patronization which impose an externally-framed, decontextualized view of forest people, creating the image of a 'mythical *seringueiro*'.

In the PLANAFLORO negotiations and in other contacts with the government this alliance has provided the means by which effective participation of the rubber tappers in the programme has been guaranteed. The rationale is simple. On the one hand, the rubber tappers have legitimate reasons for demanding the creation of extractive reserves, which incidentally are also thought to be an effective way of protecting the forest, but they lack the technical and legal resourcefulness to achieve this. Environmental NGOs, on the other hand, are interested in forest preservation and have the know-how to pressure the government into adopting environmental protection measures, but they have a fragile base for their demands. Thus, by supporting the rubber tappers, the NGOs' position is strengthened by claims that they are defending human rights and the rights of oppressed, politically disempowered people, rather than the preservation of flora and fauna. This is important since the empowerment process here is, if not reversed, at least working in both directions. In other words, this alliance empowers environmentalists as much as it empowers rubber tappers.

The participation of the rubber tappers in PLANAFLORO was a substantial achievement which might not have been possible without the the support of WWF and other NGOs. Yet, this was still not a sufficient condition in itself for the creation of extractive reserves. Demands for extractive reserves are met by strong opposition from political and economic élites, particularly those associated with cattle ranching, logging and land speculation (Hecht and Cockburn, 1989; Schwartzman, 1992), since extractive reserves contain valuable resources (wood and land) which are coveted by different actors. As shown in Table 1 these actors also have their particular strategies to further their interests.

The table distinguishes seven principal groups of actors and sets out their interests, scale of influence and means to achieve their aims. The groups are not exhaustive, nor necessarily exclusive, but they demonstrate the range of actors. Alliances have formed between a number of these different groups at various times, when common interests and aims are identified.

To overcome the political impediments outlined above (see also Silva, 1994), rubber tappers have built strategic alliances with different social and institutional actors. These alliances have been formalized by their participation in the NGO Forum of Rondônia. The Forum was created in 1991 following the commitment of the World Bank and the state government to allow a greater role for civil society in the PLANAFLORO negotiations. It is composed of non-profit organizations representing small farmers, rubber tappers, indigenous communities, local rural unions, researchers, educators, environmentalists, and groups involved in the defence of human rights. The main activity of the Forum has been to monitor and co-ordinate the participation of its members in PLANAFLORO (Millikan, 1995).

Table 1. Actors and Agents in Forest Management in Rondônia

Group	**Position in the political economy**	**Scale of influence**	**Source of power**	**Interests/aims**	**Means to achieve aims**
Rubber tappers	integrated into the national and international economy but in disadvantaged position; marginalized and ignored by the government	local and regional	limited but increasing	livelihood maintenance; secure access to forest resources through the creation of extractive reserves	coalition formed with other social groups to increase influence beyond the 'local'; use of environmental language to capture external support
Rondônia NGO Forum	very diverse	regional	strategic alliances; drawing on each others' power resources	to ensure that PLANAFLORO meets its social and environmental objectives; enable the participation of disadvantaged groups in PLANAFLORO	lobbying; networking with national and international NGOs; legal and technical advice to marginalized groups
Logging industry	flourishing business; important source of revenue in Rondônia	local, regional and national	economic importance; govt. support; ability to earn foreign exchange	profit; easy access to areas rich in high value timber often within extractive reserves	illegal logging; lobby govt. officials; offering bribes to prevent law enforcement; buy timber illegally from some tappers
Cattle ranchers	own large areas of land; important activity in Rondônia; links to the logging industry	local, regional and national	many land-owners hold important positions in local, state and national governments	profit; expanding pasture to areas occupied by extractive reserves	pressure the government to delay or prevent the creation of extractive reserves

State and national government	includes politicians with direct or indirect interests in logging and ranching	local, regional and national	political and administrative	mixed attitudes regarding the establishment of reserves; safeguard vested interests of élites; economic growth, national development	votes for legal provisions for creation of reserves to satisfy conditions imposed by the World Bank; shows little political will to create reserves; creates bureaucratic barriers for the establishment of reserves
International conservation NGOs	self appointed defenders of the world's ecological integrity	international	financial support from individuals; business and govts.; scientific knowledge; large network of support	conservation of biodiversity and ecological processes	support local struggles to protect forests; financial and technical support for local groups; lobbying World Bank and national governments
World Bank	multilateral financial institution	international	economic; institutional	economic development of region	imposes conditions on the Brazilian government for the protection of environment and livelihoods

By joining the Forum, the rubber tappers have been able to take advantage of the fact that a project funded by a multilateral financial institution is bringing a major capital influx into Rondônia. PLANAFLORO provided the political leverage through which the government could be pressured into taking notice of the rubber tappers' claims for the creation of extractive reserves. In fact, the Forum, through its large network of support, has been able to channel the *seringueiros'* claims directly to the World Bank. Since many of the contract provisions of PLANAFLORO were not being implemented by the government, the Forum submitted a request to the World Bank Inspection Panel asking for an investigation of the project. Among the contractual arrangements of PLANAFLORO was the establishment of several extractive reserves which the government had covertly tried to stall. The protests of the Forum led the World Bank to review the project and urge the Brazilian government to establish the extractive reserves without delay (Forum de ONGs, 1996).

In July 1995, after years of delay, the creation of a number of extractive reserves totalling almost 900,000 hectares was finally announced. The claims of the rubber tappers would almost certainly have gone unnoticed before the World Bank if they had not been integrated into a wider protest coming from diverse sectors of the Rondônian civil society represented by the NGO Forum. Once more, the rubber tappers were able to enrol others in their particular 'projects'. However, whether the establishment of extractive reserves has improved the livelihoods and welfare of the rubber tappers and their communities is uncertain. The following sections examine these economic dimensions of empowerment in the context of rubber tappers' livelihoods and their interactions with other NGOs in establishing extractive reserves.

ECONOMIC EMPOWERMENT AND EXTRACTIVIST LIVELIHOODS

This section first outlines the main components of extractivist livelihoods; it then identifies the key factors which act as constraints to extractivism and reviews the impact of intervention, including that by WWF with the OSR.

Extractivist Livelihoods

The livelihood of a rubber tapper family in Rondônia depends on a range of economic and subsistence activities that usually include harvesting forest products, hunting, fishing and farming. A similar system is described by Assies (1997) in Acre state. The main source of household income is derived from the sale of specific extractive products: rubber, brazil nuts and, to a lesser extent, copaíba oil.

The extraction of latex from the rubber tree (*Hevea brasiliensis*) is the main economic activity for rubber tapper households. A typical family produces an average of 900 kg of rubber annually, although this varies according to the abundance of rubber trees in each landholding (*colocação*), the number of household members dedicated to rubber extraction, and the time allocated to other activities. The income derived from rubber sales is not constant throughout the year. In the wet season rubber harvests decline considerably since the greatest concentration of *Hevea* tend to be on floodplains (*várzeas*); when these are flooded, rubber extraction becomes difficult or impossible.

Brazil nuts (harvested from *Bortholletia excelsa*) are another significant marketable forest product in Rondônia. Very importantly for the households' economic strategies, brazil nuts are harvested and marketed from December to February, when rubber extraction drops sharply or becomes impracticable. They are also consumed by the household and constitute an important component of its diet. However, since *Bortholletia excelsa* is only found in upland forest (*terra firme*), this source of income is not available in all reserve areas.

Copaíba oil (*Copaifera spp.*) is a forest product that rubber tappers have long used as a medicinal oil for treating wounds, influenza and coughs, or as fuel for lamps. In recent years the demand for copaíba oil as an homeopathic product has grown in Brazil, making it an alternative source of income for some extractivist communities. Copaíba grows in both floodplain and upland but the density of trees is low. In addition, the copaíba oil can only be harvested every few months since more frequent harvests may kill the tree. Copaíba oil is thus unlikely to be a major source of income for many households.

Forest products have not only an important economic value for rubber tapper households but also a significant subsistence value. Households use a wide range of products from the forest as food, fuelwood, building and fencing materials, and medicines. Rubber tappers hunt and fish and, for many, these are their main sources of protein. When asked about the advantages of living in the forest instead of urban areas, rubber tappers often point out that in the city they are totally dependent on cash earnings for food, whereas in the forest they can usually fish, hunt and plant manioc, rice or beans. As they say, 'in the city if you do not have work you go hungry'.

Swidden agriculture is another important component of the livelihood strategies of rubber tappers. The tappers call themselves agro-extractivists rather than just extractivists. Many households farm a swidden plot (*roça*) averaging 1.4 hectares in which manioc, beans, rice or maize are cultivated. Sugar cane, pineapple, banana and papaya may also be intercropped. The harvest is for the household's own consumption but in some cases a surplus is produced and sold, therefore constituting an additional source of income. The heterogeneity of the Amazonia region means that the suitability of soils for agriculture varies from one area to another even within the same extractive reserve. While some households are self-sufficient in food crops, others can only produce a fraction of their needs. Rubber tapper livelihoods thus have to be seen in the context of these activities, and not just rubber extraction.

Constraints on Extractivism

There are important constraints on the livelihood opportunities for agro-extractivist households in Rondônia, including dependence on one or two key products; the limitations imposed on extraction by seasonality and the ecological variability of the forest; the lack of access to reliable markets; poor prices for extractive products; and lack of integration with other aspects of livelihoods. Examination of the livelihood strategies of the rubber tappers demonstrates that extractivism is the main income-generating activity in extractive reserves; at present levels, however, this income does not meet the rubber tappers' needs. While most households rely on rubber for the greatest proportion of their income, findings suggest that on a monthly basis rubber earnings are usually less than expenditure on food and other essential items. In effect, many tappers assert that income obtained from rubber sales alone can only fully meet the basic needs of a family of five during the peak months of rubber harvesting.

This is illustrated in Figure 2 which compares average household monthly rubber production with the quantity of rubber needed to buy a bundle of basic supplies.[4] This diagram shows that in twenty-three out of thirty-four months, households did not harvest sufficient rubber to generate income to purchase basic supplies. Rubber tappers also suffer because rubber has been supported by an indirect subsidy which has gradually been reduced, bringing rubber prices down (see also Assies, 1997). The price of food and other essential items has increased disproportionately in relation to prices obtained for rubber and other extractive products, further eroding the rubber tappers' ability to meet their basic needs.

A significant share of the profits generated by forest products is captured by intermediaries. Extractivism in Amazonia has historically been characterized by unequal relations of production. For many years rubber tappers were exploited by landowners (*seringalistas*) through the *aviamento* system. When the rubber trade became less profitable landowners gradually abandoned the rubber estates and the marketing of extractive products was taken over by middlemen (*marreteiros*). At present, *marreteiros* continue to dominate the extractivist economy in Rondônia with the consequence that producer margins remain low.

---

4. These data derive from a survey undertaken in 1994 by the Institute of Amazonian Studies (IEA) in the Ouro Preto Extractive Reserve, Rondônia. IEA gathered data on household rubber production and incomes over time and compared it to the cost of a bundle of basic supplies needed to maintain an average family for thirty days, the components of which were identified by the rubber tappers themselves.

Figure 2. Rubber Production Required to Buy Basic Supplies

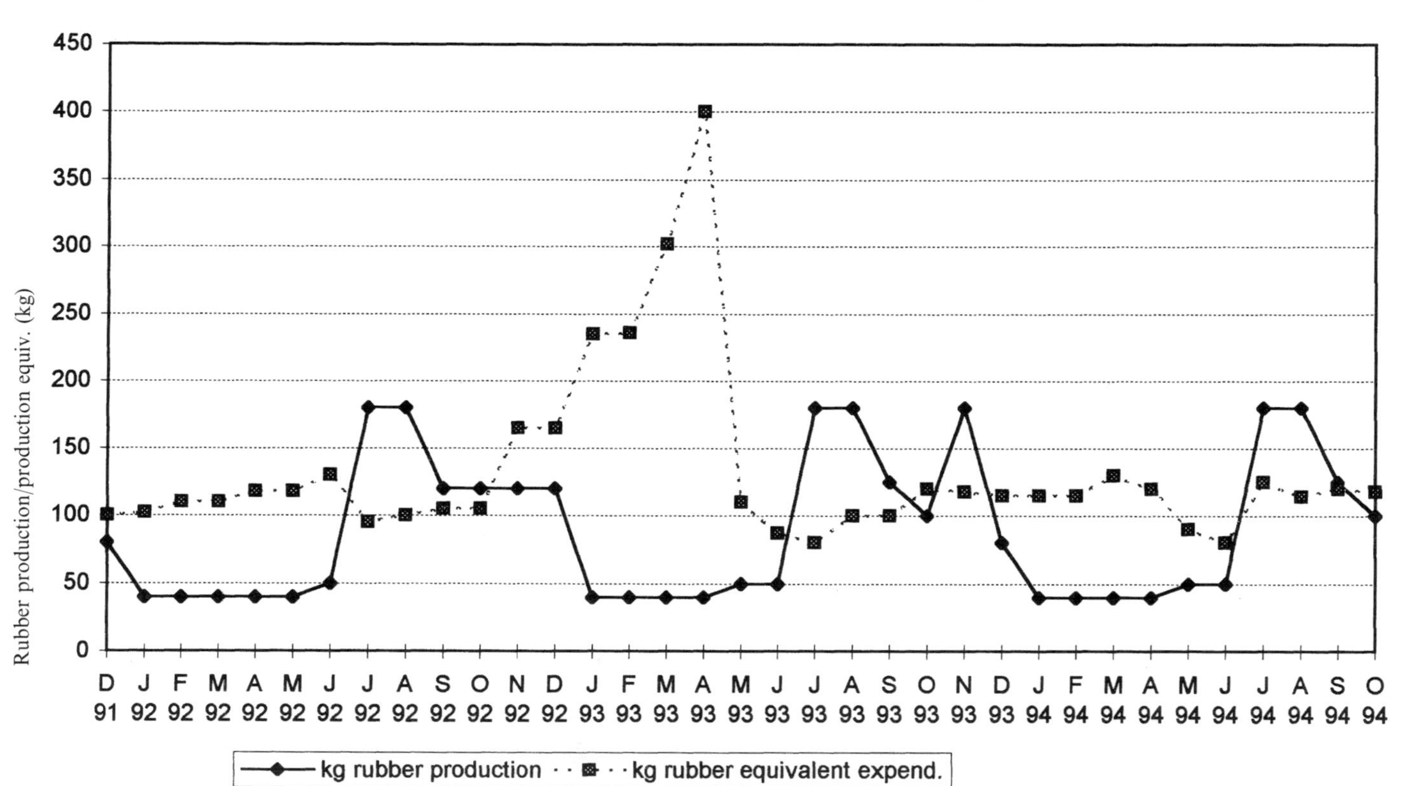

Interventions to Strengthen Livelihoods

Although extractive reserves are supported by a range of different actors, and are regarded by many as an important strategy for integrating conservation and development, reserve inhabitants are poor, lead isolated lives, and have inadequate access to health and education. They hope for better lives and identify higher incomes and improved access to social services as their most pressing needs. In recent years efforts have been made to improve and sustain livelihoods within extractive reserves. These initiatives include the G7 Pilot Programme for the Protection of the Brazilian Rain Forest (PP-G7),[5] and the PLANAFLORO project. WWF has also developed activities to enhance the incomes of reserve inhabitants either directly or by helping to strengthen the capability of the rubber tapper organizations to implement PLANAFLORO related actions. Direct initiatives include expanding the range of marketable forest products and improving marketing channels. These activities have the potential to empower the rubber tappers in economic terms and to contribute to 'sustainable livelihood security'.

WWF is attempting to address the first constraint highlighted above, the economic dependence of rubber tapper households on key extractive products. It has initiated efforts to diversify the range of commercially valuable products that can be sustainably extracted from the forest. Rubber tappers use a remarkably wide range of forest products, but only a small proportion are marketed and most products are harvested for their consumptive value as food, fuelwood, building and fencing materials, and medicines. An estimate in the field identified thirty different plant species regularly used by rubber tapper households in Rondônia. These represent only the most frequently harvested and the total number is thought to be significantly greater. Many studies provide evidence of a larger number of forest products used by other groups in Amazonia including colonist farmers (Muchagata, 1997).

Many of the products harvested by the rubber tappers have commercial potential and some already have established regional markets. Among the most important are the açaí fruit (*Euterpe oleracea*) and palm hearts extracted from a variety of palms including *Euterpe*. Açaí palms have the advantage that they can be easily managed on a sustainable basis for both fruits and palm hearts due to their multi-stemmed self-regenerative habit. Recently, WWF has initiated a study to analyse the economic and ecological viability of useful palm species such as *Euterpe* in the Jaci Paraná reserve. More information is necessary before appropriate measures for improving markets and income generation can be implemented.

---

5. The PP-G7 supports a number of extractive reserves in Brazilian Amazonia, one of which is the Ouro Preto Extractive Reserve in Rondônia.

Timber is the most valuable product from the forest. WWF is currently establishing a pilot project in two extractive reserves in Rondônia for the extraction of timber on a sustainable basis. Extractive reserves contain valuable timber resources which can be harvested sustainably to provide an additional income source for rubber tappers. The project has already generated some controversy among reserve inhabitants, in particular concerning labour inputs and distribution of benefits. Concerns have also been raised about its economic viability. The costs of implementing this kind of scheme are substantial: they include the elaboration of a management plan; adoption of harvesting methods that have a minimal impact on surrounding vegetation; and regeneration of harvested species. The economic return from timber after extraction costs are deducted is uncertain. The first stages of the project are now completed and the first timber harvests are expected to begin soon.

Given the dominance of intermediaries in the regional forest product trade WWF has helped the rubber tappers to establish alternative marketing arrangements that enable them to secure a better price for extractive products. This is the only WWF activity aimed at improving the economic welfare of rubber tappers that has been fully implemented. Its objective is to eliminate the need for marketing intermediaries in extractive reserves. However, this is not straightforward to implement. Although they exploit the rubber tappers, the middlemen form the bridge between extractive products and far-away markets. They also supply extractivist communities with staple foods and other essential goods, transporting them to the remote extractive reserves. The middlemen cannot simply be eliminated without a reliable and efficient way to supply communities with essential goods also being created. One solution is to establish community trading posts run by the local Associations (*cantinas*) in strategic locations within the extractive reserves, allowing residents to purchase basic supplies without having to travel great distances.

Setting up a marketing and supply structure requires great investment. Firstly, a workable transportation infrastructure for both goods and extractive products is needed. Secondly, storage facilities have to be built in order to keep products until a sufficient amount has been gathered for selling to a large buyer. Thirdly, there has to be initial capital to buy extractive products from producers, and to purchase goods from a wholesaler. In this case, transport and storage facilities were provided by PLANAFLORO and the initial capital was supplied by WWF.

Problems with these new marketing structures have arisen as a result of the local Associations' lack of experience in management. The local Associations have gradually lost their initial capital and this has made the sale of extractive products through the new system difficult. Furthermore, the *cantinas* have not been able to replace their stocks. This initiative has not had a great impact on the economic well-being of the rubber tappers, some of whom continued interacting with middlemen in accordance with relationships of patronage developed over many years (see Hugh-Jones, 1992, on the complexity of these interactions).

WWF has helped to strengthen the organizational and administrative capability of the rubber tappers organizations, especially of the OSR, to implement PLANAFLORO related initiatives so that these organizations can collaborate as active participants in developing and implementing the programme. More specifically, WWF has helped the rubber tappers to establish the physical infrastructure of the OSR and local Associations (acquisition of headquarters; office equipment including computer, printer, and fax machine), and it assists with maintenance expenses such as phone and electricity. WWF also provides financial resources to staff the rubber tapper organizations. This support has allowed the OSR and local Associations to pay a salary to the rubber tappers' leaders so that they can spend more of their time working in extractive reserve related activities. It has also enabled ECOPORÉ to appoint two technical and legal advisers to work with the OSR. In addition, the tapper organizations have been able to employ a secretary to help with administration and accountancy. WWF has thus contributed to building the capacity of the rubber tapper organizations, with the result that they are better prepared to negotiate, administer and implement PLANAFLORO initiatives. The exact contribution of this type of support to the economic welfare of extractivist communities is difficult to establish. To date, PLANAFLORO has provided substantial infrastructure support to extractive reserves with the construction of storage facilities and the acquisition of vehicles and machinery (rice peelers, sugar cane threshers, outboard engines). There are also plans to establish a factory to process rubber and a series of small processing units to shell, grade, and package brazil nuts. How rubber tappers will benefit from these PLANAFLORO initiatives depends greatly on their ability to manage and take control of them. Improving the organizational capability of grassroots organizations may be a decisive factor determining the extent to which PLANAFLORO can generate long-term economic benefits for extractivist communities.

In summary, there are a number of ways to address the disadvantages of extractivism so as to improve the socio-economic well-being of the rubber tappers. The WWF support for the Rondônian rubber tappers has attempted to do this by providing financial and/or technical assistance for the development of several economic activities. These include the establishment of alternative marketing networks; development of income-generating activities based on new products; search for and expansion of markets for rainforest products; and local processing of extractive products. Some of these activities are still at a very early stage of development, and thus it is difficult to anticipate their impact on the economic well-being of extractivists. Where results are already visible, namely in the establishment of alternative marketing networks, significant management problems have arisen and as a consequence, objectives have not been met. This is partly because the activities supported by WWF are related more to the creation of extractive reserves than to the development of income-generating activities that improve the socio-economic well-being of reserve inhabitants. So far, the alliances with different external agencies have not resulted in the significant economic empowerment of rubber tappers.

CONCLUSION

This review suggests that intervention by WWF and strategic alliances with other NGOs have enhanced the empowerment of the rubber tappers in political terms, but have not been so successful in economic terms. The rubber tappers have won recognition of their rights to certain resources through the establishment of large tracts of land designated as extractive reserves. On the other hand, they continue to lead precarious and impoverished lives and struggle to generate income to meet basic needs.

The case reviewed here reveals some of the complexities and dynamics of the alliances or relationships between grassroots organizations and outside agencies and NGOs. Where the objectives are mutually beneficial and are not in conflict then such alliances can be successful in furthering multiple interests. Such alliances may empower environmental groups, as well as grassroots organizations. Alliances may form for specific purposes — such as the legal establishment of extractive reserves — although the objectives and interests of the various groups and actors are quite different and they will be impacted upon in different ways.

This research raises questions about the viability of extractivism as a means of providing a secure livelihood for rubber tapper communities, and extractive reserves as sustainable conservation designations. Evidence presented here highlights both the extreme poverty of some rubber tapper households, and the lack of a sustainable livelihood security, and suggests that much needs to be done before extractivism can provide rubber tappers with secure incomes. Many rubber tapper families feel economically and socially disempowered. Rubber tappers perceive two development priorities: first, higher incomes in order to satisfy physical and material needs; secondly, access to social services such as health and education. So far, alliances with outside agencies have been unable to meet these aspirations to a significant degree. There may be reasons for this beyond the control of the current alliances, such as prices set by world markets, and structural impediments which require alliances with much more powerful groups to influence and effect change.

As yet there is the little evidence of how the successful implementation of the extractive reserves and improvements in rubber tapper livelihoods and welfare are linked. We cannot deduce whether political empowerment will lead to economic empowerment or whether assured rights of access to natural resources in the forest will lead to more secure livelihoods. Much of the available literature might lead us to expect this, although highly disadvantageous terms of trade, dependence on a few products, and unstable markets may prove too difficult to overcome, even given alliances with international organizations. Schmink and Wood (1987) have highlighted the unstable nature of extractivist economies, and a growing body of literature indicates that extractivism is unlikely to provide an adequate income for forest communities (Browder, 1992a, 1992b; Salafsky et al., 1993).

In addition to the dynamics of the different dimensions of empowerment, it is also important to recognize the dynamics of the alliances between actors and the changes in their interests, power and coalitions over time. Silva (1994) has emphasized the conditions which influence and enable the formation of alliances, and again these conditions will change over time. We have highlighted this with regard to the changing emphases of the World Bank, and with regard to the development of POLONOROESTE and PLANAFLORO, and now perhaps PP-G7, as providing alternately very conflicting and more enabling institutional and programmatic frameworks for the evolution of alliances and establishment of the extractive reserve system in Rondônia. As Conkin and Graham (1995) point out, international alliances are potentially fragile and may pose political risks domestically for grassroots organizations. This may in turn affect the sustainability of the rubber tappers' way of life if the wider political and institutional context changes drastically.

In conclusion, our study finds that strategic alliances between grassroots organizations and environmental NGOs and other agencies can empower grassroots organizations in certain circumstances, and can contribute to meeting conservation and development objectives. In the case of the Rondônian rubber tappers these interactions have helped to politically empower rubber tappers to gain legal rights to extractive reserves. These interactions have facilitated the process by which extractive reserves have become politically viable spaces in terms of being legally recognized and established; however, these areas have yet to be proven viable in terms of providing secure and sustainable livelihoods for forest dwellers like the rubber tappers.

Acknowledgements

The authors gratefully acknowledge the support of WWF International. We would also like to thank the OSR in Rondônia and many of the rubber tappers who have been generous with their time and help in the course of our research. This project has also been supported by the UK Department for International Development (DFID), which supports policies, programmes and projects to promote international development. DFID provided funds for this study as part of that objective but the views and opinions expressed are those of the authors alone. We also acknowledge the insightful comments of two anonymous reviewers. Views expressed in this chapter, and any errors and omissions, remain the responsibility of the authors.

REFERENCES

Allegretti, M. (1990) 'Extractive Reserves: An Alternative for Reconciling Development and Conservation in Amazonia', in A. Anderson (ed.) *Alternatives to Deforestation: Steps Toward Sustainable Use of the Amazon Rain Forest*, pp. 252–64. New York: Columbia University Press.

Anderson, A. (1992) 'Land-use Strategies for Successful Extractive Economies in Amazonia', *Advances in Economic Botany* 9: 67–77.

Assies, W. (1997) *Going Nuts for the Rainforest: Non-timber Forest Products, Forest Conservation and Sustainability in Amazonia.* Amsterdam: Thela Publishers.

Batliwala, S. (1994) 'The Meaning of Women's Empowerment: New Concepts from Action', in A. Sen, A. Germain, L. C. Chen (eds) *Population Policies Reconsidered: Health Empowerment and Rights*, pp. 127–38. Boston, MA, and New York: Harvard University Press.

Benchimol, S. (1989) *Amazonia: planetarização e moratória ecologica.* São Paulo: Centro de Recursos Educacionais.

Browder, J. O. (1992a) 'Social and Economic Constraints on the Development of Market-oriented Extractive Reserves in Amazon Rainforests', *Advances in Economic Botany*, 9: 33–41.

Browder, J. O. (1992b) 'The Limits to Extractivism', *Bioscience* 42: 74–181.

Browder, J. O., E. A. T. Matricardi, W. S. Abdala (1996) 'Is Sustainable Production of Timber Financially Viable? A Comparative Analysis of Mahogany Silviculture among Small Farmers in the Brazilian Amazon', *Ecological Economics* 16: 147–59.

Chambers, R. (1993) *Challenging the Professions: Frontiers for Rural Development.* London: Intermediary Technology Publications.

Cleary, D. (1991) 'The Greening of the Amazon' in D. Goodman and M. Redclift (eds) *Environment and Development in Latin America*, pp. 116–40. Manchester and New York: Manchester University Press.

Conkin, B. A. and L. R. Graham (1995) 'The Shifting Middle Ground: Amazonian Indians and Ecopolitics', *American Anthropologist* 97(4): 695–710.

Forum de ONGs (1996) 'Um novo Planafloro', *Notícias do Forum* 4(8): 3.

Hecht, S. (1989) 'Chico Mendes: Chronicle of a Death Foretold', *New Left Review* 173: 47–55.

Hecht, S. and A. Cockburn (1989) *The Fate of the Forest: Developers, Destroyers and Defenders of the Amazon.* London and New York: Verso.

Homma, A. K. O. (1993) *Extrativismo vegetal na Amazonia: limites e oportunidades.* Brazil: EMBRAPA.

Hugh-Jones, S. (1992) 'Yesterday's Luxuries, Tomorrow's Necessities: Business and Barter in Northwest Amazonia', in C. Humphrey and S. Hugh-Jones (eds) *Barter, Exchange and Value*, pp. 42–74. Cambridge: Cambridge University Press.

IEA (Instituto de Estudos Amazonicos e Ambientais) (1994) *Levantamento socio-economico da reserva extrativista do Rio Ouro Preto.* Porto Velho: IEA.

Johnson, H. (1992) 'Rural Livelihoods: Action from Below', in H. Bernstein et al. (eds) *Rural Livelihoods: Crises and Responses*, pp. 274–300. Oxford: Oxford University Press.

Long, N. (1992) 'From Paradigm Lost to Paradigm Regained? The Case for an Actor-oriented Sociology of Development', in N. Long and A. Long (eds) *Battlefields of Knowledge: Interlocking Theory and Practice in Social Research and Development*, pp. 16–43. London: Routledge.

Long, N. and J. D. van der Ploeg (1994) 'Heterogeneity, Actors and Structure: Towards a Reconstitution of the Concept of Structure' in D. Booth (ed.) *Rethinking Social Development: Theory, Research and Practice*, pp. 62–89. Essex: Longman.

Mahar, D. (1989) 'Government Policies and Deforestation in Brazil's Amazon Region'. Washington, DC: The World Bank.

Martine, G. (1990) 'Rondônia and the Fate of Small Farmers', in D. Goodman and A. Hall (eds) *The Future of Amazonia: Destruction or Sustainable Development?*, pp. 23–48. London: Macmillan.

Melone, M. (1993) 'The Struggle of the *Seringueiros* : Environmental Action in the Amazon', in J. Friedmann and H. Rangan (eds) *In Defence of Livelihood: Comparative Studies on Environmental Action*, pp. 107–26. West Hartford, CT: Kumarian Press.
Millikan, B. H. (1995) 'Pedido de investigação apresentado ao Painel de Inspeção do Banco Mundial sobre o PLANAFLORO'. Porto Velho: Forum de ONGs e Movimentos Sociais que Atuam em Rondônia/Friends of the Earth..
Muchagata, M. (1997) *Forests and People: The Role of Forest Production in Frontier Farming Systems in Eastern Amazonia.* DEV Occasional Paper OP36. Norwich: School of Development Studies, University of East Anglia.
Nelson, N. and S. Wright (1995) *Power and Participatory Development: Theory and Practice.* London: Intermediate Technology Publications.
Pimbert, M. P. and J. N. Pretty (1994) 'Parks, People and Professionals: Putting Participation into Protected Area Management'. Draft discussion paper. Geneva: UNRISD/IIED/WWF.
Rich, B. (1994) *Mortgaging the Earth: The World Bank, Environmental Impoverishment and the Crisis of Development.* London: Earthscan.
Rondônia, Governo de (1994) *Programa de trabalho PLANAFLORO: primeiro semestre de 1996.* Porto Velho: Governo de Rondônia.
Rowlands, J. (1995) 'Empowerment Examined', *Development in Practice* 5(2): 101–7.
Salafsky, N., B. L. Dugelby and J. W. Terborgh (1993) 'Can Extractive Reserves save the Rainforest? An Ecological and Socioeconomic Comparison of Non-Timber Forest Extraction in Peten, Guatemala, and West Kalimantan, Indonesia', *Conservation Biology* 7(1): 39–52.
Schmink, M. and C. H. Wood (1987) 'The Political Ecology of Amazonia' in P. D. Little, M. Horowitz and A. E. Nyerges (eds) *Lands at Risk in the Third World: Local Level Perspectives*, pp. 38–57. Boulder, CO: Westview Press.
SEDAM (Secretaria do Desenvolvimento Ambiental) (1986) *Diretrizes Ambientais para Rondônia.* Brasilia: Ministerio do Desenvolvimento e Meio Ambiente
Shwartzman, S. (1992) 'Social Movements and Natural Resource Conservation in the Brazilian Amazon', in S. Counsell and T. Rice (eds) *The Rainforest Harvest: Sustainable Strategies for Saving the Tropical Forests*, pp. 207–17. London: Friends of the Earth.
Silva, E. (1994) 'Thinking Politically about Sustainable Development in the Tropical Forests of Latin America', *Development and Change* 25(4): 697–721.
Southgate, D., M. Coles-Richie and P. Salazar-Canelos (1996) 'Can Tropical Forests be saved by Harvesting Non-Timber Products?'. CSERGE Working Paper GEC 96-02. Norwich and London: Centre for Social and Economic Research on the Global Environment, University of East Anglia and University College London.
WWF (1995) 'BR0087 Brazil: Social and Environmental Support for Rondônian Rubber Tappers', in *WWF Latin America/Caribbean Programmes*, pp. 4230–1. Gland, Switzerland: WWF.

10 Maintaining Centralized Control in Community-based Forestry: Policy Construction in the Philippines

Richard Gauld

INTRODUCTION

Many of the basic ideas which have been used to justify the maintenance of a state monopoly over forest management in the developing world have become progressively more difficult to sustain. Top-down approaches to forest management have been subject to increasing criticism in the wake of the apparent failure of scientific forestry and state ownership of forest lands to prevent continuing deforestation. Conversely, research into common property management regimes has revealed that community-based resource management systems, which the introduction of state ownership claims has tended to disrupt, can be more efficacious than centralized control in some circumstances (Bartlett, 1992; Broad, 1994; Gasgonia, 1993; Ghai, 1994; Hafner, 1995). In the context of growing social and political conflicts over dwindling forest resources, a number of researchers, planners, policy makers, NGOs, and development agencies are recommending increasing local community involvement in forest management (Colchester, 1993; Kumar and Kaul, 1996; Lynch and Talbott, 1995; Peluso, 1995). According to its proponents, community-based forestry can reconcile the goals of social justice, equity, development, empowerment and environmental sustainability by transferring forest management into the hands of local communities. It is cited as being more environmentally sustainable than top-down development planning (Colchester, 1993), since it alleviates pressures on remaining resources (Lynch, 1993), while simultaneously promoting social justice and the socio-economic elevation of rural communities (Bacalla, 1993).

Community-based management is thus emerging as a potentially important strategy within forest management planning in much of the developing world (Colchester, 1993; Hobley, 1996; Lynch and Talbott, 1995; Silva, 1994). A number of high-profile community-based policies and programmes have started to emerge (see Poffenberger, 1996; Lynch and Talbott, 1995), and these are being promoted as an important element in the ubiquitously cited but ill-defined concept of sustainable development.[1] The promotion of

1. See Lele (1991) for a critical review of the concept of sustainable development.

community-based forestry as a mainstream management strategy represents a major shift in forest management planning by the state. Contrasting such approaches with the historic hegemony of scientific forestry and state control over forest management within policy making (see Blaikie and Brookfield, 1987; Bryant and Bailey, 1997; Guha, 1989; Peluso, 1992), the move towards community-based forestry is possibly one of the most important developments in forest policy in the developing world since the adoption of scientific forestry in the latter half of the last century.

This transition in state forest policy is particularly significant in the Philippines as community-based forestry has recently been adopted as the national strategy to achieve 'sustainable forestry' and 'social justice' (Office of the President, 1995: 1). The Philippine community-based forestry policy has been described as one of the most innovative in the region (Lynch, 1993) and the proposed transfer of forest resources to communities as 'truly impressive' (Colchester, 1994). Based on a study of policy discussions and administrative procedures surrounding community-based forestry in the Philippines,[2] this chapter reveals, however, that the apparent transition in forest policy from top-down towards community-based approaches is not reflected in the way in which community-based forestry is discussed and operationalized by policy makers. Firstly, community-based forestry policy bears many of the hallmarks of scientific forestry in which technical and productivity aspects rather than social and wider environmental considerations are emphasized. Secondly, strong state control over forest management is understood as being a necessary feature of community-based forest policy. A third major issue is the predominance of a reductionist understanding of 'community' among policy makers. These characteristics of community-based forestry policy discourse and design have resulted in the marginalization of the social and environmental concerns which purportedly differentiate this policy from previous approaches. The research findings show that, in consequence, while there has been a shift in the general rhetoric of policy, the changes associated with the adoption of community-based forestry policy, even within its own terms, are not as substantive or innovative as they may initially appear.

POLICY DISCOURSE AND POLICY DESIGN

Although the relationship between state policies and deforestation represents a primary agenda within research into the political economy of

---

2. This chapter is based upon data gathered from interviews conducted over two months in the Philippines during the summer of 1995, along with analyses of supporting policy documents. Data were gathered primarily through key-informant interviews with policy officials, project-officers, and NGO representatives, supplemented by attending government data-gathering exercises and community-consultation meetings conducted in Luzon as part of a multi-purpose dam project assessment.

environmental degradation (Bryant and Bailey, 1997), the details of policy itself have often remained neglected. Even within those studies which argue that the state introduces its own self-defined interests into the logic of decision making (see, for example, Khator, 1991; Peluso, 1992), research into the role of the state in environmental management has tended to emphasize the political economy of the state while neglecting the political economy of 'truth' within state institutions (Watts, 1993). Concomitantly, social scientists conducting research into social and community-based forestry have tended to focus on understanding the perceptions and experiences of rural communities while neglecting detailed analysis of the perceptions of those who govern them, despite the fact that 'government regulations, preferences and preconceptions affect the success of almost any project involving people and trees' (Dove, 1992: 15; see also Barraclough and Ghimire, 1990).

Policy discourses, and the categorizations, problem definitions and administrative procedures which develop within and from these discourses, promote and justify particular concerns, strategies and solutions at the expense of others (Dove, 1994; Fairhead and Leach, 1995; Fortmann, 1995; Hoffman, 1995; Long and van der Ploeg, 1994; Peters, 1994). Similarly, the persistence of particular policy narratives, ideas and 'orthodoxies', as they exist in a dialectic with political and economic interests (Blaikie and Jeanrenauld, 1996; Moore, 1996), influences the way in which society–environment interaction is constructed and understood within policy (Guthmann, 1997; Hoben, 1995; Leach and Mearns, 1996; Roe, 1994). The discourses surrounding environmental management are thus intimately bound up in power relations (Banuri and Marglin, 1993; Fischer and Forester, 1993; Long and van der Ploeg, 1994; Schram, 1993). If community-based forestry policy is to be fully understood in terms of how and why policy has been designed the way it has, and its potential to create solutions to poverty and environmental degradation, it is necessary to examine how community-based forestry is articulated within policy discourse. Analyses of the way in which policy makers articulate community-based forestry, and the language used in policy documents, reports and papers, can reveal linkages between the design of community-based forestry policy and particular ideas and 'orthodoxies' as well as political and economic forces (Myerson and Rydin, 1996; Peet and Watts, 1996; Roe, 1994). This is particularly important given the significant attention and funding being directed towards the promotion of community-based forestry in many countries in the developing world.

FOREST POLICY IN THE PHILIPPINES

The Philippines represents a particularly striking example of the transition from state-centred 'scientific' forestry towards community-based approaches in general policy rhetoric. To understand and contextualize this new policy orientation, some historical appreciation of the development of forest policy is necessary. In the latter half of the nineteenth century, while under Spanish colonial rule, a centralized framework for forest management in the Philippines was first developed by the state (Chinte, 1985; Makil and Reyes, 1982). The transfer from Spanish to United States control in 1898 witnessed increasing efforts to centralize forest management. Forest policy emphasized the application of 'scientific' approaches to exploit the commercial potential of forests using 'rational' management plans (Anti-Slavery Society, 1983). Driven by an increasing desire to exploit the commercial potential of the Philippine forests, this approach resulted in increased state control over large tracts of forest land and the application of management models imported from the United States (Anti-Slavery Society, 1983; Makil and Reyes, 1982; Poffenberger, 1990a). Indigenous practices were labelled backward and destructive (cf. Dove, 1983) and traditional land-use rights among indigenous upland communities were significantly eroded. The mandate of the Bureau of Forestry was essentially to protect trees from upland communities for the purpose of commercial exploitation (Gibbs et al., 1990; Makil and Reyes, 1982; Oposa, 1992; Putzel, 1992).

This often inappropriate application of models of forest management developed in Europe and the United States, which purported to offer methods for manipulating and maximizing the growth, harvesting and regeneration of commercially valuable species, characterized many forest policies elsewhere in the developing world (see, for example, Guha, 1989; Peluso, 1992). The application of scientific forestry also tended to go hand in hand with increasing state control over forest resources and the marginalization of indigenous resource management practices.

However, a number of significant national and international ecological, political and economic processes and events following the cessation of American colonial rule after the Second World War created the context for the adoption of a more socially attuned approach to forestry. Political problems such as the growing significance of the communist insurgency movement, ecological events such as the great flood in central Luzon in 1972, which highlighted the serious effects of deforestation, and apparent success in leasing forest lands to indigenous communities in the mid-1970s (Gibbs et al., 1990; Poffenberger, 1990a) culminated in the launch of an official social forestry programme in the early 1980s. This programme was ostensibly designed to combine socio-economic development in the uplands with forest protection and development (Agaloos, 1993). However, it received little political support (Castro, 1984; Contreras, 1991).

In 1989, a community-based forestry programme (CFP) was initiated as a potentially radical new development in state policy. This policy proposed transferring the responsibility for forest management into the hands of local communities based on principles of 'social justice and resource sustainability' (DENR, nd[a]: 3). While the introduction of the programme was influenced by the historically inculcated desire to protect trees from shifting cultivators (see DENR, 1989), the inclusion of alternative livelihood opportunities offers the potential for social, economic and political empowerment (DENR, 1995c). This community-based forestry programme is seen as being an important development in forest policy in Southeast Asia (Colchester, 1994; Lynch, 1993), while the Philippine Department of Environment and Natural Resources (DENR) has been credited for facilitating community empowerment (Fox, 1993; Poffenberger, 1990b) and for being one of the most assertive environmental departments in Southeast Asia (Rush, 1991). However, this new strategy has emerged against a historical legacy of state efforts to centralize forest management and a political economy based on commercial timber exploitation and the exclusion of community access to forest lands. Through an analysis of the discourse and administrative procedures of community-based forestry policy, the following sections demonstrate the continuing influence of this historical legacy on community-based forestry policy.

RATIONALIZING COMMUNITY-BASED FOREST POLICY

There is considerable debate surrounding the efficacy and institutional design of community-based resource management within the general literature. Disagreements exist at both a theoretical level (compare, for example, Barber, 1989; Cernea, 1993; Ghai, 1994; Kirchhofer and Mercer, 1986; Ostrom, 1990; Pulhin, 1985; Reyes, 1984; de Saussay, 1987), and at a practical level (cf. Contreras, 1992; Dove, 1992; Eder, 1990; Hausler, 1993; Makil and Reyes, 1982; Rebugio, 1985). This may partly reflect the fact that the re-assessment of community-based management systems which occurred during the 1980s emerged in the form of revisions and critiques of Hardin's (1968) 'tragedy of the commons' thesis and 'prisoner's dilemma' models rather than offering coherent theoretical alternative models of community behaviour (Knudsen, 1995). Support for community-based management rests largely on empirical case-studies of such schemes in particular situations and not on any definitive new framework concerning the design and efficacy of community-based resource management (for a useful review, see Knudsen, 1995).

This confusion is mirrored within Philippine policy. According to official documentation, community-based forestry is being pursued as a policy option in the Philippines to make access to forest resources more equitable, to empower local forest communities and to achieve both 'productivity' and 'sustainability' in forest management (DENR, 1993b, nd[a], nd[b]; Office of the President, 1995). These notions saturate the discourse of contemporary development and environmental policy, but remain imprecise and conceptually complex (Adams, 1990; Barber, 1989; Ekins, 1993). Interviews with planners and policy makers revealed vague understandings and contradictions about why community-based forest management is being promoted and what fundamental principles such as 'sustainability' and 'empowerment' mean. Perceptions of community-based forestry, as described by planners and policy makers, ranged from universally suboptimal, relative to corporate management, through contingently optimal in specific circumstances, to representing the only way in which sustainability can be achieved. Sustainability emerged as a central theme, but remained largely undefined. The National Co-ordinator of the CFP, for example, stated that it was 'not possible' to define or explain how sustainability might be defined at this point in time.

Despite considerable ambiguity, the 'sustainability' of community-based forestry was frequently related to 'efficiency' and timber-productivity targets. Conversely, policy planners claimed that corporate management remained more sustainable in some areas where 'market inefficiencies' existed. These efficiencies were related to physically getting timber to markets, indicating a continuing emphasis on economies of scale upon which commercial 'scientific management' is based. This emphasis on efficiency and productivity is actively promoted by the major funding agencies who are backing the introduction of community-based forestry. Planners within the United States Agency for International Development (USAID), for example, explicitly related the introduction of community-based forestry in the Philippines to efforts to shift the production of wood away from capital intensive towards labour intensive methods. Capital intensive methods of timber extraction were described as being poorly suited to tropical forests as they cause considerable environmental damage. Conversely, small-scale labour intensive production was suggested to create 'virtually no damage'. The need to move towards free-market based 'partnerships' between communities and corporations to harvest timber was also emphasized. The Asian Development Bank (ADB) also related the introduction of community-based forestry to market efficiencies and competitiveness (ADB, 1995; DENR, 1995a). These perspectives have an important conditioning function as funding agencies exert a significant influence on domestic forest policy.[3]

This emphasis on timber production and free-market efficiency remains within the discursive boundaries of neo-liberalism which continues to dominate contemporary development paradigms. However, the emphasis on free markets and individuality (see, for example, World Bank, 1992), rests uneasily with the notions of 'community' and 'equity' with which the community-based approach is associated. Community-based forestry is

---

3. The ADB explicitly states that forestry sector lending to the Philippines is being used to promote policy and institutional reforms, including the introduction of community-based forestry (ADB, 1995; see also Salazar, 1993).

purported to engender sustainability through empowering weaker groups, but accompanying free-market reforms may counteract associated gains among weaker groups (Atumpugre, 1993). Whereas discourses concerning resource *redistribution* generally do not play a strong role in promoting policy change, Majone (1993) argues that ideas about *efficiency* can act as strong catalysts for change. It is possible that community-based forestry is being pushed in response to efficiency considerations emerging from neo-liberal theory relating to decentralization and 'rolling back' the state rather than as a response to concerns about the need for redistribution. In this regard the release of USAID funds for the CFP was conditional upon the Philippine government introducing policy reforms to promote, *inter alia*, efficiency and competitiveness in forest products industries (DENR, 1995e).

At the same time, there is a conflict between efficiency according to the neo-liberal emphasis on decentralization and rolling back the state, and efficiency as understood within the discourse of scientific forestry. As noted earlier, some DENR planners claimed that corporate-based extraction and the retention of state control may still be appropriate in some areas. This emphasis on scientific forestry even permeates community-based forestry training in the Philippines. Claims were made by an associate professor of social forestry at the main Forestry College in Los Banos, who also advises the DENR on the implementation of community-based forestry, that capital intensive corporate forest management is considerably less environmentally damaging than small-scale community forestry and, in economic terms, considerably more efficient. The primary reason for initiating community-based forestry, according to this view, is not sustainability but equity — while socially desirable, it is alleged to be an economically and environmentally sub-optimal approach to forest management.

However, it would be unrealistic to suggest that community-based forestry is purely a product of discourse(s) of efficiency. Recognizing the tensions between efficiency as understood within the DENR, with its long-standing emphasis on centralized control, and efficiency as understood by funding agencies promoting neo-liberal reforms based on decentralization, the adoption of community-based forestry by the DENR can be understood partly as a repackaging of the earlier social forestry programme, with all its attendant problems, to take advantage of international funding. In doing so, the DENR received substantial funding, as well as potentially saving face with farmers through the provision of a high-profile 'solution' to social and environmental problems in the uplands[4] and responding to populist environmentalism. This was explicitly admitted by one senior official involved in co-ordinating international funding within the DENR. Many funding agencies, on the other hand, are concerned with opening up markets and increasing efficiency in timber production. The DENR's policy

4. By 1992, the DENR was one of the main recipients of international funding among government agencies in the Philippines (Korten, 1994).

is tempered by the continuing influence of scientific forestry and state ownership while many donors place a strong emphasis on markets.

Empowering Communities

A central rationalization and assumption underlying the introduction of the CFP is that communities who are provided tenurial security over a tract of public forest land will 'develop a stake in the resource' and will be motivated to manage it sustainably (DENR, nd[a]: 4; DENR/USAID,1992: 1). Without state-sanctioned tenurial security, communities are perceived as having no incentive to adopt sustainable practices. This understanding has been used to define the need for and administrative design of 'empowerment', one of the central tenets of the CFP. Empowerment is discussed in terms of transferring tenurial rights, which are regarded as being central to the achievement of the goals of community-based forestry (see, for example, CFP Assessment Team, 1995a; DENR, nd[a]). Field officers suggested that claims over forest lands made by the state have served to confuse and break down traditional resource management practices (see also Prill-Brett, 1993) while the ownership claims of the state and the imposition of penalties for contravention of state regulations have created disincentives for community investment in natural resource management (CFP Assessment Team, 1995a; Loenen, 1993; Lynch, 1985).

Interviews with NGOs indicated, however, that some local communities treat areas of land officially designated as being under state ownership as communally and/or privately owned within the community, and manage these lands as such. DENR provincial and community officers working in areas inhabited by indigenous groups also noted that indigenous groups often have their own resource management systems which are being practised on state lands. The environmental basis of many local rural communities' livelihoods, particularly among indigenous groups living in their ancestral domain, means that many of these communities already have a strong stake in environmental management, despite the competing claims of the state.[5] Resistance to a proposed dam project in Luzon by an indigenous people called the Ibaloi illustrates the point. Resistance to the dam has rested on both environmental/livelihood criteria (concern over siltation created by the dam affecting the local environment, and hence their livelihoods), and on a reference by many of the village members to their historical relationship with their ancestral land. Further, community mapping conducted during a DENR data-gathering exercise revealed that they have established and are maintaining their own small forest planta-

5. Loenen (1993) argues that indigenous communities in the Philippines have always been concerned with environmental sustainability (see also Broad with Cavanagh, 1993).

tions. Although the area is claimed by the state, community members recognize it as being their homeland and the source of their livelihood which must be protected from degradation. This community's actions suggest that they do in fact have a long-standing stake in local environmental management, despite the absence of state-sanctioned tenurial security.

While a number of NGOs stated that they had encountered communities of recent migrant farmers who were primarily interested in 'getting rich quick', the assumption that indigenous communities and long-term occupants do not have a stake in resource management was challenged by some DENR field-workers and NGO personnel. A lack of tenurial security does not necessarily coincide with a lack of interest in environmental management among long-term occupants. Correspondingly, the granting of tenurial security does not necessarily encourage short-term occupants to develop a stake. While corporate Timber Licence Agreement holders enjoyed twenty-five years of tenurial security, their actions did not indicate that this created a stake in the environment. One of the few sociologists employed by the DENR suggested that it is necessary to distinguish between indigenous traditional communities, long-term migrants, and recent migrant settlers when considering the relationship of forest communities to their lands (see also Broad, 1994).

The actual mechanism for the creation of 'tenurial security' is the Community Forestry Management Agreement (DENR, 1993b). This tenurial instrument is valid for twenty-five years and is potentially renewable for a further twenty-five years. It grants utilization privileges, and is cited as offering 'the necessary long term security for utilisation of natural resources ... on a sustainable basis' (DENR, 1993b: 3). In evaluating what this instrument represents, appropriate points of reference are the stated objectives of 'returning the forests to the people' (DENR, nd[a]: 1) and providing 'tenurial security and ownership of public forest land' (DENR, 1995c: 1).

The first point to consider is whether a twenty-five year leasing agreement represents a suitable foundation for the promotion of ongoing community-based forest management. Long-term occupancy is viewed as being preferable to ownership as 'CFP sites are on state-owned "public" forest-land' (CFP Assessment Team, 1995a: 14). Yet elsewhere in the same document it is stated that the underlying concept of community-based forestry is that communities should be in charge of forest management (CFP Assessment Team, 1995a). Through the implementation of community-based forestry, communities become 'the ultimate resource managers' (Guiang, 1993: 8). Despite these statements, a somewhat contradictory position has been adopted in which 'the DENR will maintain the ultimate responsibility for the public domain even when the management and utilisation rights are contracted to communities under the CFP'[6] (CFP Assessment

6. 'Public' is defined by DENR planners as belonging to the state.

Team, 1995b: 27). Leasing is thus intended to encourage communities to manage forests on lands 'which they do not own' (CFP Assessment Team, 1995a: 44).

These policy statements suggest that communities cannot manage natural resources independently of state controls and that the state possesses privileged knowledge in environmental management which justifies the retention of state ownership/control over natural resources. This was supported by evidence from interviews. Planners within the DENR argued that leasing rather than ownership agreements represent the most appropriate instrument for the implementation of community-based forestry as the DENR is ultimately the most able environmental manager. Similarly, despite claiming to recognize ancestral land rights (DENR, 1993a),[7] a DENR internal working paper stated that the Department does not support removing ancestral lands from under the jurisdiction of the state, since these areas contain important natural resources which 'should be under a holistic management approach and sustainable management' (DENR, 1995d: 9). The DENR equates the retention of ownership with an environmental rationale. However, this rationale apparently contradicts the postulates underlying the community-based approach, and rests uneasily with the historical relationship between state control over natural resources control, politics and deforestation (cf. Rush, 1991). The other explanation offered by officials was that the adoption of leasing is a function of the Constitution which precludes the release of state control over classified forest lands.[8]

The leasing instrument itself is constructed in the form of a management contract which authorizes designated communities to utilize forest products (the contract 'payment') in return for managing forest lands (the contract 'service'). It includes provisions for the exclusive occupation of the site by the legally identified community, yet the ensuing provisions of leasing agreements refer primarily to tree utilization rather than land ownership (see Appendices in Dolom, 1995; see also Singh and Khare, 1993). The DENR retains authority over the identification of trees that may or may not be harvested and sets the guidelines for the development of the community management plan (DENR, 1993b). Leases transfer management powers, but 'subject to existing rules and regulations', whereas 'the DENR has the authority and jurisdiction over the CFP'. Accordingly, communities are obliged to follow 'all duly promulgated laws, rules and regulations pertinent to forest products utilisation, forest development and forest management' while the 'harvesting of timber and minor forest products shall be regulated by the DENR pursuant to existing rules and regulations' (see Dolom, 1995: 1–2).

---

7. This document does not grant *ownership* rights and, being an administrative rather than legislative order, it can and is being contested in the courts (from interviews).
8. Ancestral domain legislation was passed in autumn 1997. However, it remains to be seen what effect this will have in practice.

Leasing agreements contain many similar provisions. While a contract is required to establish a legal basis for the transfer of control over natural resource management, the provisions and conditions contained within these contracts severely limit the control that communities have over forest management *vis-à-vis* the DENR. The DENR has restricted the decision-making parameters available to communities, and it retains ultimate control over management plans since it retains the right to reject proposals. Again, the parameters guiding policy design appear to be that the state possesses privileged knowledge in natural resource management and remains the ultimate resource manager, and that communities require legally sanctioned regulation by an external leviathan if they are to manage natural resources sustainably. Policy discourse retains assumptions of local ignorance and destructiveness.

While the relationship between communities and the DENR is supposed to be a 'partnership' (CFP Assessment Team, 1995a: 9), community-based policy retains a strong emphasis on controlling communities which creates an asymmetry in the 'partnership'. The stated goal is to transfer responsibility for forest management into community control (DENR, nd[a]: 3). There is, however, no reference to a commensurate transfer of authority which is arguably a prerequisite for the effective administration of responsibilities. The restrictive nature of leasing agreements devalues their participatory element while the delegation of responsibility without authority limits the extent to which communities can be empowered to manage forest lands. Although the community and the DENR supposedly have an equal right to terminate management agreements, communities must make an agreement with the DENR if they are to obtain legal access to classified forest land. Termination converts the communities' status into that of illegal occupancy, creating an imbalance between the rights and obligations of the DENR and the rights and obligations of the community.

The potential for empowerment is further weakened by the DENR's understanding of the rights of communities over forests. While the DENR (nd[a]: 4) refers to the granting of 'rights' in promotional documents, this is transformed in policy into transferring utilization 'privileges' (DENR, 1993b: 3). The discourse and design of policy presents utilization as being a privilege granted in return for managing the resource. As outlined in leasing agreements, the DENR retains the right to withdraw privileges upon the (undefined) 'non-performance' of the community or upon violation of any provision of the agreement (Dolom, 1995: 4). This may also serve to weaken perceived tenurial security. Thus, community management is constituted as a temporary gift rather than as a right. As Vandergeest (1991) notes, gifts may empower the giver more than the receiver since conditions can be imposed on gifts (the provisions of the leasing agreements), and gifts can be withdrawn (see Dolom, 1995). The contractual obligations of the DENR in the area of socio-economic development, as per the leasing agreement, are limited to the provision of incomes through contract reforestation, timber stand improvement and assisted natural regeneration 'subject to the availability of funds' (Dolom, 1995: Appendices). Although socio-economic development is a focal concern of the CFP, socio-economic targets have not been set. On the other hand, there are clear timber production targets.

The head of the DENR's Social Forestry Division suggested that the state's role in the implementation of community-based forestry should be that of information broker. In other words, the forest should be managed by communities rather than the state. However, policy design does not position the state as an information broker and catalyst of community management but as an information controller and a community manager. Regulations and contractual obligations pre-determine the decision-making space afforded to communities, thus limiting the actual jurisdiction they have over the management of forest resources. This is based on an environmental rationale which says that the state possesses the most accurate knowledge, ability and authority to devise the parameters for forest management. Conversely communities, lacking full access to the state's supposed information monopoly and vision, are less able to manage resources and, consequently, need to be controlled through state regulations to ensure that they follow the state's conception of sustainable management practices.

Community-based Forestry as a Socio-Ecologically Sustainable Strategy

The initial injection of capital in conjunction with proceeds from utilization are presented by policy makers as ensuring the socio-economic and ecological sustainability of community-based forestry. Pump-priming activities are purported to start the 'economic ball rolling' (CFP Assessment Team, 1995a: 45), leading to sustained flows of economic and ecological benefits. However, it cannot be assumed that the granting of utilization 'privileges' will necessarily ensure the sustainability of community-based forestry. As one DENR project officer observed, many small farmers could not meet their immediate subsistence needs from sites designated under the earlier social forestry programme and abandoned them (see also Porter and Ganapin, 1988). The national co-ordinator claimed that the 'dynamism' of CFP sites prevented attempts to make generalized statements about the sustainability of community-based forestry. Planners admitted that no comprehensive socio-economic assessments or reporting systems have been established (see also Fairman, 1995). In contrast, reporting systems on financial and timber production targets are considerably more developed, and are being used as proxies to assess the effectiveness of social aspects of project implementation.

Despite the lack of clear guidelines, certain criteria are associated with the sustainability of community-based forestry. Empowerment is assessed against the capacity of officials of people's organizations (see below), specifically the president, vice-presidents and treasurers, to perform designated tasks. These tasks are timber stand improvement, agroforestry, and reforestation. In other words, the rehabilitation and protection of forests. It is unlikely that production/rehabilitation targets are appropriate indicators for long-term sustainability given the influence of social and political factors on forest degradation. Quantitative, production-orientated targets do not represent an appropriate basis for empowering communities and sustaining community-based management systems. The goal of community-based forestry is argued to be the improvement of socio-economic conditions through proceeds from forest product utilization. However, policy neglects forest development as a means towards rural development, while emphasizing forest production as an end in itself. The history of forest management in the Philippines, and elsewhere in the developing world, suggests that this may prove to be self-defeating.

In the process of incorporating only selective elements of community-based forestry into policy discourses and design, community-based forestry is being divested of many of the aspects which could challenge the *status quo* in a manner similar to the incorporation of 'sustainable development' into mainstream development planning (Adams, 1990; Ekins, 1993; Lele, 1991). Instead of resulting in a departure from previous trajectories, community-based forestry in the Philippines is being incorporated into, rather than changing, existing frameworks. This is further highlighted by the manner in which the identification and organization of community groups for participation in state-sanctioned community-based forest management projects have been constituted in policy.

DEFINING COMMUNITY

Capturing the essence of 'community' remains an unfulfilled project even within the social science literature (see Bell and Newby, 1971; Cohen, 1985; Wilkinson, 1986). Like 'sustainability', 'community' is a pervasively used but semantically elusive concept. The problem of identifying community boundaries emerges: where do they begin and end, what form do boundaries take — spatial, social, ethnic, ideological? Who is inside and who is outside community boundaries? Can a group of individuals be selected to represent the community in interactions with the state? How, indeed, can an initial interface between a state department and a community be established? It is, none the less, vital to develop some form of working definition if the transfer of forest management into the hands of local communities is to be operationalized. It has been suggested that, in practice, policy makers usually rely on various images and assumptions of what constitutes a community, reducing these to a notion of a generic community upon which policy making can be based (see Pigg, 1992). Blaikie and Jeanrenauld (1996) argue that such generic communities, as understood and used by environmental policy makers, are imaginary entities which cannot be found in the 'real world'. None the less, the manner in which 'community' is represented conceptually, socially, politically and geographically in community-based forestry policy shapes the way in which relationships and administrative procedures are constituted and enacted (see Fortmann and Roe, 1993; Li, 1996).

In addition to displaying patterns of common bonds and patterns of social interaction (Bernard, 1973; Wilkinson, 1986), communities are typically, although not universally, defined on the basis of their geographical foundations, as occupying a particular geographical space (Cohen, 1985; Dalby and Mackenzie, 1997; Elias, 1974; Fortmann and Roe, 1993). However, understandings which rely on purely geographical representations of community may erroneously conflate territoriality with deep horizontal commonalities (Anderson, 1991). Such 'imagined' communities ignore those dimensions which are not amenable to purely geographic definition such as those based upon class, gender, ethnic, tribal or ancestral identities (Anderson, 1991; Dalby and Mackenzie, 1997). The manner in which people become part of or influence a community through kinship, property ownership, or other economic, political, cultural and social relations (Fortmann and Roe, 1993) and the potentially uneven nature of intra-community relations (Elias, 1974) must be considered in juxtaposition to any geographical foundations. This is particularly important in relation to a policy concerned with social justice, equity and community empowerment.

In practice, policy makers in the Philippines have created a generic conception in which any given community is being defined almost entirely by its occupation of a particular geographical space. Within the discourse and administrative procedures of policy, these geographically-defined communities are assumed to be characterized by political homogeneity and, within the context of policy implementation, need to be organized by an external agent and encouraged to develop a stake in forest management, as the following seeks to illustrate.

The centrality of the spatial approach is highlighted in the definition of community within policy as consisting of those 'persons/individuals residing within or adjacent to the CFP site' (DENR, 1993b: 2) who have a 'shared interest' (CFP Assessment Team, 1995a: viii). What this 'shared interest' is remains unspecified. Interviews revealed that planners within the DENR focus almost exclusively on this spatial identity: internal stratification along socio-economic, political, religious, ethnic, gender or other such lines does not emerge as an important issue. According to policy planners, community members enjoy an equal political voice and spatial proximity manifests social and political homogeneity. This is not based on a pluralist conception of local political behaviour. Rather, intra-community conflict among different interest groups is simply not considered as being particularly relevant. As Guggenheim and Spears (1991) point out, however, the socio-economic, gender, age, ethnic or religious status of different community groups may limit the opportunities available to weaker members to participate in community decision making. Given the more likely scenario of political heterogeneity within spatially-defined community groupings (Eder, 1987; Loenen, 1993), participation in community decision making can be skewed in favour of more powerful subgroups. In the context of community-based forest management, Hisham et al. (1991) argue that community-based projects are typically controlled by a small number of powerful individuals.

This is particularly important in the implementation of community-based forestry in the Philippines as there are a number of project sites where different ethnic groups co-exist. Indigenous peoples have historically been the most politically, socially and economically marginalized group in the Philippines (Anti-Slavery Society, 1983; Eder, 1987; Loenen, 1993). It is acknowledged that there are a number of sites in which indigenous and migrant communities co-exist, but communities are left to 'fight it out among themselves' with no DENR intervention in this process. The leaders emerging from this process are assumed to be representative of the community as a whole. This view is evident in documentation (CFP Assessment Team, 1995a) and was generally supported by planners and policy makers interviewed within the DENR. Although policy makers and implementers suggested that local cultural traditions should and would be respected in the promotion of community-based forestry (see also CFP Assessment Team, 1995a), there are no institutionalized mechanisms to ensure the political representation of weaker groups (see Wignaraja, 1993).[9]

In preparatory assessments of potential sites, the importance of establishing interdisciplinary teams composed of foresters, agriculturalists, economists and social scientists has been recognized (Mariano, 1995). However, priorities are demonstrated by the suggestion that 'if [an economist and sociologist] are unavailable because of limited financial and human resources, a forester or an agriculturist with a strong economics or social science background may substitute for the needed expertise' (ibid.: 4). This is an institutionalized prioritization — most DENR personnel are technically trained foresters and project planning and evaluation are discussed largely in terms of the achievement of physical targets such as seedlings planted and timber production.

In defining communities as spatial entities, there is a real danger that the process of transferring resource management into the hands of local communities may simply replicate the imbalance in resource access and control on a micro-scale. In other words, the locus of many conflicts is transferred from between the state and the community to within the community itself, or community representatives may become an extension of the state (Hirsch, 1989). Community management cannot be re-created simply by transferring resource control into the hands of representatives from a spatially-defined community.[10]

9. See also Hafner (1995) for a review of a similar problem in Thailand.
10. See, however, Vandergeest (1991) for a critique of the conceptual merits of the state corrupted community versus the traditional autonomous dichotomy.

The secondary importance attached to understanding community institutions and intra-community relations is an important problem within the DENR's community-based forestry policy. The success of community-based forestry is discussed in terms of aggregated outputs (timber production and seedlings planted) and it is according to such criteria that a community is defined as 'a viable and economic entity' (CFP Assessment Team, 1995a: 21). Policy does not give any substantive consideration to intra-community dynamics, and has generally avoided many of the complex issues involved in defining and dealing with communities. This could lead to a reliance on standardized project 'packages' which may not be appropriate in some situations, particularly where intra-community politics affects access to decision making and the equitable distribution of benefits.

Bureaucratizing Communities

Once selected for inclusion in the community-based forestry programme, communities must create a 'people's organization' which operates as an economic entity with a legal personality (CFP Assessment Team, 1995a). Subsequent dealings between community members and the DENR are mediated through this organization. Since the community-based forestry policy discourse equates spatial proximity with political homogeneity, the people's organization and its leaders, however selected, are seen to represent the interests of the entire community. This assumption, in combination with the traditional orientation towards dealing with corporations in forest management, has predetermined the nature of the relationship between the state and communities.

Interviews with project co-ordinators revealed that people's organizations are treated as being synonymous with communities. This was justified in terms of a circuitous tautology — the people's organization is representative of the community because it is the community representative. Many local project implementers simply could not grasp the idea that a people's organization established through community organizing activities may be distinct from the wider community in terms of agenda, priorities, needs and wants. Yet, as Atumpugre (1993) notes, unless political leaders are committed to protecting weaker groups, community organizations may primarily serve the interests of wealthier members. There are, however, no institutionalized mechanisms in place to establish whether the people's organization is in any way representative of the community.

Within the policy discourse, communities are defined as economic and legal institutions. The difficulties of incorporating community management into rigid systems of bureaucracy has been overcome in the discourse by conceptualizing communities as not just having, but *being* concrete legal and economic personalities. In other words, the DENR has relied on the bureaucratization of communities. This process allows hierarchical and functional lines of control and authority to be identified and control mechanisms to be established. Policy makers have assumed, or claim to assume, that community bureaucracies necessarily represent and can mobilize the collective interests of the social community. Yet a community bureaucracy can never *be* the community; it only represents one possible medium through which a relationship with the community may potentially be established. In defining the community as an economic entity synonymous with an organization representing the community, the potential for empowering the wider community is considerably weakened. While the local community is supposed to have full control, researchers within the Los Banos College of Forestry claimed that the exercise of power is often vested in project leaders, and benefits tend to be concentrated in the hands of a few people.

Once a community has been identified in the manner described above, the next stage is the organization of that community into an institution receptive to a leasing agreement. The technical definition of community, in combination with the equation of community bureaucracy-based with community-based, is reflected in the assumptions underlying community organizing activities.

Organizing Communities

The idea that communities can be socially organized in the absence of a formal bureaucracy is not seriously entertained in policy or among planners. The basic premise of the policy discourse is that communities are unorganized, disorganized, or need to be reorganized (CFP Assessment Team, 1995b; DENR, 1993c). Another assumption within the discourse is that community organizing has to be exogenously stimulated. Accordingly, community organizing always represents the 'main development phase' of the CFP (CFP Assessment Team, 1995a: 13), and has been identified as one of the main strategies towards achieving sustainable development (DENR, 1993c).

Community organizing is defined by the DENR as 'a systematic, planned and liberating change process of transforming a community into an organised, conscious, empowered, self-reliant, just and humane entity and institution' (DENR, 1993c: 2). Three broad stages can be identified: establishing contacts, institution building and final withdrawal. There is, however, little operationalization of these stages into planned measures beyond vague notions such as 'establishing rapport' and 'identifying potential leaders' (DENR, 1993c: 7). There are no detailed guidelines (written or unwritten) for implementing or evaluating community organizing. A supporting DENR discussion guide on assessing community organizations explicitly states that evaluators are assumed to already possess the skills to identify and obtain relevant data, but does not specify what these skills are. It does, however, add some useful reminders such as 'be courtesy personified' and 'don't get drunk' (DENR, 1995b: 1).

The details of community organizing thus remain unspecified, partly reflecting the fact that such activities are contracted out to NGOs on the assumption that they possess the necessary expertise. However, several respondents, including an ex-director of a community-development NGO, stated that the NGOs contracted to carry out community organizing have varying interpretations and approaches. While an element of flexibility may be beneficial, problems have emerged. Many community organizations have been set up not because they are needed, but simply to satisfy regulations, and many members are often unaware of why people's organizations are needed. Claims made by NGO workers and personnel within the DENR involved in contracting NGOs,[11] that external evaluations of NGOs were 'practically non-existent', were supported by DENR field officers who described situations where community organizing had been forced upon communities where well-organized institutional structures already existed, simply because the regulations required the application of DENR-defined community organizing activities.

Given that community organizing has been identified as the main development phase of community-based forestry, it is unclear why this area remains weak in policy, while technical and productivity measures have been emphasized. The criteria that do exist emphasize technical input–output factors such as numbers attending meetings and the duration of meetings, mixed with undefined, yet crucial, terms such as the 'capability of members to manage forest sustainably' (see Mariano, 1995). These goals and techniques have strong similarities to the goals of scientific forestry. Field officers stated that they rely on assessing community organizing activities in terms of numerative outputs with no qualitatively-based assessments. If outputs such as number of seedlings planted have reached or exceeded target levels, they assume that community organizing has been successful. Thus, establishing whether or not a community is organized is based largely on an assessment of the capability of people's organization officials to plan timber stand improvement, agroforestry and reforestation.

This production-technical, input–output approach is replicated in determining the payments to NGOs contracted to carry out community organizing. The budgets for these activities are established on the basis of site hectorage — a fairly irrelevant criterion upon which to base funding if communities are organized not just through spatial relations, but also through social, cultural, economic and political relations. The most likely reason for adopting this criterion relates to the generic homogeneous community assumption within the policy discourse. The above findings lend

11. NGO evalution is supposed to be conducted by an 'NGO desk'. This proved to be a literal description — the department consisted of one desk in an open-plan office.

considerable support to claims made by a number of interviewees that social indicators are largely ignored in the DENR's community-based forestry projects while productivity measures are emphasized.

After community organizing has been carried out, the responsibility for forest management is then transferred to this organized community bureaucracy. As noted previously, however, the continuing influence of scientific management and state control serves to counteract the objectives of 'returning the forests to the people'.

CONCLUSION

Given the existence of certain commonalities in the history of forest management in many countries in the developing world (such as the development of colonial state controls, the application of scientific forestry, commercialization, the recent emergence of community-based forestry policies) the findings from this study may have implications for other countries in which community-based forestry policies are being developed and implemented.

The study has revealed that improving efficiency in forest management and market operations are central to community-based forestry within Philippine policy. It may appear surprising that in a policy potentially involving the redistribution of resources, efficiency rather than equity takes central stage. This may be understood partly in terms of Majone's (1993) suggestion that policy changes are typically precipitated by ideas about efficiency rather than redistributive concerns. The promotion of community-based forestry is not necessarily a reflection of increasing concern with the well-being of forest-dwelling communities or directly environmental concerns. Rather, it may represent part of a wider neo-liberal drive to improve efficiency through decentralization. There are, of course, a range of other ecological, political and populist influences on the adoption and design of community-based forestry policy, and the policy itself is not built upon a single unified discourse. While terms such as 'community-based forestry', 'empowerment' and 'social justice' are commonly used referents, the use of a common language does not indicate that different institutions or individuals are operating from a common frame of reference. There are no clear definitions of what community-based forestry and key concepts such as 'sustainability' and 'empowerment' mean within policy. These may be laudable aims, but without clear and shared definitions, policy objectives may be subject to contradictory interpretations. This increases the potential for political conflicts about the direction of community-based forestry and the extent to which it can be regarded as successful. Without clearly defined tactical and strategic objectives, money and effort could be expended in inappropriate directions. Consequently, underlying conflicts may not surface until later, when the effects of policy become apparent.

One potential source of conflict lies between neo-liberal and scientific forestry understandings of efficiency. Community-based forestry is still regarded by many planners as being environmentally and economically suboptimal since it is argued that transferring forest management into community control may lead to economic inefficiencies. This has manifested itself in a policy design which supports the maintenance of centralized state control over forest management. Foresters remain bound by their traditional perspectives — most are technically trained in tree management and not in community development (cf. Dove, 1992; Peluso, 1992). Thus, perhaps unsurprisingly, there is a continuing emphasis on centralized control and the production and technical aspects of forest management. Productivity indicators are used as a proxy for socio-economic development and as a proxy for social, economic and political empowerment. Yet such measures cannot represent the social targets of community-based forestry. A bureaucratized conception of community has been adopted in an effort to maintain clear lines of control, but by doing so, policy has positioned communities as essentially spatially-definable monoliths.

Historical, institutional, economic and political considerations create barriers to the implementation of policies involving the transfer of power over natural resources from within the state into the hands of local communities. However, strong financial incentives in the form of substantial international funding can encourage the formal adoption of community-based forestry policy. As Peluso (1992) notes, the challenge created by conceptual developments such as community-based forestry is to re-orientate state forest bureaucracies away from traditional models which emphasize trees and production towards the new paradigms which emphasize people and the socio-economic development of rural communities living in forest lands. The historical legacy of state control over natural resources in combination with the continuing production-orientation of forest management are evident within community-based forestry policy. The assumption that the state remains the most privileged and the ultimate environmental manager continues, while the belief that community management practices are 'backward' still shapes policy design. Consequently, the state has chosen to lease forest lands rather than releasing ownership rights, and policy still emphasizes controlling communities rather than devolving forest management authority.

Community-based forestry in the Philippines is being constructed in a manner not entirely distinct from earlier strategies built on corporate-based timber production. With temporary leasing to community bureaucracies through a contractual agreement, there is no guarantee that community-based forestry will not simply reproduce what happened under earlier timber licensing systems — more powerful groups exploiting natural resources on an unsustainable basis at the expense of weaker groups.

As Rush (1991) notes, forest management remains inherently political since it revolves around control over resources. Policy reflects the prominence of particular orthodoxies and received wisdom, as well as reflecting the historical legacy of state control over forest lands introduced during the colonial period, and the political economy of commercial timber production. Timber production rather than community development has historically dominated the political economy of forest use in the Philippines, and elsewhere in the developing world, and this continues to shape policy planning.

Acknowledgements

This chapter is based upon data gathered from interviews in the Philippines during the summer of 1995, along with analyses of supporting policy documents (see footnote 2). I wish to express thanks to all those who facilitated my research in the Philippines, particularly Sharon Quitzon, Romeo T. Acosta and Robert Belen, and to Ray Bryant in the UK and two anonymous referees who provided helpful comments on an earlier draft of this chapter. I must also gratefully acknowledge the financial support of the ESRC who provided funding for the research project from which this chapter is drawn.

REFERENCES

Adams, W. M. (1990) *Green Development: Environment and Sustainability in the Third World*. London: Routledge.

Agaloos, B. C. (1993) 'Social Forestry in the Philippines: An Overview', *Regional Development Dialogue* 14(1): 183–99.

Anderson, B. (1991) *Imagined Communities: Reflections on the Origin and Spread of Nationalism*. London: Verso.

Anti-Slavery Society (1983) *The Philippines: Authoritarian Government, Multinationals and Ancestral Lands*. London: Anti-Slavery Society.

Asian Development Bank (ADB) (1995) *The Bank's Policy on Forestry*. Manila: ADB.

Atumpugre, N. (1993) *Behind the Lines of Stone: The Social Impact of a Soil and Water Conservation Project in the Sahel*. Oxford: Oxfam.

Bacalla, D. T. (1993) 'Policy Changes and Upland Management in the Philippines', in K. Warner and H. Wood (eds) *Policy and Legislation in Community Forestry*, pp. 69–80. Bangkok: RECOFTC.

Banuri, T and F. A. Marglin (1993) 'A Systems-of-Knowledge Analysis of Deforestation, Participation and Management', in T. Banuri and F. A. Marglin (eds) *Who Will Save the Forests? Knowledge, Power and Environmental Destruction*, pp. 1–23. London: Zed.

Barber, C. V. (1989) 'The State, the Environment and Development: The Genesis and Transformation of Social Forestry in New Order Indonesia'. PhD dissertation, University of California.

Barraclough, S. and K. Ghimire (1990) 'The Social Dynamics of Deforestation in Developing Countries: Principal Issues and Research Priorities'. UNRISD Discussion Paper no. 16. Geneva: UNRISD.

Bartlett, A. G. (1992) 'A Review of Community Forestry Advances in Nepal', *Commonwealth Forestry Review* 71: 95–100.

Bell, C. and H. Newby (1971) *Community Studies: An Introduction to the Sociology of the Local Community*. London: George Allen and Unwin.

Bernard, J. (1973) *The Sociology of Community*. Glenview, IL: Scott, Foresman.

Blaikie, P. M. and H. Brookfield (1987) *Land Degradation and Society*. London: Methuen.

Blaikie, P. M. and S. Jeanrenauld (1996) 'Biodiversity and Human Welfare'. UNRISD Discussion Paper no. 72. Geneva: UNRISD.

Broad, R. (1994) 'The Poor and the Environment: Friends of Foes?', *World Development* 22(6): 811–22.

Broad, R. with J. Cavanagh (1993) *Plundering Paradise: The Struggle for the Environment in the Philippines*. Berkeley, CA: University of California Press.

Bryant, R. and S. Bailey (1997) *Third World Political Ecology: An Introduction*. New York: Routledge.

Castro, C. (1984) 'Look Again, Forester, at this Thing they call Social Forestry, Now', *Tropical Forests* 1(1): 34–6.

Cernea, M. M. (1993) 'Strategy Options for Participatory Reforestation: Focus on the Social Actors', *Regional Development Dialogue* 14(1): 3–31.

Chinte, F.O. (1985) 'Legacies of Pioneers in Philippine Forestry', *Philippines Lumberman* 31(1): 50–4.

Cohen, A. P. (1985) *The Symbolic Construction of Community*. London: Routledge.

Colchester, M. (1993) 'Forest People and Sustainability', in M. Colchester and L. Lohmann (eds) *The Struggle for the Land and the Fate of the Forests,* pp. 61–95. London: Ecologist/Zed.

Colchester, M. (1994) 'Sustaining the Forests: The Community-based Approach in South and South-East Asia', *Development and Change* 25(1): 69–100.

Community Forestry Program Assessment and Planning Team (1995a) 'National Community Forestry Program Implementation plan Vol I: Background, issues, options and planning directives'. Quezon City: Jaakko Poyry Consulting AB (Interforest)/DENR.

Community Forestry Program Assessment and Planning Team (1995b) 'National Community Forestry Program implementation plan Vol II: The Plan'. Quezon City: Jaakko Poyry Consulting AB (Interforest)/DENR.

Contreras, A. P. (1991) 'Genealogy and Critique of the Upland Development Discourse', *Philippines Journal of Public Administration* 35: 55–69.

Contreras, A. P. (1992) 'The Discourse and Politics of Resistance in the Philippines Uplands', *Kasarinlan* 7(4): 34–50.

Dalby, S. and F. Mackenzie (1997) 'Reconceptualising Local Community: Environment, Identity and Threat', *Area* 29(2): 99–108.

Department of the Environment and Natural Resources (DENR) (no date [a]) 'Community Forestry Program'. Quezon City: DENR.

Department of the Environment and Natural Resources (DENR) (no date [b]) 'Highlights of the Community-based Forest Management Program (CBFMP)'. Quezon City: DENR.

Department of the Environment and Natural Resources (DENR) (1989) 'Community Forestry Program's Manual of Operations Appended to DAO 123, Series of 1989 Dated Nov 28, 1989'. DENR Policies, Memoranda and Other Issuances on the National Forestation Program 3: 13–41. Quezon City: DENR.

Department of the Environment and Natural Resources (DENR) (1993a) 'Rules and Regulations for the Identification, Delineation and Recognition of Ancestral Lands and Domain Claims'. DAO 02, Series of 1993. Quezon City: DENR.

Department of the Environment and Natural Resources (DENR) (1993b) 'Revised Guidelines for the Community Forestry Program'. DAO 22, Series of 1993. Quezon City: DENR.

Department of the Environment and Natural Resources (DENR) (1993c) 'Institutionalising Community Organising as an Approach and Strategy in the Planning and Implementation of Programs and Projects of the Department of Environment and Natural Resources'. DAO 62, Series of 1993. Quezon City: DENR.

Department of the Environment and Natural Resources (DENR) (1995a) 'DENR's GATT-Related Measures'. Memorandum from the Assistant Secretary, Planning and Policy Studies to the Head Executive Assistant dated 11 August 1995. Quezon City: DENR.

Department of the Environment and Natural Resources (DENR) (1995b) 'Discussion Guide'. Working Paper. Quezon City: DENR.

Department of the Environment and Natural Resources (DENR) (1995c) 'The Community Forestry Program'. Internal Status Report. Quezon City: DENR.

Department of the Environment and Natural Resources (DENR) (1995d) 'Untitled Paper Relating to DENR Policy on Ancestral Domain'. Working Paper. Quezon City: DENR.

Department of the Environment and Natural Resources (DENR) (1995e) 'A Compilation of DENR Foreign Assisted Projects: Profile'. Quezon City: DENR.

Department of the Environment and Natural Resources (DENR)/USAID (1992) 'The Community Forestry Program: Vision and Implementation Strategy', *NRMP News* 11(4): 1–7.

Dolom, B. L. (1995) 'Technical Assistance Completion Report of the Community Forest Management (CFM)'. Quezon City: Development Alternatives Inc.

Dove, M. R. (1983) 'Theories of Swidden Agriculture, and the Political Economy of Ignorance', *Agroforestry Systems* 1: 85–99.

Dove, M. R. (1992) 'Foresters' Beliefs about Farmers: A Priority for Social Science Research in Social Forestry', *Agroforestry Systems* 17: 13–41.

Dove, M. (1994) 'The Existential Status of the Pakistani Farmer: Studying Official Constructions of Social Reality', *Ethnology* 33(4): 331–51.

Eder, J. F. (1987) 'Tribal Filipinos, Ancestral Lands, and Cultural Survival: Or the "Negrito Case"', *Pilipinas* 9: 1–9.

Eder, J. F. (1990) 'Deforestation and Detribalisation in the Philippines: The Palawan Case', *Population and Environment* 12: 99–115.

Ekins, P. (1993) 'Making Development Sustainable', in W. Sachs (ed.) *Global Ecology: A New Arena of Political Conflict*, pp. 91–103. London: Zed.

Elias, N. (1974) 'Towards a Theory of Community', in C. Bell and H. Newby (eds) *The Sociology of Community: A Selection of Readings*, pp. ix–xii. London: Frank Cass.

Fairhead, J. and M. Leach (1995) 'False Forest History, Complicit Social Analysis: Rethinking Some West African Environmental Narratives', *World Development* 23(6): 1023–35.

Fairman, D. (1995) 'Proposal for Mid-Term Evaluation of CFP and RRMP Project Impact at the Household Level: Draft Household Questionnaire'. Memorandum to NCCO Co-ordinator, Obien.

Fischer, F. and J. Forester (1993) 'Introduction', in F. Fischer and J. Forester (eds) *The Argumentative Turn in Policy Analysis and Planning*, pp 1–17. London: UCL.

Fortmann, L. (1995) 'Talking Claims: Discursive Strategies in Contesting Property', *World Development* 23(6): 1053–63.

Fortmann, L. and E. Roe (1993) 'On Really Existing Communities — Organic or Otherwise', *Telos* 95: 139–46.

Fox, J. (1993) 'The Tragedy of Open Access', in H. Brookfield and Y. Byron (eds) *Southeast Asia's Environmental Future: The Search for Sustainability*, pp. 302–15. Tokyo: United Nations University Press.

Gasgonia, D. Z. (1993) 'Development Assistance and Property Rights in the Philippines Uplands', in J. Fox (ed.) *Legal Frameworks for Forest Management in Asia: Case Studies of Community/State Relations*, pp. 53–71. Honolulu, HI: East-West Center.

Ghai, D. (1994) 'Environment, Livelihood and Empowerment', *Development and Change* 25(1): 1–11.

Gibbs, C., E. Payuan and R. del Castillo (1990) 'The Growth of the Philippine Social Forestry Program', in M. Poffenberger (ed.) *Keepers of the Forest: Land Management Alternatives in Southeast Asia*, pp 253–65. West Hartford, CT: Kumarian Press.

Guggenheim, S. and J. Spears. (1991) 'Sociological and Environmental Dimension of Social Forestry Projects', in M. M. Cernea (ed.) *Putting People First: Sociological Variables in Rural Development*, pp. 304–39. New York: Oxford University Press.

Guha, R. (1989) *The Unquiet Woods: Ecological Change and Peasant Resistance in the Himalaya*. Oxford: Oxford University Press.

Guiang, E. S. (1993) 'Community-Based Forest Management: Its Evolution, Emerging Prototypes, and Experiences from the Philippines'. Paper presented during the Fourth Annual Common Property Conference of the International Association for the Study of Common Property, Manila (16–19 June).

Guthmann, J. (1997) 'Representing Crisis: The Theory of Himalayan Environmental Degradation and the Project of Development in Post-Rana Nepal', *Development and Change* 28(1): 45–69.
Hafner, J. A. (1995) 'Beyond Basic Needs: Participation and Village Reforestation in Thailand', *Community Development Journal* 30: 72–82.
Hardin, G. (1968) 'The Tragedy of the Commons', *Science* 162: 1243–8.
Hausler, S. (1993) 'Community Forestry: A Critical Assessment; the Case of Nepal', *Ecologist* 23: 84–90.
Hirsch, P. (1989) 'The State in the Village: Interpreting Rural Development in Thailand', *Development and Change* 20(1): 35–56.
Hisham, M. A., J. Sharma, A. Ngaiza and N. Atumpugre (1991) *Whose Trees: A People's View of Forestry Aid.* London: Panos.
Hoben, A. (1995) 'Paradigms and Politics: The Cultural Construction of Environmental Policy in Ethiopia', *World Development* 23(6): 1007–21.
Hobley, M. (1996) *Participatory Forestry: The Process of Change in India and Nepal.* London: ODI.
Hoffman, J. (1995) 'Implicit Theories in Policy Discourse: An Inquiry into the Interpretations of Reality in German Technology Policy', *Policy Science* 28: 127–48.
Khator, R. (1991) *Environment, Development and Politics in India.* Lanham, MD: University Press of America.
Kirchhofer, J. and E. Mercer (1986) 'Putting Social and Community Forestry in Perspective in the Asia-Pacific Region', in S. Fujisaka, P. Sajise and R. del Castillo (eds) *Man, Agriculture and the Tropical Forest: Change and Development in the Philippines Uplands,* pp. 323–35. Bangkok: Winrock International.
Knudsen, A. J. (1995) *Living with the Commons: Local Institutions for Natural Resource Management.* Bergen: Chr. Michelsen Institute.
Korten, F. F. (1994) 'Questioning the Call for Environmental Loans: A Critical Examination of Forestry Lending in the Philippines', *World Development* 22(7): 971–81.
Kumar, A. and R. N. Kaul (1996) 'Joint Forest Management in India: Points to Ponder', *Commonwealth Forestry Review* 75(3): 212–16.
Leach, M and R. Mearns (1996) 'Challenging Received Wisdom in Africa', in M. Leach and R. Mearns (eds) *The Lie of the Land: Challenging Received Wisdom on the African Environment,* pp. 1–33. Oxford: International African Institute.
Lele, S. M. (1991) 'Sustainable Development: A Critical Review', *World Development* 19(6): 607–21.
Li, T. M. (1996) 'Images of Community: Discourse and Strategy in Property Relations', *Development and Change* 27(3): 501–27.
Loenen, M. (1993) 'The Philippines: Dwindling Frontiers and Agrarian Reform', in M. Colchester and L. Lohmann (eds) *The Struggle for the Land and the Fate of the Forests,* pp. 264–90. London: Ecologist/Zed.
Long, N. and J. D. van der Ploeg (1994) 'Heterogeneity, Actor and Structure: Towards a Reconstitution of the Concept of Structure' in D. Booth (ed.) *Rethinking Social Development: Theory, Research and Practice,* pp. 62–89. Harlow: Longman.
Lynch, O. J. (1985) 'Ancestral Lands: The Case of IP's Forest Dwellers', *Tropical Forests* 2(2): 4–11.
Lynch, O. J. (1993) 'Securing Community-Based Tenurial Rights in the Tropical Forests in Asia: An Overview of Current and Prospective Strategies', in K. Warner and H. Wood (eds) *Policy and Legislation in Community Forestry,* pp. 27–34. Bangkok: RECOFTC.
Lynch O. and K. Talbott (1995) *Balancing Acts: Community-Based Forest Management and National Law in Asia and the Pacific.* Baltimore, MD: World Resources Institute.
Majone, G. (1993) 'When Does Policy Deliberation Really Matter?' EUI Working Paper SPS No 93/12. Florence: European University Institute.
Makil, P. Q. and R. E. Reyes (1982) *Toward a Social-Forestry Oriented Policy: The Philippines Experience.* Quezon City: Ateneo de Manila University.
Mariano, S. M. (1995) 'Terminal Report'. Quezon City: Development Alternatives Inc.

Moore, D. S. (1996) 'Marxism, Culture and Political Ecology', in R. Peet and M. Watts (eds) *Liberation Ecologies: Environment, Development and Social Movements,* pp. 125–64. London: Routledge.

Myerson, G. and Y. Rydin (1996) *The Language of Environment: A New Rhetoric.* London: UCL.

Office of the President (1995) 'Adopting Community-Based Forest Management as the National Strategy to ensure the Sustainable Development of the Country's Forest Resources and Providing Mechanisms for its Implementation'. Executive Order 263. Malacanang Palace: Office of the President.

Oposa, A. A. (1992) 'The Legal Arsenal for Forest Protection'. Natural Resources Management Program No 492–0444 document no. pl 10–92. Quezon City: Development Alternatives.

Ostrom, E. (1990) *Governing the Commons: The Evolution of Institutions for Collective Action.* Cambridge: Cambridge University Press.

Peet, R. and M. Watts (1996) 'Liberation Ecology: Development, Sustainability, and Environment in an Age of Market Triumphalism', in R. Peet and M. Watts (eds) *Liberation Ecologies: Environment, Development and Social Movements,* pp. 1–45. London: Routledge.

Peluso, N. L. (1992) 'Traditions of Forest Control in Java: Implications for Social Forestry and Sustainability', *Natural Resources Journal* 32: 883–918.

Peluso, N. L. (1995) 'Whose Woods are These? Counter-Mapping Forest Territories in Kalimantan, Indonesia', *Antipode* 27(4): 383–406.

Peters, P. E. (1994) *Dividing the Commons: Politics, Policy and Culture in Botswana.* London: University Press of Virginia.

Pigg, S. L. (1992) 'Inventing Social Categories through Place: Social Representations and Development in Nepal', *Comparative Studies in Society and History* 34(3): 491–513.

Poffenberger, M. (1990a) 'The Evolution of Forest Management Systems in Southeast Asia', in M. Poffenberger (ed.) *Keepers of the Forest: Land Management Alternatives in Southeast Asia,* pp. 7–26. West Hartford, CT: Kumarian Press.

Poffenberger, M. (1990b) 'Introduction' in M. Poffenberger (ed.) *Keepers of the Forest: Land Management Alternatives in Southeast Asia,* pp. 163–6. West Hartford, CT: Kumarian Press.

Poffenberger, M. (1996) *Communities and Forest Management: Report of the IUCN Working Group on Community Involvement in Forest Management.* Cambridge: IUCN.

Porter, G. and D. F. Ganapin (1988) *Resources, Population and the Philippines Future.* Washington, DC: World Resources Institute.

Prill-Brett, J. (1993) 'Common Property Regimes Among the Bontok of the Northern Philippines Highlands and State Policies'. Cordillera Studies Center Working Paper 21. Baguio: University of the Philippines College.

Pulhin, J. M. (1985) 'The Profits and Perils of Social Forestry in the Philippines', *Philippines Lumberman* 31(3): 24–37.

Putzel, J. (1992) *A Captive Land: The Politics of Agrarian Reform in the Philippines.* London: Catholic Institute for International Relations.

Rebugio, L. L. (1985) 'Social Forestry: For What and for Whom?', *Tropical Forests* 2(3): 28–35.

Reyes, M. R. (1984) 'Perpetuating the Dipterocarp Forest in Productive Condition', in *The Key to Philippine Forest Conservation: The Defence of the Dipterocarps,* pp. 28–60. Manila: Columbian Publishing Corporation.

Roe, E. (1994) *Narrative Policy Analysis: Theory and Practise.* Durham, NC: Duke University Press.

Rush, J. (1991) *The Last Tree: Reclaiming the Environment in Tropical Asia.* Boulder, CO: Westview Press.

Salazar, R. C. (1993) 'Policy Reform and Community Forestry: Selected Experiences From the Philippines', in K. Warner and H. Wood (eds) *Policy and Legislation in Community Forestry,* pp. 205–10. Bangkok: RECOFTC.

de Saussay, C. (1987) 'Land Tenure Systems and Forest Policy'. FAO legislative study 41. Rome: FAO.

Schram, S. F. (1993) 'Postmodern Policy Analysis: Discourse and Identity in Welfare Policy', *Policy Sciences* 26(3): 249–70.

Silva, E. (1994) 'Thinking Politically about Sustainable Development in the Tropical Rainforests of Latin America', *Development and Change* 25(4): 697–721.

Singh, S. and A. Khare (1993) 'People's Participation in Forest Management', *Commonwealth Forestry Review* 72: 279–83.

Vandergeest, P. (1991) 'Gifts and Rights: Cautionary Notes on Community Self-help in Thailand', *Development and Change* 22(3): 421–43.

Watts, M. (1993) 'Development I: Power, Knowledge, Discursive Practices', *Progress in Human Geography* 17(2): 257–72.

Wignaraja, P. (1993) 'Rethinking Development and Democracy', in P. Wignaraja (ed.) *New Social Movements in the South: Empowering the People,* pp. 4–35. London: Zed Books.

Wilkinson, K. P. (1986) 'In Search of the Community in the Changing Countryside', *Rural Sociology* 51(1): 1–17

World Bank (1992) *Strategy for Forest Development in Asia*. Washington, DC: The World Bank.

11 Unpacking the 'Joint' in Joint Forest Management

Nandini Sundar

INTRODUCTION

India's Joint Forest Management (JFM) programme is an attempt to forge partnerships between forest departments and the rural users of forest resources, in order to regenerate degraded forest land. In formal existence since the passage of a resolution by the Government of India in 1990, it has involved the creation of village-based forest committees, known variously as forest protection committees (FPC) or Van Suraksha Samitis (VSS). In return for their protection activities, the villagers are promised some intermediate benefits and a share of the regenerated timber when it is finally harvested. There are estimated to be some 10,000-15,000 committees protecting over 1.5 million ha of state forest land in India (Agarwal and Saigal, 1996) — although these figures conceal great differences in the degree of participation and protection. In the 1990s, JFM has received an enormous amount of attention from development practitioners, planners, donors, and academics alike.[1] In these circles, the idea that JFM represents a resurgence of civil society is a popular one. Kamala Chowdhury, the chairperson of the Society for Promotion of Wastelands Development (SPWD), a nodal non-governmental organization (NGO) working on natural resource management in India, speaks of NGOs as the 'new bankers' and the 'new social capitalists', because they create social capital, and the 'new abolitionists' because they eliminate poverty, and talks about the creation of a new civil society through joint forest management.

According to forest officials, faced with rapidly accelerating forest degradation and worn out by constant conflict between recalcitrant villagers and beleaguered forest staff in which both sides had been known to lose lives and limbs,[2] there was no alternative but to turn from coercion to consent,

---

1. In donor-funded forestry projects, approximately 28 per cent of funding is allocated to participatory components. Within academia, there is a burgeoning number of PhD theses on the subject, both in India and abroad.
2. Pathan et al. (1990) record that in Gujarat between 1985 and 1989, 383 forest officers were injured and four died in 376 cases of dispute.

at least in certain areas. [3] In the activist/NGO version, on the other hand, JFM represents the culmination of a long struggle to gain control over natural resources by and for local communities (Chattre, 1996: 113). This struggle has been waged at different levels and in different forms: by forest-dependent communities, in the form of large-scale rebellions or Chipko-type movements;[4] by loose coalitions of activist groups such as those which came together to frame alternative forest legislation; and by professional NGOs like SPWD or the Aga Khan Rural Support Programme (AKRSP). In any case, as one development practitioner put it, the JFM resolution of 1990 and the setting up of a National Support Group for JFM by SPWD released 'tremendous energies' among individuals, researchers and NGOs, who all went out to document what was happening. [5] They found widespread instances of 'community' protection of forests, especially in the state of Orissa, but also in parts of Bengal, Bihar, Gujarat and elsewhere.

Some critics see JFM as a form of co-optation by the state and a concession of power to the forest department (Rahul, 1997). The National Alliance of People's Movements (NAPM) and the Bharat Jan Andolan — both umbrella organizations of activist groups — have coined the slogans *Jal, Jungle Jameen Hamara Hain* (The Water, Forest and Land belong to us) and *Hamare Gaon me Hamara Raj* (Our Government in Our village). These challenge the state's ownership of water, forests and land, and the system of centralized governance that currently prevails, and argue for total community control. Other organizations, generally the more corporate NGOs, think that JFM represents the thin end of a wedge which might lead into wider possibilities (Fernandes, 1996: 50). Both views have some truth, and it would perhaps be better to see community management of resources today as a double-edged sword. On the one hand, it makes villagers responsible for afforestation when they may not have been the ones responsible for deforestation in the first place. On the other hand, coming at a particular historical moment following a long-standing critique of state exclusivity, it represents a partial victory and leaves open room for expanding local control to all types of forest land, not just the degraded forest land that the JFM resolution currently allows.

Whatever the differences among NGOs and activists, there is a common tendency to treat community forest protection as a manifestation of civil society at work. In general 'the community', especially the supposedly homogeneous tribal community, is seen as that autonomous remnant which the colonial and post-colonial state and the incursion of the market could not smash, and community management of resources therefore a step towards

---

3. In one of Gramsci's renderings, civil society is identified with consent and the exercise of ideological hegemony, while the state is identified with coercion (Anderson, 1977).
4. A famous movement in Uttarakhand, North India, where women saved their local forests from being felled by contractors by 'hugging' the trees.
5. Interview with Arvind Khare, former Executive Director of SPWD, 1996.

reversing the alienation introduced by state appropriation of resources. In this celebration of community and civil society, what is perhaps equally important for NGOs is that the JFM policy of 1990 represents a recognition of their role as legitimate brokers with villagers. Given the prevailing context of political and economic pressure towards decentralization and disinvestment, they find common cause with donors keen to reduce the powers of a centralized state.

The assumption upon which 'joint' forest management is predicated is of a historical dichotomy between civil society and the state, as if villagers and forest departments previously existed in water-tight compartments, locked into mutual antagonism. This article seeks to challenge that view. Although the predominant thrust of the colonial forest department was towards appropriation and centralization of forest resources to serve the needs of an expanding capitalist economy, there were also pockets in which village level institutions were allowed to manage resources. By examining the commonalities between older examples of joint or co-management of resources between villagers and the bureaucracy, and current practices of joint forest management, we might get a clearer perspective of the situation today and the potential limits to its long-term sustainability. Creating a sharply defined and delimited space for community management of resources has often been a politically and economically more viable way of managing those resources for the state than taking on the tasks of administration itself. The conditions which make this option preferable for the state over centralization include a loss of legitimacy, financial stringencies, and the ability to appear flexible and participatory while at the same time retaining the deciding vote. Consequently, community management initiatives have tended to be like desert mirages — false images of an oasis of civil society, quickly covered by the shifting sands of the state.

THE STATE AND CIVIL SOCIETY: THE MIRAGE OF MUTUAL INDEPENDENCE

Constructed Communities

In describing the range of local forest-protecting institutions across India, it is common to differentiate between community forest management and joint forest management, largely on the basis of who initiated them. Madhu Sarin lists three categories of institutions (Sarin, 1996: 168–73). First are those which emerged out of local initiatives, such as the many committees in Orissa and Bihar which are managed by village youth clubs or village elders, in many cases protecting village forest land, but also reserve forests. The second category are those promoted by the forest department, especially in states with large donor-funded forestry projects like Madhya Pradesh and Andhra Pradesh. Finally there are the committees initiated by NGOs such as the AKRSP-funded Gram Vikas Mandals in Gujarat, which have been undertaking forest protection in addition to their other functions. However, the distinction between externally-initiated and autonomous groups is often one of degree rather than of rigid category. For instance, many of the Van Panchayats in Uttar Pradesh (UP) which were formally initiated by the forest department became defunct and were later revived by local NGOs (Agarwal, 1997: 16–17). Moreover, it is also important to consider other factors, such as the parameters within which each of these groups operates, and the consequences of particular interventions.

When thinking of 'community', the most common associations that spring to mind are small, homogeneous, harmonious, territorially-bound, ascriptive units in which people enjoy face-to-face interactions, or what Partha Chatterjee sums up as 'an inherited network of social obligations' (Chatterjee, 1998: 278). Benedict Anderson's well-known rendering of nations as imagined communities picks up on precisely this sense of 'fraternity' as central to communities (Anderson, 1991:7), although he points out that 'all communities ... are imagined. Communities are to be distinguished, not by their falsity/genuineness, but by the style in which they are imagined' (ibid.: 6). However, the fraternity factor — particularly in the case of nations but also in the case of village communities — often seems more prominent in its absence than its presence. What is important, according to Sabean, is not 'shared values or common understanding so much as the fact that members of a community are engaged in the same argument ... the same discourse, in which alternative strategies, misunderstandings, conflicting goals and values are threshed out' (Sabean, 1988: 28; see also MacIntyre, 1981).

Not only are communities often hierarchical and conflict-ridden rather than homogeneous and harmonious, but individuals are involved in a multiplicity of different relationships. In any one Indian village, for example, villagers have links with members of their caste within and outside the village, with political parties, religious groupings, marital circles, and so on. In other words, there are several 'little traditions' of local community life (Chatterjee, 1998: 278). What may be seen as dividing lines — such as race, class, gender, religion, ethnicity or age — may also be forms of community in their own right. Each of these forms of community and the categories themselves have to be constructed, and often make sense only in opposition to others. For instance, 'Hindus' are not a community — it has required a great deal of mobilization, propaganda, and political invective to create a Hindu identity out of multiple others, assisted by decades of imperial historiography (see Breckenridge and Van der Veer, 1993; Ludden, 1996). The role of administrative instruments like the census has often been crucial in fixing identity (Cohn, 1990).

Although India's villages have been famously described as institutions that have outlived the rise and fall of successive states, much recent work has shown how, on the contrary, these 'village communities' are intrinsically affected by state formation or disintegration, and administrative practice (see Breman, 1997). According to Minoti Chakravarty-Kaul (1996), both village 'communities' and their 'commons' in Kangra (North India) were created in the early-mid nineteenth century. The joint landlord tenures of certain parts of the Punjab, like Delhi and Karnal, were replicated in places where they had not previously existed, such as Kangra and Hissar, in order to ensure joint revenue responsibility. Village commons were created by enclosing long grazing fallows and vesting them in particular village communities, thus destroying other 'communal' relationships between farmers in the highlands and lowlands.

There were a variety of ways in which the governments informally abrogated their powers in favour of village management. In many parts of eastern and central India, especially in the former princely states, land revenue administration took place through village lessees called *malguzars* or *thekedars*, or through *gaontias*, who were responsible for the management of the village forests. In the words of the Sambhalpur gazetteer, the *gaontia* 'acted as a trustee on behalf of the government responsible to ensure that the village forests were used for community needs' (Government of Orissa, 1971: 347). In other words, the system of revenue administration (which included the administration of village forest land) was 'joint', between village communities and the state, the person of the *gaontia* providing the link by acting as agent of both. Where the 'community' itself was settled by the *gaontia* — who was usually granted a rent-free period in order to extend cultivation into hitherto unbroken tracts — the distinction between 'community' and 'state' becomes even more slippery.

What appear to be rules set in place by autonomous community protection groups in fact reflect a history of state intervention and differentiation between different categories of people. For instance, in village Lapanga of Sambhalpur, where protection is reputed to have been going on for about a hundred years, initial settlers and later immigrants pay different rates for use of produce from the protected patch (see Sundar et al., 1996). In many *zamindari* areas under rules indirectly introduced by the colonial administration, landless labourers were made to pay higher dues for grazing and non-timber forest produce than agricultural tenants, on the grounds that they did not contribute through land revenue. Those engaged in non-agricultural occupations also had to pay according to fuel consumed (Kamath, 1941; Ramadhyani, 1942).

Thus, communities are not a natural excrescence of the soil, but come into being through a variety of historical processes. Old laws, however hotly contested at the time, eventually become custom and gain new legitimacy. Similarly, communities which have been settled by the state or institutions which have their rules framed by government intervention, eventually solidify into seemingly natural associations which are then counterpoised to the state. Instead of expecting to find ready-made communities which can be mobilized for different causes, including management of forests, we need to examine the ways in which communities are constructed for specific purposes. As Appadurai (1995: 207) notes:

> Much that has been considered 'local' knowledge is actually knowledge of how to produce and reproduce locality under conditions of anxiety and entropy ... Drawn into the very localisation they seek to document most ethnographic descriptions have taken locality as ground not figure, recognising neither its fragility nor its ethos as a property of social life.

We might thus read the communities that the JFM policy assumes as partial and ongoing products of the procedures specified by the resolutions themselves, rather than as inherent attributes of the people and the place. The JFM resolutions conceive of the community as settled (excluding migrants, pastoralists and shifting cultivators and any links the settled villagers may have to these categories); as one that consists of stable married households (with its criterion of membership and shares by household rather than on the basis of all adults); as one that privileges the male links of property within the village as against the multiple other links that individual households share with their affines; and as one that has an identifiable relationship to an identifiable resource (with its allocation of particular patches to particular FPCs). In practice, there may be multiple violations of these requirements. As the policy takes root, however, and people acquire stakes in the new dispensation, these artificially constructed attributes of the community begin to take on a life of their own.

Early Forms of Co-Management

Within the nascent forest department in the late nineteenth century, there were differences of opinion about the degree to which peasant access to forests should be recognized. Highlighting the debate between 'annexationists', 'pragmatists' and 'populists' in the making of the 1878 Forest Act, Gadgil and Guha (1992: 123–34) show that the outcome of the conflict between state right over forests and the proprietary right engendered by customary peasant use was not a foregone conclusion. The annexationists argued for complete control by the state over all forests, while the populists argued for those rights to be entirely vested in village communities. Pragmatists, represented by the first Inspector General of Forests, Brandis, argued for the creation of different types of forests — state forests, village forests and private forests. Peasant needs were to be met from village forests which would be under the management of village communities. The Deputy Superintendent of Bangalore, B. Krishnaiengar, suggested 'that a council of men (one elected from each village) form a forest court to settle disputes arising out of the management of village forests' (Guha, 1996: 92). This was evidently a prescient suggestion, given the problems caused by inter-village conflicts arising out of the distribution of land under JFM and the inter-village federations that emerged in response. Eventually, however, whether due to internal politics between the forest and revenue departments and different provincial governments (Pathak, 1999), or the presumptions of conquest, the system of reserved and protected forests which was set up was kept firmly under the control of the forest service. The rights of peasant and tribal users were classified merely as 'privileges and concessions'. As Guha (1996: 94) notes: 'while the Act did have a provision for the constitution of village forests, the option was not exercised by the government except in a few isolated instances'.

The exceptions are perhaps more numerous and significant than Guha's formulation suggests, since forest tenures varied considerably across provinces. In practice, there has often been a complex mixture, ranging from complete state control to village control. Within state-owned forest land, villagers have had varying degrees of right, such as the *mafikat* (free timber) system in Gujarat and *nistar* in the former Central Provinces (now the state of Madhya Pradesh). As a revenue term, *nistar* connotes the right to take forest produce for non-commercial household use, either free or at concessional rates, from both protected and reserved forests. Conversely, forest departments exercise control over the management of village lands: certain trees like teak cannot be cut, even on private revenue land, without forest department permission. In Madhya Pradesh, Orissa and Bengal, *zamindari* forests covered substantial areas, and were subject to different rules (Sivaramakrishnan, 1996).

Among the better known examples of early co-management of forests are the Van Panchayats in Uttar Pradesh (UP) and the Kangra Forest Co-operatives in Himachal Pradesh, set up in the 1930s and 1940s respectively. Although both are often cited as examples of village-based institutions with their own rights and obligations (see, for example, Agrawal, 1994: 10), one of the more striking features about them is the limited degree of autonomy under which they functioned. Only certain classes of land were specified as under the control of community management (in UP, Class I and Civil Forests), while other valuable timber areas were left within the purview of the state. In both cases, the administration exercised control over different aspects of management. In UP, for instance, while the revenue department was responsible for the overall framework of the Panchayat Act, individual van panchayats were supposedly free to frame their own rules regarding specific rights of access, grazing, punishment of offenders, finances, election of office bearers etc. However, the van panchayats have always functioned in the shadow of the state, and have rarely acted without the approval or even the instigation of government officials. New rules passed in 1976 extended this dependence on the state, for instance requiring the district administration's permission before the van panchayat could engage a watchman, sell timber seized from an offender, or get their share of funds from the sale of pine resin released. Yet state support is lacking precisely because these forest areas have been ostensibly given over to the control of panchayats (Ballabh and Singh, 1988; Saxena, 1995, 1996: 55).

In Kangra there were, by 1953, seventy co-operative societies covering 58,000 acres of forest land (Agarwal and Singh, 1996: 40). They faced dual control, by the forest and by the co-operative departments, and membership was limited to heads of households and permanent residents owning land in the village, thus depriving several categories of users (such as pastoralists, women, and poorer residents) from decision making (Agarwal and Singh, 1996: 38–48; Sood, 1996: 6–7). In 1973, the forest department stopped grants-in-aid to the co-operatives, drastically reducing their income. What both these cases illustrate is the manner in which community management of forests served as the occasion for the state to abdicate serious responsibility without abdicating overall control (see Mosse, 1999 for a similar argument with respect to community irrigation management in South India).

There are other examples in which existing community management of forests was recognized by the administration and formalized, only for the process to be later reversed in a changed political climate. Gadgil and Guha cite the case of three village forest councils in Uttara Kannadda district of Karnataka — Chitragi, Kallabe and Halakar — which had their own systems of forest management. This involved keeping a watchman and regulating the harvest. These village forest councils were given formal recognition in 1926, and continued till the 1960s when they were disbanded by the Karnataka forest department. Attempts by two of the villages to challenge this dissolution were suppressed and a timber contractor was given permission to extract wood from the Kallabe forest (Gadgil and Guha, 1995: 41).

A parallel process can be seen in the working of the Ulnar *nistari* scheme in Bastar, a former princely state in Madhya Pradesh. In 1937, the Chief Forest officer came across a system of forest protection in Ulnar. The villagers protected their *nistari* or village forests by appointing jungle watchers, paying them 30–60 kg of paddy per year and exempting them from village *begar* (compulsory labour collectively performed by villagers for landlords or state officials). This was then formalized into a Working Plan by S. R. Daver, which ran from about 1937/8 to 1952. The Ulnar *nistari* jungle, primarily sal, comprised about ten square miles and was surrounded by twenty-three villages. As part of the working scheme, the forest was divided into seven or eight felling series. Each felling series was assigned to a set of villages, which were then responsible for its management and the payment of watchers. The felling series were further divided into forty 'coupes', one of which was closed every year for felling, the produce being distributed among the relevant villages. Certain trees, such as mahua (*Bassia latifolia*), tamarind (*Tamarindus Indicus*), hurra (*Terminalia chebula*), mango (*Mangifera Indica*) and trees forming the sacred grove around the local deity's shrine, were not to be cut. For each felling series, the villagers appointed a *panch* or council, which met weekly at the bazaar in Bajawand. The council had to be approved and confirmed by the administration, who also had powers to override the council's judgement. The council was vested with powers to impose fines of upto Rs 25 for offences connected with illicit felling, or excessive removal of timber, fuel, grass, and non-timber forest products (NTFP). The money went to 'the furtherance of the Ulnar forest conservancy'. Individual councillors were responsible for appointing the watcher for their felling series and collecting contributions in grain from each household, which went towards paying the watchman, buying uniforms and axes, the construction and repair of the grain depot, and so on. The rights of the villages were recognized and upheld in any dispute over the use of *nistar* from these forests (Daver, 1938). In short, while operating through the 'traditional' institutions of village panchayats, a new system of a forest panch was created, subject to the overall approval of the administration.

The whole system was eventually disbanded once upper caste immigrants settled in the area and began cutting wood illegally and smuggling it out of the district. The prevailing emphasis on nationalization of forests in the 1952 Forest Policy further undermined support for the Ulnar system — neither villagers nor forest staff felt committed to improving the condition of a particular forest, the former because it was no longer theirs, and the latter because they were prone to transfer and had no long-term interest in the area (pers. comm., Divisional Forest Officer, Bhopalpatnam, 1988–90).

These instances of community forest management or systems of co-management in the colonial period have striking implications for our understanding of joint forest management in the current context and the debate over its significance. In the first place, several of them, such as the UP Van Panchayats and the Ulnar Nistari scheme, followed a struggle by communities to regain lost rights in forest access. In 1910, a violent rebellion had swept Bastar state, one of the chief causes of which was the enclosure of approximately two thirds of the forest land of the state, excluding *zamindaris* (Sundar, 1997). Similarly, the joint forest management policy comes after sustained critique by environmentalists.

Secondly, in recognizing community management as a particular form of forest management, the state drastically refashioned those communities. By specifying the contours of the community (who had what rights), and demarcating the extent to which community institutions would manage resources, it reduced the ability of communities to function as a real alternative to bureaucratic management. There are implications here for the situation in the state of Orissa, where in the name of JFM, the forest department is superimposing new committees (VSS) on existing informal village committees.

Thirdly, by retaining its leading role, the state limited the ability of the community to manage its own affairs on a sustained basis. Community forest management committees become susceptible to the overall imperatives of the forest department. When market and other pressures impinge, forest departments have often disbanded or ignored village committees in favour of revenue considerations, or under pressure from more powerful commercial interests. As will become clear, there are obvious similarities between this and the uncertain legal status of the FPCs discussed in the second part of this article.

The Parallels Between 'Custom' and Participation

There is a striking parallel between the practice of community participation in the management of forests, and the concept of 'customary law'. Within the terms of current development discourse, there are several levels at which people's participation may be summoned:

> as a presence, as objects of a theoretical process of economic and political transformation; as expected 'beneficiaries' of programmes with pre-set parameters; as contributors of casual labour to help a project achieve its ends; as politically co-opted legitimisers of a policy; or as people trying to determine their own choices and direction independent of the state. (Nelson and Wright, 1995: 6)

As we shall see in the following section, participation in JFM schemes does not extend to the last category. The overall policy is set, and popular participation is simply expected to make it work better. Similarly, customary law operates within a larger legal framework. According to Fallers (1969): 'Customary law is not so much a kind of law as a kind of legal situation which develops in imperial or quasi-imperial contexts in which dominant legal systems recognise and support the local law of politically subordinate communities'. Others have extended this observation to show how local law itself was created anew under such a situation (Starr and Collier, 1989; Vincent, 1990: 383).

In at least one place, Bastar, and at two different points in history, the two experiments of popular participation in judicial and in natural resource management seemed to occur simultaneously. The first occurrence was in the 1930s and 1940s, when sensitive anthropologist-administrators were developing a notion of tribal exceptionalism. The second is found in the 1990s, when ideas of participation that have been touted since the 1970s finally began to find expression in laws and government programmes.

Indirect rule in princely Bastar was initially dictated by considerations of distance, the dense jungle and the reputation of its inhabitants for savagery. As Fields (1985: 32–41) puts it in a parallel context, this was making a virtue of the necessity of 'rule on the cheap'. Gradually, as forest resources gained in value, this indirect control gave way to more direct rule. However, after the 1910 rebellion, and with the need to keep the nationalist Congress party at bay, the value of isolating Bastar as a tribal oasis emerged once again. Colonial administrators believed in a sharp distinction between Hindus and tribals and cast themselves as paternalist protectors of tribals in the face of a predominantly Hindu Congress. It was during this period that there were a number of attempts at reserving spheres for custom — both in the assignment of certain judicial powers to newly regulated village panchayats and in schemes such as the Ulnar *nistar* forest scheme, discussed above. In the judicial sphere, village and *pargana* panchayats were given control over seventeen sections of the Indian penal code, and the power to impose fines up to Rs 25. At the same time, certain anomalies had to be ironed out, such as the absence of *pargana* heads in some places, the mismatch of judicial and revenue boundaries in others. In other words, while jurisdiction over some cases was given over to 'customary institutions', these institutions were themselves transformed as far as possible to resemble the structure of the formal legal system.

In any case, since the purview of customary law is so limited, the village panchayats have been unable to function as significant alternatives to the formal legal system. Even cases deliberated in a village panchayat are influenced by considerations of what the punishment would be in the formal system. Yet, the seduction of the concept of customary law continues to be such that it is enshrined in a recent act which is applicable to all areas governed under the 5th schedule of the Indian constitution, that is, those with predominantly tribal populations, apart from the North East.[6] Under this Act, tribal communities will have the right to manage their local natural resources and settle their disputes.

However, the concept of community that is invoked is not problematized in either customary law (where women are routinely excluded from village councils) or in official pronouncements on natural resource management. As Michael Anderson (1990: 165) put it in the context of customary law, but which could as well apply to the notion of community developed in joint forest management schemes:

> the judiciary seems to have developed a view of custom that occluded local servitudes of gender, age and status. Custom was frequently described in terms of organic social harmony, which permitted its validation as it was incorporated into the protocols of state administration ... colonial legal administration carried out a brand of juridical homogenisation whereby certain indigenous terms were adopted and deployed to signify a diverse group of social practices.

The crucial link being made here is that 'participation' (of which jointness is one form) in current development discourse has much the same function as 'custom' did in the discourse of the colonial state. Both often follow a long struggle by subaltern groups to regain lost ground in terms of rights or to gain real control or participation; both are motivated, in part, through the financial and administrative exigencies of the state;[7] both carry positive connotations that are intended to legitimate whatever function they are put to, whether it is in the implementation of afforestation programmes or the implementation of criminal justice; both are carefully circumscribed and reworked to resemble or fit the overall framework of formal state structures. In other words, both represent a fight-back by the state to regain its control over processes, when individuals or groups have begun to take management into their own hands. Finally, both work with a flattened sense of

6. The Provisions of the Panchayats (extension to the Scheduled Areas) Act, No. 40 of 1996.
7. Several foresters interviewed in 1995–7 argued that JFM is dictated by the state's dependence on donors who insist on community and NGO involvement. Others deny this, but clearly JFM is premised on the notion that community involvement is a cheaper and more effective way of regeneration than old style forest department plantations.

community that ignores differences in interests between groups. This last point, especially, will be illustrated in the following sections.

THE 'JOINT' IN JOINT FOREST MANAGEMENT

Several features distinguish joint forest management schemes from earlier social forestry schemes which aimed to provide for villagers' fuel and fodder needs from private land and village commons. These include government resolutions that facilitate community sharing of final harvests and intermediate benefits from state-owned forest lands in return for protection, and that specify membership rules, duties and benefits; a stress on microplanning to be conducted through participatory rural appraisal exercises (PRA); a preference for planting villagers' choice of species, wherever natural regeneration is not enough; the provision, in some places, of alternative development inputs to wean people away from dependence on the forest or at least to build up trust in the forest department; and the involvement of appropriate NGOs to motivate the people. Although not a direct part of JFM, donor pressure has made restructuring of the forest department a parallel condition for aid. Thus in the minds of reluctant foresters, popular participation and departmental reorganization and training are linked within a single ensemble which is seen as being imposed from above.

The following sections focus on two aspects of JFM: first, how participation has been defined, shaped and limited, and second, how forest usage patterns and needs are constructed within JFM. They are based on material from a comparative study of JFM in four Indian states — Gujarat (Western India), Madhya Pradesh (Central India), Andhra Pradesh (South India) and Orissa (Eastern India) — conducted by Edinburgh University in collaboration with the Indian Council for Forestry Research and Education from 1994 to 1997.[8]

The Scope for Participation in Joint Forest Management

In most state resolutions, people's protection is limited to forest land with less than 40 per cent crown density. The exceptions are Bengal and Madhya Pradesh where two types of committees, FPCs and VFCs, have been set up to protect good forest and degraded forest respectively. Given the degree of

---

8. The research was carried out by Abha Mishra and Neeraj Peter in Orissa, Prafulla Gorada in Andhra Pradesh, Nabarun Sengupta and Shilpa Vasavada in Madhya Pradesh and Ajith Chandran and Monika Singh in Gujarat. I refer to the researchers wherever I draw on their field notes. Other examples are drawn from my own field notes, and are not individually referenced.

excitement and the number of workshops that JFM has generated it is easy to forget that the area covered by JFM constitutes such a limited part of the national forestscape. On the other hand, faced with 30 million ha of degraded land, even this limited amount of afforestation is likely to benefit villager users (Mr Rizvi, SPWD, pers. comm., 1995). However, this in turn raises three further issues. First, villagers' participation is mobilized for a specific purpose, namely the afforestation of degraded land, not for any agenda they might have set themselves, such as getting more NTFPs and timber for local needs. Secondly, not all villages have equal measures of degraded forest land which they can protect: exclusion from the protected patch, as well as from the development benefits and wages for plantation work that are associated with JFM, aggravates conflict between villages. Third, while villagers may be active in protecting their degraded forest land, their needs do not disappear and pressure to fulfil these needs is often merely shifted to alternative, good forest land.

The domain of participation, moreover, is strictly limited: while certain questions may be jointly decided, other issues of far-reaching importance are not. In at least three of the four sites studied, the major threats to forest land are commercial interests or large-scale development projects, such as tourism and bauxite mining in Borra in Paderu division of Visakhapatnam (Andhra Pradesh), proposed power plants in Sambhalpur (Orissa), and a planned paper mill near Kevdi village in Mandvi (Gujarat). While geological and geographical factors restrict the number of sites appropriate for power projects, there has been no move to allow villagers participation in the management of, or shares in, industries in lieu of displacement. Whilst 'jointness' in forest management is a step to be applauded, without necessarily also expecting jointness in coal sector management or tourism, people's lives are not divided into separate compartments — what happens in one sphere affects their ability to participate effectively in others. According to local estimates, there have been 200,000 migrants from Koraput district in Orissa to Visakapatnam district in Andhra Pradesh across the border, largely due to displacement by the Machkund river valley project and other development projects. The migrants have settled on reserve forest land, resulting in intensified pressure on resources, and conflict and inequities between earlier settlers and later oustees. While shifting cultivation is seen by the forest department as the prime cause of deforestation in the area, and while villagers tend to blame each other,[9] Michael Dove (1993: 19) points out that the main cause of deforestation may not be 'shifting cultivators' but 'shifted cultivators'.

I have selected four aspects from the case material to illustrate the sphere assigned to participation: the legal status of committees, the benefits

9. The Andhra villagers complain that it is the migrants from Orissa who practise shifting cultivation, while the Oriyas say that they are unfairly blamed, and that it is the Andhra villagers, who have permanent land titles, who nevertheless encroach on forest land.

villagers can expect, membership rules, and the selection of JFM villages.[10] In most states, the committees are simply registered with the forest department, except in Gujarat where the village protection committees have an independent existence as co-operatives registered under the co-operative society act. Forest departments reserve the right to dissolve committees if they perform unsatisfactorily, or at least to deny them the shares expected (Poffenberger and Singh, 1996: 71). Forest departments have also been known to refuse to register existing committees, especially if the forests under their protection have changed status from degraded into good forests, as happened in the Panchmahals in Gujarat (Agarwal and Saigal, 1996: 71). In Orissa, the forest department has attempted to make use of existing committees while denying the legitimacy of rules they may have framed earlier which do not fit into state resolutions. For instance, in village Kenaloi in Sambhalpur district, an existing committee was formally given the status of a VSS in 1994, but a new president and executive committee members were appointed in place of the existing ones. Villagers who were cutting fuelwood from their protected patch in accordance with the rules of the original committee were caught and fined by a new assistant conservator, since — according to JFM rules — villagers can only collect dry and fallen wood (Mishra and Peter, field notes). The role of the committee in silvicultural management in JFM is limited to assisting forest officials in carrying out operations in accordance with the management plan and in the distribution of forest produce among the villagers. Not surprisingly, the Kenaloi villagers were loath to give up the control they had formerly enjoyed in exchange for the limited 'participation' offered to them under JFM: they decided to scrap the formal JFM committee.

The share of the forest protection committee in the final harvest is contingent on its performance. Given that there is no benchmark for performance, this could make the committee dependent on the goodwill of the forest department. There is also the danger of the rules being changed midway, as in Madhya Pradesh. A village which began protection in 1991 in expectation of a 20 per cent share of timber after five years, might find after 1995 that all it could expect under revised JFM rules is continued access to *nistar*. Even this right becomes attractive only because it is now being taken away from other villages: those beyond a radius of 5 km from good forests will be required to pay market rates for small poles, etc.[11] Of course, changes in rules may not be a bad thing in themselves, especially if they are the result of negotiation by the state and affected groups or those representing them. The fact remains, however, that participation then becomes a rule-bound exercise, being used in different measure and for different purposes according to the current rule.

---

10. There are, of course, other features through which the same point could have been made, especially in terms of silvicultural issues (see Ravindranath et al., 1996).
11. Order No. F-7-22/93/10/3, dated 26 December 1994.

The question of who participates is also specified by state government resolutions. Madhu Sarin points out that by restricting membership in the FPC to one person per household, the rules systematically exclude women (Sarin et al., 1998). Even where the rules are revised, as in West Bengal, Orissa and Madhya Pradesh, to include one man and one woman per household, important sections, such as widows or younger daughters-in-law, are still left unrepresented. In several villages, especially in Madhya Pradesh and Andhra Pradesh, women were unaware that they were members of a general body, let alone of the executive committee (Gorada, field notes; Vasavada and Sengupta, field notes). It is not just that women have been excluded from decision-making bodies — this is an old feature of village politics — but that JFM rules, in the name of protection, give further power to men (especially certain classes of men) to exclude women (especially certain classes of women) from the forests. Sarin et al. (1998) give several examples of women suffering severe deprivation due to closure of the forests, being forced to resort to inferior fuels and coming into confrontation with members of the patrol committee. This picture is corroborated by evidence from all the four states studied.

Well-off villagers who have alternative sources of fuelwood on their farms and can avoid using the forest for a while often stand to gain through JFM. They acquire shares in timber that can be grown because of the abstinence exercised by poorer villagers, who suffer heavy opportunity costs through giving up head-loading or even through the labour time spent in protection. Membership of the executive committee of the FPC also gives people a certain leverage in village politics. In Kilagada and Seekari villages in Andhra Pradesh, for instance, forestry works sanctioned under JFM such as cleaning and planting are widely regarded as patronage opportunities for the chairperson.

There are limits on the number of forest protection committees that can be set up, which are circumscribed by targets, funding (especially in the states where there are large donor projects) and the person-power available to the forest department. Under the Madhya Pradesh World Bank project, nine divisions were to be taken up in the first year, with three ranges in each division and three villages in each range. One criterion for village selection is the presence of at least 300 ha of degraded land, without any prior plantation. In Dewas, for example, some villages were selected by range staff on the basis of their perception of which would be 'good, responsive' villages or which had some major problem to be addressed like timber smuggling. In one range, there were only three villages which satisfied the criterion of 300 ha of degraded land. In Andhra Pradesh, in 1995–6 the department had a target of eighty VSS. Choices are generally made on the basis of visibility and accessibility, but the contacts of individual villagers with the administration also make a difference. Given that JFM is at some level a privatization of land into the hands of individual villages, the consequences of not being selected for JFM can be quite serious. There are no overall mechanisms for choosing between villages in a division in which villagers can also exercise their opinions, for instance through a district council meeting. Villages which are not selected naturally have their own opinions as to why one village is chosen over another, ranging from semi-serious allegations of corruption to acrimonious aspersions on private morality. Gorada cites the example of a dispute between two villages, Kothauru and Kunchappally, after the latter was excluded from a protected patch. In an attempt to solve the dispute, a woman from Kothauru said that they had written permission from the Divisional Forest Officer (DFO) to protect the forest and punish the intruders. Before she could complete her sentence, a woman from Kunchappally responded 'Yes, yes, DFO has given a very big letter authorizing you. Maybe you are sleeping with him, he is so pleased to give the hill' (Gorada, field notes).

There are thus several important aspects of joint forest management which are completely non-participatory, or which are fragile, such as the status of the committees and the benefits expected. What the examples regarding membership and the selection process also illustrate is the manner in which the non-participatory nature of the programme affects the ability of different sections to participate within the community, in terms of access to resources or their ability to negotiate in future government programmes. In other words, the contours of the 'community' become refashioned, along with the balance of power between different communities. This in turn affects the way in which schemes are translated on the ground. Finally, the lack of participation extends to the forest department, whose hierarchical structure is perceived by insiders as both a weakness and a strength. While it enables the forest department to function as a cohesive disciplined body and to implement orders, it allows little recognition for local initiative (see Jeffery et al., forthcoming).

The Construction of Usage Patterns and Needs

Underlying the concept of JFM is a certain notion of villagers' needs and patterns of use. A critical part of the Orissa JFM resolution is the preparation of microplans, defined in the following terms: 'It would be based on a diagnostic study of the specific problem of forest regeneration of the locality and the specific cost effective solution for the same that may emerge from within the community. The views of all sections of the community, particularly the women folk, should be elicited in the PRA exercise for preparing the microplan'. Similarly, in Madhya Pradesh spearhead teams have been formed in those divisions taken up under the World Bank project. These teams have been trained in 'participatory methods' and are expected to train other staff in turn.

In theory, therefore, villagers are asked what their problems are and the best way of redressing them. Despite the growing emphasis on the uses of 'indigenous knowledge' even in World Bank circles (see Warren, 1991), this is perhaps the first time that local knowledge and action are being formally tapped within forest policy. In practice, of course, much of the work of forest departments has depended on the intimate knowledge and expertise of villagers regarding the forests in their vicinity, especially for the collection of NTFPs which have constituted an underrated but significant portion of forest department revenues (see also Grove, 1998).

In many states, while microplans are drawn up, working plans which lay out in detail the working of the forest department over a twenty-year period continue to exercise pre-eminence. In Gujarat, AKRSP had to abandon its plans for afforestation in a particular village because it was not due for plantation under the working plan. This resulted in the loss of thousands of seedlings as well as employment opportunities for villagers, many of whom were forced to migrate in search for work (Chandran and Singh, field notes). Even when microplans are given priority over working plans, they follow the same silvicultural prescriptions and no attempt is made to integrate villagers' needs or knowledge into the technical management of the forest floor.

Mapping resources and social composition of a community through the use of participatory rural appraisal techniques is one of the major components on which the brave new world of joint forest management is based. In practice, decisions are made through a process of negotiation between competing agendas. In some cases, the choice of the particular plot is left to the villagers (often the more active or visible ones), based on their usage patterns and their knowledge of their immediate environment, while in other cases, it is the forest department which does the apportioning, based on their mental maps of the forest as degraded or non-degraded. Lands which appear to foresters to be degraded may in fact yield a range of products to villagers.

Where villagers themselves initiate protection, they may choose to apportion land in informal ways that are generally ignored by forest departments in setting up their village committees. The villagers of Kenaloi in Sambhalpur district of Orissa decided to start protecting their forests sometime in the 1980s. Initially there was a village-wide committee, but after a period when protection flagged, villagers decided to overcome the problem by having a sub-committee in each hamlet to look after the protection of the forest patch closest to it. The total forest area was divided into seven parts and in each hamlet a president and a secretary were made responsible to the original committee (Mishra and Peter, field notes). Following the enactment of the state order, however, the Orissa forest department set up a formal forest protection committee in the village, with a village-wide base. In many cases, decisions regarding the closure of certain areas for protection each year are made regardless of the fact that this may affect the needs of hamlets differentially.

Villagers and foresters sometimes subscribe to the same notion of custom, though this may be violated by the forest department when it suits them.

One such customary concept is the idea that forest land within the notional boundary of a village should be given to it for protection. Yet contemporary notions of village boundaries are often determined by settlements made by earlier forest departments, and the 'customary rights' that villagers fight to retain are often the 'privileges' conferred by earlier administrations. The following example illustrates the complications caused when the forest department violates 'custom'.

In 1992, the forester responsible for two neighbouring villages in Rajpipla West Division, in Gujarat, gave 61 ha of forest land that fell within the revenue boundaries of Bharada village to Tabda village to protect. The logic behind this was that the Bharada villagers had in the past encroached on state forest land. Furthermore, Tabda was a larger village, and therefore seen as requiring more forests. On this 61 ha within the Bharada boundary, the FD had planted *khair* (*Acacia Catechu*) in 1987, as part of a standard departmental plantation, which would have totally excluded villagers' use. When Tabda received the patch to protect, they started with a five year plantation. According to the Tabda villagers, the forester told them that since the land belonged to the forest department, there was no problem in transferring some of it from Bharada to Tabda: 'Whether it is in Bharada or Tabda's revenue boundary it belongs to the Forest Department and whoever protects the area will be entitled to the timber from it'. In contrast, the Bharada villagers feel strongly that the forest land within their revenue boundary is theirs, because their ancestors put labour into maintaining it. They also say that their alleged 'encroachments' represent permanent agriculture of long standing, and that they have recently won their case to get title deeds for the land. In fact, there were several hamlets of Bharada in this area which were depopulated due to cholera, which suggests that the land did not always belong to the forest department but was taken over by it only relatively recently.

Matters are further complicated by the practical problems imposed by the type of species planted on the land, and the fresh claims that protection has generated. Khair, like teak, takes a long time to mature. The Tabda villagers say that they will return the land with the khair plantation to Bharada after the first felling, while the Bharada villagers argue that the forest department should give wages for protection for four years to Tabda and return the land to Bharada immediately. Bharada also has its own protection committee, but is protecting lands which are further off. In short, Bharada has some claim to the khair plantation because it is on their land, and because they looked after the plantation for the first five years. Tabda has a claim to it because they invested the next four years in protecting the trees. While some benefit-sharing arrangement on the trees could be worked out, the entrenchment of a territorial approach which equates benefits with land under a village's jurisdiction, makes this difficult. Given the complex interplay of 'customary' boundaries, actual usage, the definition of forest as state property and new claims as a result of a changed context — all of which are invoked at different times — the concept that site selection can be a function of 'local' knowledge is highly problematic.

Unlike the selection of forest areas, which may or may not be left to villagers to decide, the selection of species for plantation in the protected areas has been specifically assigned to them, and it is generally assumed that these species will be NTFP bearing ones. Sivaramakrishnan (1996: 19) highlights the compartmentalized nature of participation:

> Foresters are intent on preserving their control (in the final instance) through silviculture, a knowledge that through manuals and working plans is claimed as their exclusive preserve. Given the participatory framework, this leaves the awkward question of what should be conceded to the domain of local knowledge, where under the rules of JFM villagers are skilled practitioners. The answer, jointly provided by environmentalists and development specialists, is the knowledge of NTFP collection and processing.

The emphasis on NTFP in JFM arises for several reasons. In most states, villagers have some rights (or concessions) to collect NTFPs, but no rights to timber. Offering villagers an increase in NTFPs as part of JFM agreements does not normally require any change in the legal rules or existing balance of power. Secondly, timber products will take a long time to mature (forty years, in the case of teak), and villagers are assumed to need some interim incentives. Thirdly, the degraded lands that are given for JFM may never be capable of good marketable timber, whereas even the most degraded patch is probably capable of giving some NTFPs, including fuelwood, grasses, and so on.

In practice, targets and the availability of species in the forest department nursery often determine what is planted. This could mean the planting of NTFP species, but it could also mean that timber species like eucalyptus are planted, because that may be all the local forest department has. The DFO in Paderu recognized that the department 'should give priority to people's choices', but he added, 'November is the time for preparing the nurseries. If we ask the villagers they may suggest mango, tamarind etc. But at this time the department does not have the seed stock for these and they are very costly. So we prefer to plant whatever is feasible within the time limit' (Gorada, field notes).

Where NGOs are involved they may be better placed to cater to villagers' expressed demands, but there too, the process by which demands are expressed is not straightforward. For instance, in the plantation projects of the AKRSP in Gujarat, tree selection is done by village-level extension volunteers through PRA techniques such as species ranking. This is carried out with different groups of men and women and the two are then compared and a consensus formed (Ajith Chandran, ex-PO Forestry, pers. comm.). However, the very idea of ranking is something that is project-orientated and associated with the delivery of benefits, and not necessarily connected with the wide range of uses that villagers make of different plants (Mosse, 1995). In one village in Bastar, a couple of old women identified, at a casual sitting, some fifty-one species for which they had various uses. This is itself probably a small sample of their knowledge. Of these, however, only a few would be nursery species, and even fewer would be provided within a plantation programme.

In several villages in Andhra Pradesh which were practising JFM, we found that villagers did not ask for trees which had traditionally grown in their forest areas or even those species from which they collected NTFPs for sale. When asked why, a common reply was that these would regenerate naturally from protection so it made sense to ask for new species, preferably timber species or commercially valuable species like coffee or jaffra, a red dye plant. In one village, the villagers asked for teak although they had previously had mainly rosewood in their forests, because they had learnt from neighbouring villages and NGOs that teak was an important timber species. One finds villagers in diverse ecological settings asking for the same species regardless of the fact that it may not be suitable to the soil and the climate of the area. This is often despite their awareness of the suitability of particular species to particular soil types. In addition, although villagers may recognize that regeneration might be sufficient to afforest degraded land without additional plantation, they are loath to turn down the possibility of getting free seeds and wages for plantation from the forest department. Forest staff, on the other hand, sometimes complain that they have to persuade the villagers of the benefits of natural regeneration and of traditional NTFP species rather than giving into their demands for the plantation of exotics like cashew. This is in many ways an interesting reversal of roles, in a country where the forest department has long been castigated for large-scale plantations of alien species like eucalyptus.

In other words, villagers' mould their own demands according to what they feel the project can deliver (see also Mosse, 1995: 9), and the opportunity this gives them — or at least some among them — for entering into the commercial sphere, as against continuing with their old subsistence economy. The specific context also matters. In Koraput, Orissa, villagers wanted eucalyptus because of the proximity of various paper mills. Just over the border in Paderu, Andhra Pradesh, the primary demand was for silver oak, although successive conversations have yielded different reasons for this. Initially, fresh from the 'motivation speech' delivered by the forest staff, villagers in Kilagada, Paderu, gave ecological reasons like increased rainfall for their involvement in JFM. Another reason mentioned was the fact that the forest department constructed a village meeting room and provided various other goods in order to improve their relations with the villagers. In subsequent meetings, villagers cited the increase in NTFPs as a factor for engaging in protection and later came down to coffee as the major reason why they had taken to JFM. In Paderu, the forest development corporation and the Integrated Tribal Development Authority, which together constitute the major initiators of 'development' in the area, have flooded it with silver oak, which acts as a good shade for coffee. Coffee plantations are an important source of employment. Thus, even where the soil and climate is not suitable for silver oak, villagers often end up asking for it to be planted in their protection areas (Gorada, field notes).

In general, however, the request for coffee comes from richer villagers who are not so dependent on the forests for the collection and sale of NTFPs. The choice of certain species over others, or the choice of certain silvicultural methods over others, is not just a question of local knowledge or local choice but is, in addition, a gendered and class question. Commercially-valuable timber species are often associated with male élites whereas fruit and fodder-bearing trees are associated with women and lower classes. This has been documented in the Chipko case by Shiva (1988). In many cases, especially for poorer women, 'subsistence' includes the sale of NTFPs or firewood through head-loading, and unless some acknowledgement is made of this, environmental action meant to help women or poorer sections can often end up harming them (Rangan, 1993; Sarin et al., 1998). In short, what is represented as local knowledge to outsiders, or what is adopted by villagers from the outside, are not merely 'matters of instrumentalities, technical efficiencies, or hermeneutics ... but involve aspects of control, authority and power that are embedded in social relationships' (Long, 1992: 270).

CONCLUSION

This article has dealt with three main themes, all of which have implications for our understanding of jointness in joint forest management today. First, the jointness is not new — creating spaces for co-management or community management has historically been as much a part of state management of resources as centralization. What is perhaps new is the degree to which it is being promoted by other actors like NGOs or donors. Secondly, it is not really joint — the overall agenda is set by the state, and the policy is unable to answer many needs on the ground, such as the need for ongoing supplies of small timber. What it does do in many cases is strengthen certain hierarchies in society. Thirdly, in the limited spheres of participation that have been opened up by the state, there is a new and joint construction of needs. This has to be seen in the context of the different options available and not in terms of some pure unmediated concept of needs. Even where villagers do exercise initiative, it is under terms dictated by the overall framework of targets and activities prescribed by government rules, which in some sense distorts their agency (see also Mosse, 1995: 16). Further, the kind of local knowledge that is mobilized is presumed to be common to all groups within a village, regardless of their differing interests.

Participation by communities is limited to matters such as the choice of species or how to patrol forests, and cannot by itself solve the 'bigger issues'. '(T)he fashionable focus on the "community" or NGOs tends to neglect the essential role of the local state in articulating locality beyond the parochial to the generalisable interest' (Porter, 1995: 83). In the case of inter-village conflicts arising out of the apportionment of hitherto open access forest land, for instance, or in articulating the need of distant stakeholders for forest products, the state plays a crucial role. The state has also retained a role for itself in deciding the appropriate balance between national and state interests in forest resources. Hence attention to reorganizing the functioning of the forest department or NGOs is at least as important as reforming local institutions.

Of course, there is scope for change in all these aspects of JFM, but the basic structural problem remains. Participation is necessary not only in small-scale sectoral units, but also in influencing the entire direction of the political process. At the moment, ordinary people have little or no say in a whole range of critically important policies — they are limited to voting for politicians imposed from above by centralized, undemocratic political party structures. Rather than asking how the entire system of representative democracy can be transformed to give more power to people, donor institutions and development planners have, by focusing on village-based 'participatory committees', helped to create a discourse that diverts attention from the real issues. Such small-scale participatory attempts may be a necessary first step towards wider changes; they may lead to increased politicization at the local level, and give villagers control over aspects of their daily lives which matter to them. However, if there is no broad vision unifying these efforts, they too easily degenerate into multiple committees, compartmentalized into different management modes, serving no useful short-term or long-term purpose.

Acknowledgements

This paper was first presented as a seminar at Teen Murti, New Delhi in January 1997. I am grateful to Roger Jeffery, Mahesh Rangarajan and Siddharth Varadarajan for their inputs, as well as to my former colleagues in the Edinburgh University Joint Forest Management Research project, funded by the ESRC, and entitled 'Organising Sustainability: NGOs and Joint Forest Management in India'.

REFERENCES

Agarwal, B. (1997) 'Environmental Action, Gender Equity and Women's Participation', *Development and Change* 28(1): 1–44.

Agarwal, C. and S. Saigal (1996) 'Joint Forest Management in India: A Brief Review'. SPWD Working Paper. New Delhi: SPWD.

Agarwal, C. and K. Singh (1996) 'Forest Co-operatives in Kangra (HP): A Case Study', in *Wasteland News* 11(2): 38–48.

Agrawal, A. (1994) 'Small is Beautiful but could Larger be Better? A Comparative Analysis of Five Village Forest Institutions in the Indian Middle Himalyas'. Paper prepared for the

meeting of the FAO Forestry Working Group on Common Property, Oxford Forestry Institute, Oxford (15–18 December).
Anderson, B. (1991) *Imagined Communities*. London: Verso.
Anderson, M. (1990) 'Classifications and Coercions: Themes in South Asian Legal Studies in the 1980s', *South Asia Research* 10(2): 158–77.
Anderson, P. (1977) 'The Antinomies of Antonio Gramsci', *The New Left Review* 100: 5–79.
Appadurai, A. (1995) 'The Production of Locality', in R. Fardon (ed.) *Counterworks*, pp. 204–25. London: Routledge.
Ballabh, V. and K. Singh (1988) 'Van (Forest) Panchayats in Uttar Pradesh Hills: A Critical Analysis'. Research Paper 2. Anand: Institute of Rural Management.
Breckenridge, C. and P. Van der Veer (eds) (1993) *Orientalism and the Postcolonial Predicament: Perspectives on South Asia*. Philadelphia, PA: University of Philadelphia Press.
Breman, J. (1997) 'The Village in Focus', in J. Breman, P. Kloos and A. Saith (eds) *The Village in Asia Revisited*, pp. 15–75. Delhi: Oxford University Press.
Chakravarty-Kaul, M. (1996) *Common Lands and Customary Law: Institutional Change in North India over the Past Two Centuries*. Delhi: Oxford University Press.
Chandran, A. and M. Singh (1995–7) Field Notes on JFM in Rajpipla West division, Gujarat, for the Edinburgh University Project on Joint Forest Management.
Chatterjee, P. (1998) 'Community in the East', *Economic and Political Weekly* 33(6): 277–82.
Chattre, A. (1996) 'Joint Forest Management: Beyond the Rhetoric', in W. Fernandes (ed.) *Drafting a People's Forest Bill: The Forest Dweller–Social Activist Alternative*, pp. 102–120. Delhi: Indian Social Institute.
Cohn, B. (1990) 'The Census, Social Structure and Objectification in South Asia', in B. Cohn *An Anthropologist Among the Historians and Other Essays*, pp. 224–54. Delhi: Oxford University Press.
Daver, S. R. (1938) Working Plan for the Sal Areas in the Ulnar Nistari Forest, Compilation No. XXIII, 31/B Part I, Jagdalpur Record Room, Bastar.
Dove, M. (1993) 'A Revisionist View of Tropical Deforestation and Development', *Environmental Conservation* 20: 17–56.
Fallers, L. (1969) *Law Without Precedent*. Chicago, IL: Aldine.
Fernandes, A. P. (1996) *The Myrada Experience — Working with the Government in Multilateral and Bilateral Projects*. Bangalore: Myrada.
Fields, K. (1985) *Revival and Rebellion in Colonial Central Africa*. Princeton, NJ: Princeton University Press.
Gadgil, M. and R. Guha (1992) *This Fissured Land: An Ecological History of India*. Delhi: Oxford University Press.
Gadgil, M. and R. Guha (1995) *Ecology and Equity: The Use and Abuse of Nature in Contemporary India*. New Delhi: Penguin Books.
Gorada, P. (1995–7) Field Notes on JFM in Paderu division, Andhra Pradesh, for the Edinburgh University Project on Joint Forest Management.
Government of Orissa (1971) *Orissa District Gazetteers, Sambhalpur District*. Cuttack: Government of Orissa Press.
Grove, R. (1998) 'Indigenous Knowledge and the Significance of South-West India for Portuguese and Dutch Constructions of Tropical Nature', in R. H. Grove, V. Damodaran and S. Sangwan (eds) *Nature and the Orient: The Environmental History of South and Southeast Asia*, pp. 187–209. Delhi: Oxford University Press.
Guha, R. (1996) 'Dietrich Brandis and Indian Forestry: A Vision Revisited and Reaffirmed', in M. Poffenberger and B. McGean (eds) *Village Voices, Forest Choices*, pp. 86–100. Delhi: Oxford University Press.
Jeffery, R., P. Khanna and N. Sundar (forthcoming) 'Joint Forest Management: A Silent Revolution among Forest Staff?', in B. Vira and R. Jeffery (eds) *Participatory Natural Resource Management: Analytical Perspectives*. London: Macmillan.
Kamath, H. S. (1941) *Grazing and Nistar in the Central Provinces Estates: the Report of an Enquiry*. Nagpur: Government Printing Press.

Long, N. (1992) 'Conclusion', in N. Long and A. Long (eds) *Battlefields of Knowledge: The Interlocking of Theory and Practice in Social Research and Development*, pp. 268–77. London: Routledge.

Ludden, D. (1996) *Making India Hindu: Religion, Community and the Politics of Democracy in India*. Delhi: Oxford University Press

MacIntyre, A. (1981) *After Virtue: A Study in Moral Theory*. London: Duckworth.

Mishra, A. and N. Peter (1995–7) Field Notes on JFM in Sambhalpur division, Orissa, for the Edinburgh University Project on Joint Forest Management.

Mosse, D. (1995) 'People's Knowledge in Project Planning: The Limits and Social Conditions of Participation in Planning Agricultural Development'. ODI Network Paper 58. London: Overseas Development Institute.

Mosse, D. (1999) 'Colonial and Contemporary Ideologies of "Community Management": The Case of Tank Irrigation Development in South India', *Modern Asian Studies* 33(2): 303–38.

Nelson, N. and S. Wright (1995) 'Participation and Power', in N. Nelson and S. Wright (eds) *Power and Participatory Development*, pp. 1–18. London: Intermediate Technology Publications.

Pathak, A. (1999) 'Legislating Forests in Colonial India 1800–1880'. PhD dissertation, University of Edinburgh.

Pathan, R. S., N. J. Arul and M. Poffenberger (1990) 'Forest Protection Committees in Gujarat: Joint Management Initiative'. Sustainable Forest Management Working Paper no. 7. New Delhi: Ford Foundation.

Poffenberger, M. and C. Singh (1996) 'Communities and the State: Re-establishing the Balance in Indian Forest Policy', in M. Poffenberger and B. McGean (eds) *Village Voices, Forest Choices*, pp. 56–85. Delhi: Oxford University Press.

Porter, D. J. (1995) 'Scenes from Childhood: The Homesickness of Development Discourse', in J. Crush (ed.) *Power of Development*, pp. 63–86. London: Routledge.

Rahul (1997) 'Masquerading as the Masses', *Economic and Political Weekly* 32(7): 341–2.

Ramadhyani, R. K. (1942) *Report on Land Tenures and the Revenue System of the Orissa and Chattisgarh States*. Berhampur: Indian Law Publication Press.

Rangan, H. (1993) 'Romancing the Environment: Popular Environmental Action in the Garhwal Himalayas', in J. Friedman and H. Rangan (eds) *In Defense of Livelihood: Comparative Studies on Environmental Action*, pp. 155–81. West Hartford, CT: Kumarian Press.

Ravindranath, N. H., M. Gadgil and J. Campbell (1996) 'Ecological Stabilization and Community Needs: Managing India's Forest by Objective', in M. Poffenberger and B. McGean (eds) *Village Voices, Forest Choices*, pp 287–323. Delhi: Oxford University Press.

Sabean, D. W. (1988) *Power in the Blood: Popular Culture and Village Discourse in Early Modern Germany*. Cambridge: Cambridge University Press.

Sarin, M. (1996) 'From Conflict to Collaboration: Institutional Issues in Community Management', in M. Poffenberger and B. McGean (eds) *Village Voices, Forest Choices*, pp. 165–209. Delhi: Oxford University Press.

Sarin, M., with L. Ray, M. S. Raju, M. Chatterjee, N. Banerjee and S. Hiremath (1998) *Who Gains? Who Loses? Gender and Equity Concerns in Joint Forest Management*. New Delhi: SPWD

Saxena, N. C. (1995) 'Towards Sustainable Forestry in the UP Hills'. ODA Working Paper. London: ODA.

Saxena, N. C. (1996) *I Manage, You Participate*. Bogor: Centre for International Forestry Research.

Shiva, V. (1988) *Staying Alive: Women, Ecology and Survival in India*. New Delhi: Kali for Women.

Sivaramakrishnan, K. (1996) 'Forest, Politics and Governance in Bengal, 1794–1994'. PhD dissertation, Yale University.

Sood, M. P. (1996) 'New Forestry Initiatives in Himachal Pradesh'. IIED Working Paper. London: IIED.
Starr, J. and J. F. Collier (eds) (1989) *History and Power in the Study of Law*. Cornell, NY: Cornell University Press.
Sundar, N. (1997) *Subalterns and Sovereigns: An Anthropological History of Bastar, 1854–1996*. New Delhi: Oxford University Press.
Sundar, N., A. Mishra and N. Peter (1996) 'Defending the Dalki Forest: "Joint" Forest Management in Lapanga', *Economic and Political Weekly* 31(45/46): 3021–5.
Vasavada, S. and N. Sengupta (1995–7) Field Notes on JFM in Dewas division, Madhya Pradesh, for the Edinburgh University Project on Joint Forest Management.
Vincent, J. (1990) *Anthropology and Politics: Visions, Traditions and Trends*. Tucson, AZ: University of Arizona Press.
Warren, D. M. (1991) 'Using Indigenous Knowledge in Agricultural Development'. World Bank Discussion Paper 127. Washington, DC: The World Bank.

12 Community Forestry and Tree Theft in Mexico: Resistance or Complicity in Conservation?

Dan Klooster

INTRODUCTION

Participation, decentralization, and community involvement are centre-stage topics in current development debates over the management of forests and other renewable natural resources. Community-based conservation has been called a new paradigm, one that confronts the 'biggest conservation challenge of all: how to deal with the vast majority of the earth's surface, where there are no parks and where the interests of local communities prevail' (Western and Wright, 1994: 7; see also Agrawal and Gibson, 1999; Brosius et al., 1998; Castilleja, 1993; Poffenberger, 1994; Romm, 1993). One of the most compelling reasons for states to foster participatory management approaches is that they have not been able to police forests effectively (Ascher, 1995; Johnson and Cabarle, 1993; World Bank, 1996).

With their traditional woodcutting and vegetation management practices declared illegal by forest bureaucracies, forest communities 'steal' trees and commit 'arson', and in many other ways anonymously contest the restrictions imposed on them. United against the impositions of a forest bureaucracy, community members see nothing wrong in their resistance, and collude to protect their members from detection and punishment (Guha, 1989; Peluso, 1992). In the measured words of the World Bank, 'Exclusive bureaucratic control of [grazing lands, wildlife, forests, and water sources] has proved inadequate in many different institutional settings, in some cases leading to confrontation between the users of these resources and the public officials seeking to manage them' (World Bank, 1997: 118). So does community-based forestry create a situation in which communities change from being accomplices to arson and tree theft,[1] but instead convert

---

1. There are good reasons to bring the specific topic of tree theft into debates on natural resource management. Precious tree species like mahogany and Spanish cedar, in tropical America, or yew and cedar in the Pacific Northwest, can be the focus of intense cutting not only by transnational logging companies on poorly managed concessions, but also by small local businesses, peasant farmers, and other forest dwellers in national parks, national forests, and common property territories. In the United States, tree theft from public lands removes timber worth an estimated $100 million a year (Pendleton, 1998).

to become complicit in conservation? The question is complicated by the apparent incompatibility between the goals and methods of scientific forestry and peasant agroforestry systems.

This study first describes the apparently deep-seated conflicts between scientific forestry and rural communities, conflicts related to local conceptions about justice in access to forest resources. It then goes on to examine a case-study from Mexico's rich experience with community forestry, describing the imposition of scientific forestry and the evolution of tree theft in a case-study community, and identifying community processes which restrain and legitimate tree theft.

SCIENTIFIC FORESTRY AND RESISTANCE

Scientific forestry developed in the context of political centralization and nation-building in late eighteenth century Germany and France. It is clearly associated with the rise of a modernist approach to natural resource development which supplants communal control of forests with technocratic control (Pincetl, 1993; see also Scott, 1998). Forestry science stresses wood production over other forest uses, and generally requires spatial control to organize production in rotating zones of intense cuts and long fallows. It discourages agricultural clearings, burning, woodcutting, and grazing, declaring many of these practices illegal even when they are perceived as customary rights (see Marx, 1975). Scientific forestry is a political and ideological system of resource and territorial control (Banuri and Apfell Marglin, 1993; Bryant, 1994; Guha, 1989; Peluso, 1992; Scott, 1998; Vandergeest, 1996). The social confrontation accompanying forest management appears to go deeper than mere problems with centralization, however; this literature identifies inherent conflicts between scientific forest management and the livelihood goals, strategies, and world views of forest-dwelling people.

Different management systems do exist. Careful fieldwork continues to discover contemporary rural societies exercising highly sophisticated local resource management systems to grow forests and enrich them with useful ritual, subsistence, and commercial species (Alcorn, 1981; Fairhead and Leach, 1996; Padoch and Peters, 1993; Peluso, 1996). Not all forest-dwelling people have these kinds of knowledge systems, however. Furthermore, rural societies often choose modern production methods as part of their own vision of cultural survival and mediated modernization (Bebbington, 1996; Rangan, 1996). Sometimes these include commercial forestry.

New perspectives in forest ecology stress the ubiquity of environmental flux, disturbance, and the need to manage for risk. Within this perspective, 'local inhabitants are, in many cases, most able to identify the spatial and temporal heterogeneities of their biophysical environments and to help plan accordingly' (Zimmerer, 1994: 118). Disturbance ecology also suggests that many issues of forest conservation cannot be addressed without awareness of ecological scales and disturbance patterns operating well above the community level (Baker, 1992; Noss, 1993; Zimmerer and Young, 1998). Putting global interests in climate change and biodiversity conservation above the livelihood rights of local people entails great moral danger (Peluso, 1993), but these global concerns are important resource management goals, none the less. Despite the potential for conflict, scientific forestry remains indispensable in many situations.

Co-management strategies promise to integrate community participation, scientific resource management, and state resource management bureaucracies (Berkes, 1997; McCay and Jentoft, 1996; Pinkerton, 1989). But does community participation resolve contradictions with scientific forestry? Do the kinds of resistance that wreck centralized management approaches continue to undermine possibilities for sustainable scientific forestry, even with community participation? Addressing questions like these requires a better understanding of how groups of people define some forest uses as permissible and others as reprehensible, especially how they come to legitimate behaviours that are officially defined as illegal, translating tree 'theft' and 'arson' into anonymous acts of protest called 'resistance'[2] (Scott, 1985). Researchers often use the concept of moral economy[3] to clarify the poors' motivations in protest. Moral economy indicates much more than mere subsistence concerns. It includes 'conceptions of social justice, of rights and obligations, of reciprocity' (Scott, 1976: viii; Thompson, 1991). Peluso (1996) contributes the related idea of an ethic of access, a notion of justice in access to common property resources which motivates resistance to restrictions and re-assignment of the benefit from forest resources. This too encompasses much more than subsistence rights, including kinship, ritual considerations, and other aspects of the social meanings of the resource. When the restrictions of scientific forestry violate customary rights or otherwise insult an ethic of access, this disturbs a community's moral economy. In the eyes of rural society, enforcing forest laws then becomes a 'moral crime' on the part of the state (Peluso, 1992: 11).

For villagers who would never steal a neighbour's crop, this removes the moral dilemma and social stigma of taking trees from state forest lands (Peluso, 1992: 202; Pendleton, 1998). When very few in the community perceive pilfering trees as wrong, the resource management system requires a powerful monitoring and enforcement apparatus. When rural society develops a culture of resistance, actively protecting members who commit

---

2. Resistance stops short of collective outright defiance. It includes slander, arson, pilfering, foot dragging, flight. No co-ordination or planning is required and it contributes to individual self-help. Through resistance, the peasantry defends its interests from powerful outside sectors (Scott, 1985).
3. Drawing on writings from the late eighteenth century, when the term emerged in opposition to developing notions of political economy, E.P. Thompson introduced the concept into contemporary academic exchanges (Thompson, 1991).

'illegal' acts, the resource control system needs even more coercion to meet its goals. It becomes increasingly oppressive and ineffective.

In the context of forest co-management, the resistance perspective raises a number of troubling questions. Since community-based resource management systems often carry the history of repressive centralized forestry with them, how will their baggage of resistance affect outcomes? Scientific forestry prohibits many long-standing customary usages of the forest, and this especially catalyses resistance. Can community management avoid this contradiction of resource management systems? How do novel forest uses with tenuous links to customary usages become entrenched (or not) into cultures of resistance? Resistance is also a phenomonen internal to fragmented rural societies, perhaps between landless sectors and the relatively wealthy, or between men and women (Agarwal, 1994, 1998; Peluso, 1992: 16). How does internalized resistance affect co-management?

COMMUNITY FORESTRY IN MEXICO: THE CASE OF SAN MARTIN OCOTLAN

This section addresses the preceding questions through an analysis of the Mexican community forestry experience. It centres on a case-study of San Martin Ocotlán (a pseudonym), a forest community with forty years of logging experience. Despite fifteen years of community forestry, San Martin is still wracked by internal conflict over access to resources, including widespread tree theft.

San Martin is one of an estimated 8,000 *ejidos* and *comunidades agrarias* owning 80 per cent of remaining Mexican forests. These forests suffer widespread deforestation and degradation: estimated deforestation rates range from 0.8 to 2 per cent a year, although case-studies give rates as high as 12 per cent a year, while forest fires affect between 90,000 ha and 500,000 ha of forests per year (Cairns et al., 1995; Masera, 1996; World Bank, 1995). Estimates of the volume of tree theft in some regions can be four to seven times greater than the authorized cut, far above estimates of the annual growth increment (Masera et al., 1998). These highly selective and intense cutting pressures can change the species compositions of forests in ways that extirpate precious species, cause commercial degradation,[4] and undermine sustainable development possibilities.

4. Degradation is a socially-laden term, and some of the forest changes associated with woodcutting that are not co-ordinated by scientific forestry might result in beneficial environmental changes for some groups — more fodder for sheep, goats and cattle, a different array and density of mushrooms, medicinal plants, more game species and so on. At the same time, these kinds of cutting pressures can extirpate commercially-valuable species, and this *commercial* degradation can decrease the viability of the community-based commercial forestry option. It also undermines the long-term viability of timber-smuggling economies, which rely on these species.

San Martin Ocotlán, Oaxaca, is home to some 600 households and 3,300 inhabitants. One fourth of community residents speak the Mixtec language in addition to Spanish, and there is an ethnic distinction between a relatively mixed-race (*mestizo*) capital village and five outlying hamlets. Planting *milpa* — maize combined with beans, squash, potatoes, various leafy vegetables, and flowers — occupies virtually all households in the community, at least part-time. Many *milpas* are grown on plots irrigated by stream diversion, with one crop every year. Others are rainfed, grown in short rotations with fallows of one to three years. Agriculture rarely produces enough maize for year-round food self-sufficiency, however, nor does it generate cash needed to purchase food, clothing, and sundries. San Martin has a sub-subsistence agricultural system,[5] requiring inputs from livestock, off-farm labour, and especially the forest. It is a forest-dependent community.

The forest provides the most accessible supplementary activities to *milpa* production. Commercial logging generates sporadic employment opportunities for about half the community's population of working men, especially during the winter dry season when agricultural activities ebb. Work in the forestry business is not always available when agricultural activities allow other work, but woodcutting is nearly always a possibility. Currently, the most dependable and least problematic way to earn the cash needed to supplement the *milpa* is the sale of oak firewood and charcoal. The only equipment needed is an axe, a machete, and a mule, although chainsaws are becoming increasingly common. In addition to firewood, woodcutters also cut building posts highly valued in regional informal construction markets, and these fetch from eight to fifteen pesos each in Zaachila and Oaxaca markets.[6] Woodcutters also find a small market for firewood in San Martin Ocotlán village, where a mule-load of wood fetches five pesos. Local truck drivers with permits purchase larger volumes and transport it for resale in the weekly markets at Zaachila and Oaxaca, where the price nearly doubles. Charcoal makers add value to oak, and reduce its weight for transport. Like firewood cutters, charcoal makers sometimes market small quantities themselves, carrying gunnysacks to the roadside by mule and hitching a ride on a public bus. Larger quantities require a deal with a permit-holding truck owner. Truckers charge 400 pesos to transport a load of 100 gunnysacks of charcoal to Oaxaca City, where wholesalers pay roughly twelve pesos per bag. Charcoal makers can also sell their product in the forest, for about eight pesos per gunnysack.

A riskier, but more lucrative complement to the *milpa* is the preparation and sale of rough-hewn boards, roundwood posts, beams, and rafters from

5. For a more elaborated discussion of forest dependence in an agricultural community, see Klooster (forthcoming).
6. In 1995, the exchange rate averaged 6.4 Mexican pesos per 1US$.

pine. Forestry management plans favour pine, and both official and community forestry enforcement focuses on pine, hence this activity is termed tree theft.[7] After they fell the selected tree, small groups of cutters carve it into eight-foot lengths, square the logs, and mount them on a makeshift platform to provide a flat, dry, clean, space on the forest floor. Taking strings impregnated with the black lead powder from the insides of batteries, expert cutters mark the lines along which to cut boards from the squared-off log, take their chainsaws, and then skilfully slice the log into boards. Truck owners co-ordinate the activity. The pickup truck is the key to moving the boards to market. A history of commercial forestry means that the high altitude areas where most illegal cutting occurs have good communications, and old logging roads offer numerous escape routes past the patrols and roadblocks which community authorities often post.

The boards and square building timbers sell in informal markets in Oaxaca City and other Oaxaca valley urban centres. Much of this lumber enters formal construction markets, sometimes with minimal additional processing. Tree theft generates benefits much higher than other labour opportunities in the area, both to truck drivers and to cutters. One driver who fell into the hands of communal authorities told me that his pickup-load of wood took three people four hours to cut and would have sold for 1000 pesos in Oaxaca. After discounting the costs of gasoline and oil for cutting and transport, that still nets each cutter more than 200 pesos for four hours of work, in an economy where the going rate for an entire day's work weeding someone's *milpa* is only fifteen to twenty-five pesos. In the forest, a board goes for six pesos while in Oaxaca the same board sells for ten pesos, so a driver expects to earn a 400 peso return from a typical load of 100 boards purchased from other cutters. Some Oaxaca lumber merchants blame cheap lumber from the community for driving several legal lumberyards out of business.[8]

Tree theft has the potential to drastically affect the forest. Based on smuggler's estimates, in 1994 tree thieves sneaked twenty pickup-loads of rough-hewn boards past roadblocks each week. If the activity were unrestrained, there are enough trucks, chainsaws, and potential cutters in the community to remove significantly more pine wood than the forest can replace each year. Currently, cutting intensities are below foresters' estimates of the rate of forest growth, but this does not mean that cutting is benign. Even at intensities below the replacement rate, tree thieves'

---

7. The Spanish term used by both law enforcers and law breakers is *clandestinaje*. An alternative term is timber poaching, which perhaps better captures the moral ambiguity of individual appropriation of common property trees, in violation of the law. Poaching is awkwardly applied to trees, however, since it more often refers to the taking of game and fish.
8. *El Imparcial* (Oaxaca City newspaper) 4 June 1995.

selective cutting practices tend to change the genus composition and commercial quality of the forest in ways inimical to long-term community interests.

Timber thieves focus their work on the very best trees in the forest — mature pines with straight, branchless boles, close to roads. Even when cutters thin the forest by cutting young pines for posts and beams, they also choose the best trees available. The effects of this kind of high grading in Mexican pine forests are well known. 'The superior provenances (trees best adapted to a particular site and set of environmental conditions) are frequently the first to be removed. The trees left to regenerate are often of poor form, stagheaded (or rogue), and as seed trees produce genetically inferior progeny' (Styles, 1993: 415). The end effect will be genetic impoverishment of pines, lack of pine regeneration, oak dominance, and long-term decreases in available commercial volume.

The community's prescribed silvicultural method, in contrast, calls for culling commercially undesirable individuals in thinning cuts, saving the best individuals for seed trees. Tree theft removes precisely the trees the silvicultural method aims to reserve for seed trees. Timber theft subverts silviculture and decreases the long-term commercial value of the forest, diminishing the ability of the forest to meet the demands for pine that both individual woodcutters and the communal forestry business place on it.

The Co-Evolution of Tree Theft and Scientific Forestry under Concessions

Tree theft co-evolved with scientific forestry in San Martin, initially developing in parallel to a concession system, but continuing even under community forestry. Logging in the community's forest began in 1958, under the control of a small private firm that left little record of its impact in the community. Starting in 1964, a series of state-owned and parastatal concessionaire logging firms began to work San Martin's forests.

The concessionaires, together with foresters from the federal forestry department, introduced San Martin to the goals and resource use restrictions of scientific forestry. This resource control system came to Mexico in the early twentieth century,[9] brought by the French-trained civil engineer Miguel Angel de Quevedo. He founded forestry schools with French-trained instructors, drafted a legal framework for forestry, and established a forest bureaucracy based on European concepts of centralized control, sustained yield of timber, and forest conservation for flood control (Simonian, 1995).

9. This is much later than in Asia. Strong nineteenth century colonial states imposed scientific forestry there, but this was lacking in Mexico. Neither the Spanish crown nor the various nineteenth century national governments that followed it created a forest bureaucracy capable of enforcing forest laws (Bogati, 1978; Gonzalez, 1981; Moguel, 1994; Simonian, 1995).

In the four decades following the Second World War, concessions[10] were a key vehicle for making forests meet import substitution and other national development goals through the methods of scientific forestry. The federal government granted integrated logging and processing industries monopsonistic rights over extensive forest areas. Forest-owning communities maintained nominal control over forest land in these concessions, but they could only sell logging rights to the big industrial entities and they could not convert forests to other uses.

Foresters associated with these concessions[11] restricted peasants' clearing, cutting, and burning practices. Evidence of this in San Martin goes back to the earliest record of forestry in the community, when foresters began the long process of getting peasants to see the forest the way professional foresters see it, using community assemblies to explain the problems presented by forest fires, agricultural clearings, and cutting roofing shingles. The concessionaires also used these arguments in documents requesting logging permits in the area, presenting commercial logging under scientific forestry as the way to save San Martin's forests from the threat of peasant degradation:

> The major causes of forest destruction, as much natural as caused by man, are well known (agricultural clearing, fires, insect plagues, diseases, grazing, and so on). Even so, we can conclude that the principal cause is economic, due to the fact that residents of forest zones, faced with the lack of permanent or semi-permanent sources of work and faced with the necessity to survive, resort to the often-mentioned practice of clearing the forest for agriculture. This causes, first of all, a decrease in forested area that increases with time, and in the second place, it causes forest fires which, when not controlled, affect huge areas, and year after year destroy regeneration and affect the development of trees to varying degrees. We can also mention the forest destruction caused by household uses, which occur without any control or monitoring. (Escarpita, 1977: 213)

---

10. In a third of the nation's forested area (but not in the case-study area) the Mexican government established bans on all logging and clearing activities in order to control small, private logging firms, halt deforestation, and maintain reserve areas for future industrial expansion by the integrated forest industries. Bans failed to conserve forests but did create conflicts with resident communities unable to use forest lands for agriculture or wood production. Logging by individuals and small firms continued there, as did deforestation for peasant agriculture. Communities benefited little from this clandestine logging on their forested lands, however. Neither were there any of the potential benefits from organized logging, such as reforestation or sustained-yield logging techniques. Clandestine timber economies in Puebla, Chiapas and, especially, Michoacan became entrenched during this period. *Rentismo*, where logging takes place following short-term contracts between small logging firms and community forests, has also been a widespread phenomenon in Mexican forest history, and remains common to this day in certain regions of the country (Beltran, 1964; Bray and Wexler, 1996; Calva et al., 1989; Hinojosa, 1958; Moguel, 1994; Simonian, 1995).
11. At different points in time, the federal forestry department granted rights to exercise scientific forestry to the concessionaires, and at other times the state retained the provision of professional forestry services as a monopoly. Even in the latter case, however, foresters were usually dependent on the big firms for transportation, food and lodging while in the field.

As the best solution to this problem, Escarpita goes on to propose scientific, centralized, commercial forestry. 'This would be of great benefit for the protection and conservation of this resource since the creation of permanent sources of work for the people in the region, who would then have other means of meeting their needs, would eliminate one of the principal causes of destruction which are agricultural clearings and clandestine cutting'. In addition to employment, there would also be a 'campaign of persuasion' so that people would see the forest as an economic resource, and thus combat and prevent forests fires, and limit domestic use and grazing (ibid.: 214). Enforcement also played a role, with brigades consisting of a patrol chief, nine assistants, and a vehicle, and with concessionaires establishing watchtowers in strategic locations where guards could spot forest fires. 'These personnel are always ready to resolve or report any anomaly in the control of forestry, agricultural clearings, fires, and illegal taking of forest products that occur in their zone of jurisdiction' (ibid.: 216).

The company also had a cadre of workers who could be expected to help with controlling illegalized forest uses. Community members who had privileged positions with the company were there to keep tabs on burning and to call on forestry authorities when there was any illegal clearing or tree theft. In the inspections that resulted, inspectors from the concessionaire would accompany personnel from the enforcement and monitoring section of the federal forestry directorate, who had the power to levy fines. Since the community could be fined for forest clearing infractions, community authorities became part of the enforcement infrastructure, monitoring land use and helping enforce the restrictions against clearing and illegal cutting.

In San Martin Ocotlán, the restrictions and enforcement apparatus of scientific forestry — together with the development of new markets, technological change, and improved road infrastructure — transformed forest usages. Traditionally, the forest provided commoners with materials for building houses, furniture, corrals for their animals, tools, and firewood for cooking their food. Before the establishment of commercial forestry and the restrictions of scientific forestry, the forest also provided peasants with marketable products to supplement the *milpa*. Don Hilario, an elderly peasant farmer from the region, recalls that thirty years ago (before forest controls), 'with one trip to the forest you could come back with six or seven mule loads of fat wood[12] which was worth ten or fifteen pesos. With that you could buy everything you needed at the market'. Cutting wood shingles provided another way of turning trees into money. Farmers would sometimes manufacture boards and building timbers for sale as well, but before chainsaws and the penetration of roads, this was time-consuming, difficult, and therefore uncommon.

---

12. Fat wood (*ocote* in Mexican Spanish) is resin-impregnated pine wood used for kindling and lighting.

Parallel to the illegalization of peasant activities came regional economic changes. Logging roads penetrated the forest in the 1960s, an all-weather dirt road connected San Martin Ocotlán village with markets in Zaachila in 1968, and community members acquired pickup trucks in the 1970s and 1980s. Transporting wood products became much easier just as population growth and deforestation in areas closer to Oaxaca City increased demand for wood from San Martin Ocotlán. Even firewood and charcoal became commercial possibilities. Finally, a profusion of chainsaws in the late 1970s and early 1980s made wood cutting fast and efficient, and added rough-hewn boards to the collection of marketing possibilities.

Even as forestry restricted agriculture and provided wage opportunities, the role of agriculture in subsistence was decreasing. Markets for maize were becoming increasingly regional, with declining prices. Subsidized maize was more often available for farmers to purchase, if they had the money (Fox, 1993; Zabin, 1989). Meanwhile government programmes promoted chemical fertilizer and farmers began to depend on this increasingly expensive input. More than ever before, *milpa* activities required money inputs.

Amidst these regional economic changes, the restrictions of scientific forestry raised barriers to trafficking in wood. However, these have tended to reshape the activity rather than eliminate it. Commoners have long sought ways to get around restrictions and gain access to small-scale commercial resources in forestry. In the 1960s, under concessionaires, community members were already interested in the financial opportunities of commercializing firewood, but in order to transport and sell such forest products, they needed the sanction of the experts overseeing forestry in order to get the documentation required by the federal laws of the time. They requested permits in assemblies in 1966 and again in 1967, when the Delegate for Agrarian Affairs authorized commoners to sell firewood taken from the branches and tips of logged trees, with the conditions that they pay a small fee to the communal treasury, that the communal authorities co-ordinate the work, that the expert forester dictate the necessary measures to meet forestry law, and that any infraction be paid by the commoner who breaks forestry law.

The distinction within traditional usages between domestic use and petty commerce blurred, with wood for both purposes sanctioned by fees and permits. In 1979, as part of yearly negotiations for logging contracts, the concessionaire agreed to assist with processing the documentation required to extract firewood, charcoal and small timbers for domestic use, which was often a cover for petty commerce. In theory, permits only sanctioned wood gleaned from logging areas, but in practice they were available to legalize wood taken from anywhere within the community's boundaries. The permitting system helped to entrench the San Martin Ocotlán truck owners' knowledge, skills, and dependence on marketing minor wood products. Once the parameters of marketing firewood and building timbers with the permit system became entrenched, it was easy to slip into timber smuggling when permits were unavailable, enforcement slackened, and as markets developed for rough-hewn pine boards.

Community Forestry

Timber smuggling in San Martin was just one minor symptom of an increasingly dysfunctional forest management system. Throughout Mexico, payments for stumpage rights were low. Concessions marginalized peasants from control over forests and from the benefits of forestry, and so the forest became, from their perspective, marginal. Alienated from their resources (Guha, 1989), they often resisted the system with tree theft, clearing, and burning. At the same time, the concessionaire's own logging practices were poor, and reforestation lax (Beltran, 1964; Calva et al., 1989; Chambille, 1983; Hinojosa, 1958).

In 1980, suspicious that volume mismeasurements and wasteful logging practices were reducing the community's benefits from forestry, San Martin community leaders enlisted support from the Agrarian Reform agency and a government-affiliated peasant union, broke free from the monopsonistic controls of the concessionaire, and established one of the first community logging businesses in the state of Oaxaca. Community forestry in Mexico was still incipient at the time, but the conditions were ripe. Although small compensation for the late nineteenth century liberal reforms which dispossessed indigenous communities of 90 per cent of their lands, land reforms of the 1930s and 1970s had placed increasing areas of forests in the full legal control of peasant communities (Guerrero, 1988; Otero, 1989; Sanderson, 1984). Starting in the 1970s neopopulist progressives in the agrarian reform agency and the forestry agency experimented with community-owned logging businesses, especially in areas where bans had been recently removed and established logging and processing interests were not directly challenged (Mendoza, 1976). In the late 1970s, communities in areas under logging concessions began to organize, conducting production strikes by refusing to sign contracts, and applying pressure on the government to rescind concessions or allow them to expire. Following pressure from unions of forest-owning communities and from reformers within both the forestry bureaucracy and the agrarian reform agency, the 1986 Forestry Law formally recognized the right of communities to form their own logging businesses (Bray and Wexler, 1996).

Communities took advantage of the opening their organized resistance had created. They built on experience gained from working for the concessionaire logging firms and established their own logging operations. At first they sold logs to the same big firms that had previously employed them and rented their forests, but now in a competitive market with much higher prices. Some, including San Martin Ocotlán, were able to quickly capitalize and buy cranes, trucks, bulldozers, and even sawmills. By 1992,

15 per cent of lumber and 40 per cent of timber was from the organized community forest management sector (Bray and Wexler, 1996). Short-term logging contracts between small firms and community forest owners remain common, but the number of communities with community logging businesses has increased since then, as more forest-owning communities make the transition from renting their forests to running their own logging businesses.[13]

Scientific Forestry under Community Control

After San Martin established its own forestry business, the parameters of enforcement and restriction on wood cutting and forest clearing remained basically the same as under the concessionaires, but communal authorities took on a greater degree of responsibility for enforcing restrictions and issuing permits in accordance with regulations from the department of forestry. Communal authorities can give commercial transport documentation for oak firewood, oak building materials, charcoal, and even pine on the pretext that the wood is from dead trees or from the area of legal commercial logging. Enforcement consists of community forest patrols, confiscation of chainsaw-cut boards discovered in the forest, and roadblocks to inspect passing traffic and apprehend trucks carrying clandestinely-cut timber. Always aware of the potential for violence, communal authorities undertake these activities with caution, in small groups armed with pistols. Punishments available range from verbal admonishment, to confiscation of wood and trucks, to temporary imprisonment in the community's small jail. The possibility of support from Oaxaca state police, the federal environmental enforcement agency, and federal police, is an important complement to community enforcement, especially when community members resist with violence.[14]

Forestry restrictions and enforcement tend to be tightly applied to pine, but both communal and federal authorities look the other way if oak is the only species involved in an infraction, especially if the quantity involved is not a truck-load, but a mule-load. Enforcers also exercise their authority to differentiate between infractions involving dead wood, and those involving freshly-cut trees. In this way, some of the usages with which poor commoners supplement agricultural activities continue to be respected.

Some communal administrations, including one in place in 1994, pay very little attention to enforcement. In 1995, however, communal authorities

13. Personal communication, Sergio Madrid, Consejo Civil Mexicano para la Silvicultura Sostenible, 1998.

14. One particularly notorious local tree thief is now in federal prison after being convicted of murdering a member of the community's forest protection committee with whom he had personal conflicts unrelated to forest enforcement activities.

aggressively patrolled the forest to control unsanctioned cutting. They meted out fines, temporarily decommissioned vehicles and chainsaws, reported repeat offenders to federal authorities, and actively recruited official support for their internal enforcement activities.

Extent and Potential Impact of Tree Theft

During the 1995 season of aggressive patrols, roadblocks, and confiscations, timber theft was greatly reduced from previous seasons. Andrés, a young man who admits to having been a full-time clandestine cutter in the past, estimates that in 1995 only two trucks a week made it past patrols and roadblocks to Oaxaca Valley markets. In 1994, however, when enforcement was lax, he smuggled five loads a week on his own. His estimates of about twenty trips per week between ten smuggling trucks would result in an illegal cut of 4,500m^3 standing timber. In a hypothetical free-for-all, in which the majority of the community's thirty-five pickup trucks participated, smugglers could conceivably cut more than 20,000m^3 of pine per year (Table 1).

Restraint and Legitimation of Tree Theft

In spite of the profit potential of tree theft, this activity is restrained; a free-for-all does not occur. Many commoners recognize that the pressures to cut and clear could get out of hand and threaten the forest. They see individual appropriation of pine trees for sale as a form of theft against the community, because individuals cut community-owned trees, sell the lumber, but leave nothing of common benefit. There is still substantial support for restrictions on tree theft, especially when they are directed against the

Table 1. The Estimated and Potential Impact of Commercial Logging and Tree Theft on San Martin's Forest

Community enforcement scenario	Basis of clandestine estimate	Clandestine cut pine total volume m^3	Legal cut [a] pine m^3/year
Free-for-all (hypothetical)	30 trucks, 3 loads/wk ea.	20,000	16,000[a]
Lax enforcement (1994)	10 trucks, 2 loads/wk ea.	4,500	8,000
Vigorous enforcement (1995)	2 truckloads per week[b]	450	7,000

Notes:
[a] Maximum sustainable yield is 16,000 m^3 total tree volume according to the forestry management plan, assuming optimal harvesting methods.
[b] Assuming each truck can carry 2.5m^3 of fresh wood, 45 weeks per year are dry enough to allow transit in the forest, and cutters take about 50% of the volume of the trees they fell (Klooster, 1997).

minority of full-time cutters and carriers, who do it not for 'need', but for 'avarice'. A significant number of community members collaborate with communal authorities, informing them when they see or hear evidence of clandestine cutting. Tree theft not only contravenes forest laws, it insults the community's ethic of access. Paradoxically, however, timber smugglers also use the ethic of access to legitimate their activities. Resolving this apparent contradiction requires a better understanding of the social context of community forestry in San Martin.

Corruption and the Imbalance of Power in San Martin

Concessionaires fostered fragmentation of an already-stratified community with a history of feuds and strong individual leaders. They developed an internal constituency in the communities where they had contracts, channelling key jobs like winch operator and bulldozer driver to influential community members. They also fostered truck ownership and cultivated a class of truck owners dependent on forestry for work, transporting logs and acquiring permits to legally commercialize truckloads of firewood. The concessionaires also provided salaries and reimbursed expenses to members of communal government. In the short run, these practices increased the chances of favourable outcomes in the yearly community assemblies where contracts were renewed, since this group was likely to support a status quo on which their livelihood depended. In the long run, such practices reinforced the development of a clique of men accustomed to disproportionate benefits from forestry, equipped with trucks and other capital, knowledge of the forestry business and wood markets, and power in community assemblies.

The establishment of community forestry in 1980 did not change the balance of power between an élite from the central village and the predominantly Mixtec commoners from the community's five outlying settlements, nor did it correct the concentration of forestry benefits. Social investments paid for with forestry profits are concentrated in the central village, including the Catholic church, a cobblestone street, a health clinic, government buildings, and the community-owned sawmill. The outlying settlements, by contrast, consistently see their requests for funds for electrification, schools, roads, and communal pickup trucks rejected. Although forestry generates sporadic earnings for nearly half the working-age men in the community, a select few earn substantially more than average, and this group comes disproportionately from San Martin Ocotlán village.

In 1995, after six years without any profit distribution from the forestry business, commoners solicited assistance from the agrarian agency and the Oaxaca state government to conduct audits of the communal forest business. These revealed loans of money and wood to a group of wealthy commoners who owed the forestry coffers nearly 208,000 pesos (more than

40 per cent of the payroll) but who refused to acknowledge their debts. The overwhelming majority of recent debtors were from the central village, and many owed sums in excess of 10,000 pesos. A sawmill audit uncovered an additional problem with the misclassification of boards, which represented a bonanza for local truckers who could buy cheap and resell at a higher, more expensive, classification.

Dissidents also requested the federal environmental enforcement agency to conduct an audit of logging methods, and this revealed practices similar to the concessionaires that preceded the communal logging business. The forestry élite of San Martin Ocotlán pressured the professional foresters in their employ to direct logging towards sawmill-quality trees at the expense of costly thinning cuts. They failed to clean logging debris, which inhibits regeneration and causes a fire hazard, and although they conducted reforestation, they did not meet the goals of the official management plan. Most community members neither participate in decisions regarding collective use of the forest nor do they share in the jobs and economic benefits which commercial forestry produces. In effect, a forestry élite usurps the community's forest, alienating it from an important sector of the community.

The formal institutions of common property management ought to provide controls against mismanaging the communal forestry enterprise. The basis of local power in San Martin Ocotlán is supposed to be the community assembly. Comprised of mostly male heads of households, the community assembly elects community members to an executive committee that represents the community to outside authorities, selects members to oversee the logging business, appoints a supervisory committee to monitor the executive committee and the logging administrators, and appoints a committee charged with combating tree theft (Figure 1). These communal institutions ought to provide a system of checks and balances to maintain the accountability of communal authorities and forestry administrators, but the forestry élite finds ways to circumvent such checks on its power.

The élite dominates communal institutions using intimidation, rigging elections, avoiding supervision, and discouraging participation in community assemblies. Threats, violence, bribes, and the manipulation of reciprocal obligations are common tools of internal politics. 'Some threaten, others invite you to drink', is how one internal dissident put it. The élite comprise the majority on the Council of Distinguished Men, a traditional body of authority parallel to the general assembly, and this traditional institution circumvents the community assembly in decision making. These weapons of the not-so-weak reproduce the forestry élite's power and privilege, while undermining the democratic potential of the formal institutions of community management.

They also undermine the community basis for restrictions on tree theft, because the same communal authorities who enforce restrictions on clearing and cutting are associated with corruption in the logging and milling business, and this also offends the community's ethic of access. The internal

Figure 1. Political Organization of San Martin Ocotlán

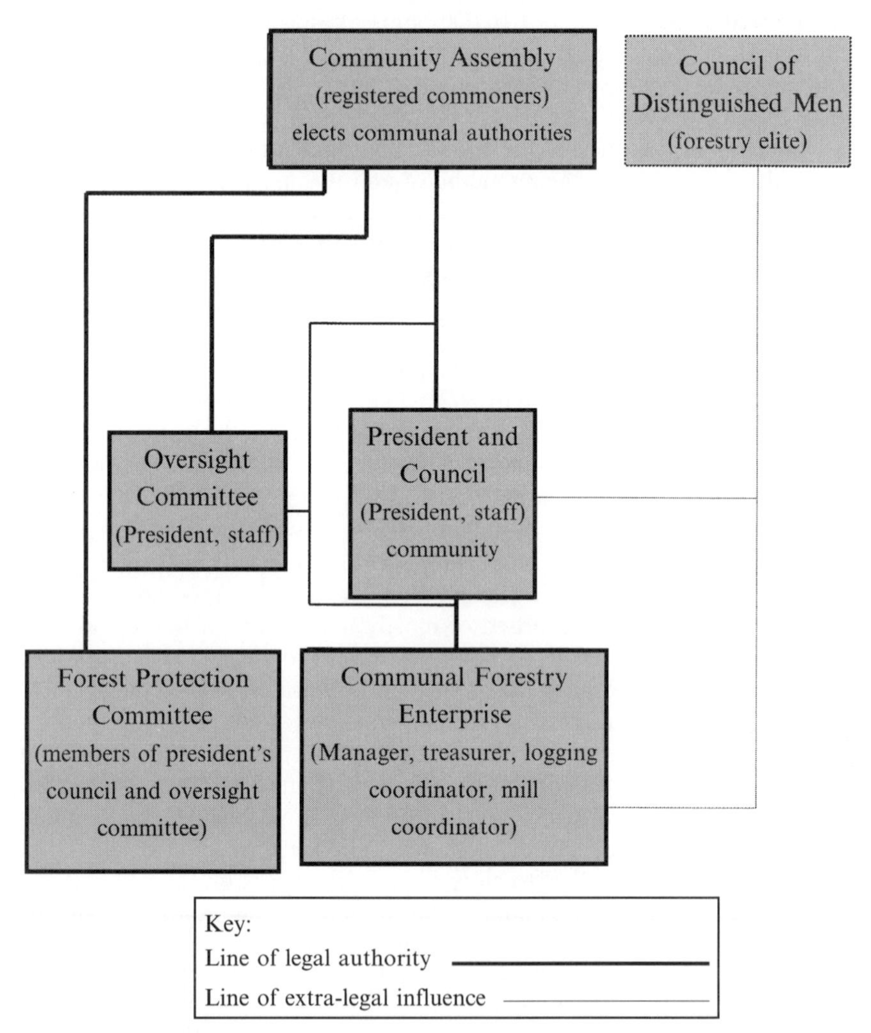

centralization of community forestry, therefore, provides tree thieves with a mantle of legitimacy. They boast of the number of families their activity feeds, in contrast with legal logging. The village authorities who are charged with controlling contraband are also the ones thought to benefit from corruption in the forestry business, so cutters consider the punishments they mete out hypocritical and unjust. In the words of Pedro Diaz, an admitted tree thief and one of the loudest critics of village authorities: 'When we are cutting one tree they want to throw us in jail, so how can they cut 50,000 trees and not produce anything? You can't say "I'm going to cut 50,000 trees, and you, two gunnysacks of charcoal". There are two kinds of tree thieves. Some have licenses but still leave nothing for the community'. Together with perceptions that enforcement against tree theft is unfair, the loan issue has been particularly galling. Constituting a clear re-assignation of forest benefits to an undeserving internal élite, it grated against the ethic of access, fuelling internal resistance. 'Sometimes the rage felt by people deprived of resource access derives not from the denial itself but from the reassignment of access to others whose claims are considered invalid' (Thompson, 1975: 9, quoted in Peluso, 1992: 14).

Contrast with Successful Communities

San Martin has not so far been successful in integrating scientific forestry in a manner congruent with values surrounding the use of forest resources. The experiences of a number of other forest communities in Mexico indicate that this is a possibility (Abardia and Solano, 1995; Alvarez-Icaza, 1993; Bray, 1998; Bray et al., 1993; Chapela, 1999; Jaffee, 1997; Klooster, 1999; Lopez and Gerez, 1993; Merino, 1997; Sanchez, 1995). Despite similar histories of battling concessions, they avoided San Martin's twin problems of internal corruption and tree theft.[15]

Compared to San Martin, community assemblies are more common and better-attended in the more successful communities, and they also have accounting procedures and institutional arrangements which allow them to detect mismanagement of funds and to replace leaders who err. Community assemblies make decisions about employment possibilities and revenue distribution that community members consider fair.

In the implementation of scientific forestry, these communities found ways to respect traditional usages, minimize the costs involved, compensate those whose rights were affected, and enforce restrictions in a graduated way. Some co-ordinate fuelwood gathering with logging activities, for example, while others maintain woodcutting reserve areas. They have internal agreements to forgo cutting pine trees for the traditional roofing shingles, but as compensation for this restriction the forestry business supplies free tin roofing materials and lumber when needed. Community members maintain their rights to take trees from communal forests for house building, but now co-ordinate with village authorities and the forestry business to select their tree. An illustrative example of the integration of customary rights with scientific forestry comes from Nuevo San Juan Parangaricutiro, in Michoacan. All of San Juan's forest lands are parcelled

15. This section summarizes information gathered during site visits to six communities in Oaxaca and one in Michoacan, literature reviews, and interviews with communal authorities. Klooster (1997, 2000) provides more detailed analysis.

out to community members for resin tapping. Although agrarian laws provide no basis for individual forest parcels in a common property forest, leaders establishing the community logging business respect their neighbour's informal usufruct rights to these forest plots. Resin collection continues in these plots in accordance with individual interests, but when the forest management plan indicates it is time for the community business to cut trees, it does so, giving forest plot possessors a stumpage payment as an incentive for having protected community trees. Nuevo San Juan once had serious problems with tree theft carried out by community members supplying the twenty-three local packing crate workshops. The community hired the best cutters, instituted armed patrols to dissuade community members and outsiders from stealing community trees, and started selling sawmill scrap at reduced prices to the family-owned workshops (Klooster, 1997; Sanchez, 1995).

Rights to land and rights to trees held by households and communities overlap in complex ways (Fortmann and Bruce, 1988; Rocheleau and Edmunds, 1997). These communities have discovered a way to navigate those rights, imposing the restrictions of scientific forestry and distributing its benefits in a way congruent to the ethic of access, resonant with the moral economy.

A Struggle for the Forest in San Martin

There is some chance that San Martin will be able to resolve its internal conflicts and establish a resource control system congruent with community values. Under conditions of co-management, the association of community leaders with corruption and inequity does more than legitimate tree theft as resistance among some members of the community: it also generates other kinds of protest. In 1995, the community assembly called for audits and then demanded accountability for the unpaid loans to the élite.

In one of the high points of internal conflict over natural resource management, members of Benito Juárez, an outlying village, put up a chain to stop logging trucks from reaching the forest zone slated for logging that year. Members of the forestry élite mobilized 200 supporters from the central village who descended on Benito Juarez, where thirty men and twenty women quickly took up positions in front of the chain. After a heated discussion failed to resolve matters to their liking, the forestry élite sent in a bulldozer to break the chain and reassert their authority over the forest. To stop them, one dissident man from Benito Juarez lay down in front of the bulldozer, but communal authorities carefully picked him up and moved him gently to one side. Then a dozen women lined up in front of the chain, standing defiant, babies in their arms, the blade of the huge bulldozer towering several feet above their heads. The communal authorities dared not move them and the bulldozer could not proceed.

These events put a stop to logging, but they do not represent a rejection of scientific forestry. Dissidents never called for a permanent ban on logging, or the closure of the forestry business. Rather, their dissent focused on the distribution of the benefits and restrictions of scientific forestry, and on the corruption of the logging business. The struggle over access to resources in San Martin might result in greater internal justice in the distribution of access to the forest. It might allow for the implementation of scientific forestry in a way that does not offend an ethic of access or other aspects of the moral economy. In late 1997, for the first time in twenty years, members of outlying settlements were elected to positions of communal authority. In 1999, however, efforts to resolve conflicts, control the continuing problem of tree theft, and return to logging under more equitable arrangements were still ongoing.

CONCLUSION

It should not be concluded from the San Martin case that community-based forest management strategies are doomed to unresolvable internal conflict, continued tree theft, and other forms of resistance that confound organized forest management. Despite the problems identified here, management in San Martin is better than it was under concessions. Reforestation began only after independence from the concessionaires, but now continues with broad community participation. Controls on agricultural clearing on hillslopes are mostly honoured, and both moral restraints and a community-level enforcement apparatus partly constrain tree theft. Despite carrying the baggage of resistance against scientific forestry under concessions, communities are sometimes able to integrate forestry with common property management systems and become complicit in conservation. Community forestry in Mexico can be successful at implementing forestry in an internally-legitimate way and generating rural development benefits including employment and revenues for social infrastructure like roads and health clinics (Abardia and Solano, 1995; Alvarez-Icaza, 1993; Bray, 1998; Bray et al., 1993; Chapela, 1999; Jaffee, 1997; Lopez and Gerez, 1993; Merino, 1997; Sanchez, 1995). It deserves greater support from Mexican policy makers and the international community (Klooster, 1999).

Creating spaces in which to reduce conflict over forest resource management requires a better understanding of how communities of resource users come to condone or condemn rule-breaking. Research to inform community-based management strategies must go beyond consideration of how communities come to create rules (Agrawal and Gibson, 1999; Ostrom, 1990). It should also address how communities come to integrate those rules into local concepts of just and legitimate natural resource use (Klooster, 2000; Ostrom, 1998). The concepts of resistance, moral economy, and the ethic of access are useful in this task, and in the case of San Martin, they help explain a relationship between resistance (tree theft), internal exclusion, and a concentration of forest wealth that offends the community's ethic of access and insults its moral economy.

It also calls into question excessively romantic readings of peasant resistance under scientific forestry. The relationship between scientific forestry and clandestinity in San Martin is not a simple story of scientific forestry criminalizing traditional forest activities and converting them into acts of resistance. Tree theft does not reaffirm some traditional cultural system of harmonious relations with nature, as Guha (1989) implies for resistance in the Garwhal Himalayas (see Rangan, 1996). Rather, it is a reflection of already-transformed production strategies. Tree theft in San Martin is a new activity. It evolved out of customary usages amidst a regulatory regime of scientific forestry, and in the context of regional economic change. More than being created through official definition (Marx, 1975), tree theft, in this case, is a by-product of scientific forestry. It evolved not just as opposition to an alien system of resource control, but also in relation to internal politics and processes of community change (Ortner, 1995; Peluso, 1992). In San Martin, these include the concentration of power and the development of an internal élite that marginalized the majority of community members from participating in legal forestry. The problem behind resource management conflicts is not scientific forestry, but injustice and a lack of democratic participation in forestry, which is why this study reaffirms calls for more attention to the implications of social equity and inclusiveness in community forestry policies (Agarwal, 1998; Andersen, 1995; Ribot, 1996; Sarin, 1996).

Furthermore, the case of San Martin identifies a number of crucial roles for the state in co-management schemes. Interventions from state agencies were important, for example, in establishing community forestry and constructing the framework of co-management.[16] Within this system, the state plays a vital backup role to support community authority against repeat, and sometimes violent, offenders of restrictions on forest access. But the state's role goes far beyond coercion in the imposition of co-management. Community dissidents, motivated by the lack of congruence between co-management and the ethic of access, called on various federal and state agencies in their attempts to hold leadership accountable for forest mismanagement and the misdirection of forest wealth.

Most importantly, however, the case of San Martin suggests that co-management offers an arena where protest can cross over from covert and individual resistance like tree theft, to the kinds of organized collective resistance which might lead to social change and a fairer distribution of resources (Agarwal, 1994). The forms of protest engendered by community management in San Martin are much more likely to integrate scientific

16. The state should also play a stronger role in providing training in forest management, business administration, and making capital available for investments in forestry so that forest communities can generate greater wealth from forestry and distribute it with greater clarity (Klooster, 1999).

forestry into locally-acceptable resource control systems than the concession system, or other forms of centralized bureaucratic control.

Acknowledgements

Fulbright-Hayes, the Inter-American Foundation, and UCLA provided support for Mexican fieldwork. This work would not have been possible without the co-operation of authorities and villagers of San Martin Ocotlán. Thanks also to the Princeton Environmental Institute, the Science Technology and Environmental Policy Program at Princeton University, and William and Jane Fortune for providing support during the writing of this article. Peter Vandergeest and three anonymous reviewers made insightful comments on an earlier draft.

REFERENCES

Abardia Moros, F. and C. Solano Solano (1995) 'Forestry Communities in Oaxaca: The Struggle for Free Market Access', in N. Forster (ed.) *Case Studies of Community-Based Forestry Enterprises in the Americas: Presented at the Symposium 'Forestry in the Americas: Community-Based Management and Sustainability' University of Wisconsin-Madison, February 3–4, 1995*, pp. 99–120. Madison, WI: University of Wisconsin Madison Land Tenure Center and Institute for Environmental Studies.

Agarwal, B. (1994) 'Gender, Resistance and Land: Interlinked Struggles Over Resources and Meaning in South Asia', *Journal of Peasant Studies* 22(1): 81–125.

Agarwal, B. (1998) 'Environmental Management, Equity, and Ecofeminism: Debating India's Experience', *Journal of Peasant Studies* 25(4): 55–95.

Agrawal, A. and C. C. Gibson (1999) 'Enchantment and Disenchantment: The Role of Community in Natural Resource Conservation', *World Development* 27(4): 629–49.

Alcorn, J. B. (1981) 'Huastec Noncrop Resource Management: Implications for Prehistoric Rain Forest Management', *Human Ecology* 9(4): 395–417.

Alvarez-Icaza, P. (1993) 'Forestry As a Social Enterprise', *Cultural Survival Quarterly* 17(1): 45–47.

Andersen, K. E. (1995) 'Institutional Flaws of Collective Forest Management', *Ambio* 24(6): 349–53.

Ascher, W. (1995) *Communities and Sustainable Forestry in Developing Countries*. San Francisco, CA: Institute for Contemporary Studies.

Baker, W. L. (1992) 'The Landscape Ecology of Large Disturbances in the Design and Management of Nature Reserves', *Landscape Ecology* 7(3): 181–94.

Banuri, T. and F. Apfell Marglin (eds) (1993) *Who Will Save the Forests? Knowledge, Power and Environmental Destruction*. London: Zed Books.

Bebbington, A. (1996) 'Movements, Modernizations, and Markets: Indigenous Organizations and Agrarian Strategies in Ecuador', in R. Peet and M. Watts (eds) *Liberation Ecologies: Environment, Development, Social Movements*, pp. 86–109. London and New York: Routledge.

Beltran, E. (1964) *La Batalla Forestal: Lo Hecho, Lo No Hecho, Lo Por Hacer*. Mexico City: Editorial Cultura.

Berkes, F. (1997) 'New and Not-So-New Directions in the Use of the Commons: Co-Management', *Common Property Resource Digest* 42: 5–7.

Bogati, Andres G. (1978) 'Algunos Aspectos Historicos Del Recurso Forestal Desde La Conquesta Hasta El Principio De La Legislacion Moderna', *Bosques y Fauna* 1: 19–30.

Bray, D. B. (1998) 'Mexican Community Forestry and Its Global Significance'. Unpublished manuscript. Miami: Florida International University.

Bray, D. B. and M.B. Wexler (1996) 'Forest Policies in Mexico', in L. Randall (ed.) *Changing Structures of Mexico: Political, Social and Economic Prospects*, pp. 217–28. Armonk, NY: M. E. Sharpe.

Bray, D. B., M. Carreon, L. Merino and V. Santos (1993) 'On the Road to Sustainable Forestry', *Cultural Survival Quarterly* 17(1): 38–41.
Brosius, J. P., A. L. Tsing and C. Zerner (1998) 'Representing Communities: Histories and Politics of Community-Based Natural Resource Management', *Society and Natural Resources* 11(2): 157–68.
Bryant, R. L. (1994) 'Shifting the Cultivator: The Politics of Teak Regeneration in Colonial Burma', *Modern Asian Studies* 28(2): 225–50.
Cairns, M. A., R. Dirzo and F. Zadroga (1995) 'Forests of Mexico: a Diminishing Resource', *Journal of Forestry* 93(7): 21–4.
Calva Tellez, J. L., F. Paz Gonzalez, O. Wicab Gutierrez and J. Camas Reyes (eds) (1989) *Economia Politica De La Explotacion Forestal En Mexico: Bibliografia Comentada 1930–1984*. Mexico City: Universidad Autonoma Chapingo, Universidad Nacional Autonoma de Mexico.
Castilleja, G. (1993) 'Changing Trends in Forest Policy in Latin America: Chile, Nicaragua and Mexico', *Unasylva* 44(175): 29–35.
Chambille, K. (1983) *Atenquique: Los Bosques Del Sur De Jalisco*. Mexico: Instituto de Investigaciones Economicas-UNAM.
Chapela, F. (1999) *Silvicultura Comunitaria En La Sierra Norte De Oaxaca: El Caso De La Union Zapoteco-Chinanteca*. Mexico City: Red de Gestion de Recursos Naturales y Fundacion Rockefeller.
Escarpita Herrera, J. L. (1977) *Proyecto General De Ordenacion: Estudio Forestal Fotogrametrico*. Unidad de Ordenacion Forestal 'Silvicola Magdelena, S. de R.L.' Filial de Fabricas de Papel Tuxtepec, S.A., Mexico.
Fairhead, J. and M. Leach (1996) *Misreading the African Landscape*. Cambridge: Cambridge University Press.
Fortmann, L. and J. W. Bruce (1988) *Whose Trees? Proprietary Dimensions of Forestry*. Boulder, CO: Westview Press.
Fox, J. (1993) *The Politics of Food in Mexico: State Power and Social Mobilization*. Ithaca, NY: Cornell University Press.
Gonzalez Pacheco, C. (1981) 'La Explotacion Forestal en Mexico', *Revista Del Mexico Agrario* XIV(4): 11–40.
Guerrero, G. (1988) 'Estado, Madera y Capital', *Ecologia Politica/Cultura* 2(4): 5–13.
Guha, R. (1989) *The Unquiet Woods: Ecological Change and Peasant Resistance in the Himalaya*. Berkeley, CA: University of California Press.
Hinojosa Ortiz, M. (1958) 'Los Bosques De Mexico: Relato De Un Despilfarro y Una Injusticia'. Mexico City: Instituto Mexicano de Investigaciones Economicas.
Jaffee, D. (1997) 'Restoration Where People Matter: Reversing Forest Degradation in Michoacan, Mexico', *Restoration and Management Notes* 15(2): 147–55.
Johnson, N. and B. Cabarle (1993) *Surviving the Cut: Natural Forest Management in the Humid Tropics*. Washington, DC: World Resources Institute.
Klooster, D. J. (1997) 'Conflict in the Commons: Conservation and Commercial Forestry in Mexican Indigenous Communities'. Dissertation. University of California, Los Angeles, Department of Geography.
Klooster, D. J. (1999) 'Community-Based Conservation in Mexico: Can it Reverse Processes of Degradation?', *Land Degradation and Development* 10(4): 363–79.
Klooster, D. J. (2000) 'Institutional Choice, Community, and Struggle: A Case Study of Forest Co-Management in Mexico', *World Development* 28(1): in press.
Klooster, D. J. (forthcoming) 'Forest Conservation and Degradation in a Sub-Subsistence Agricultural System: Community and Forestry in Mexico', in C.B. Flora (ed.) *Agroecosystems and Human Communities*. New York: CRC Press.
Lopez Arzola, R. and P. Gerez Fernandez (1993) 'The Permanent Tension', *Cultural Survival Quarterly* 17(1): 42–44.
Marx, K. (1975) 'Debates on the Law on Thefts of Wood', in *Collected Works Karl Marx and Frederick Engels*. London: Lawrence and Wishart; Moscow: Progress Publishers.

Masera, O. R. (1996) 'Desforestacion y Degradacion Forestal en Mexico'. Documentos De Trabajo No 19. Patzcuaro, Michoacan: Grupo Interdisciplinario de Tecnologia Rural Apropiada (GIRA) AC.

Masera, O. R., D. Masera and J. Navia (1998) *Dinamica y Uso De Los Recursos Forestales En La Region Purepecha: El Papel De Las Pequenas Empresas Artesenales*. Patzcuaro, Michoacan: Grupo Interdisciplinario de Tecnologia Rural Apropiada (GIRA) AC.

McCay, B. J. and S. Jentoft (1996) 'From the Bottom Up: Participatory Issues in Fisheries Management', *Society and Natural Resources* 9: 237–50.

Mendoza Medina, R. (1976) 'La Politica Forestal En El Sector Ejidal y Comunal', *Revista Del Mexico Agrario* 9(2): 35–70.

Merino, L. (ed.) (1997) *El Manejo Forestal Comunitario En Mexico y Sus Perspectivas De Sustentabilidad*. Cuernavaca: Centro Regional de Investigaciones Multidisciplinarias, UNAM.

Moguel Santaella, E. (1994) 'Legislación Forestal Mexicana: De Tenochtitlan a Nuestros Dias', in J. C. Ayala Sosa and C. F. Romahn de la Vega (eds) *Reunion de Analisis de la Iniciativa de Ley Forestal 1992: Memoria*, pp. 1–24. Chapingo, Mexico: Division de Ciencias Forestales Universidad Autonoma Chapingo.

Noss, R. F. (1993) 'Sustainable Forestry or Sustainable Forests?', in G. H. Aplet, N. Johnson, J. T. Olson, and A. V. Sample (eds) *Defining Sustainable Forestry*, pp. 17–43. Washington, DC: The Wilderness Society.

Ortner, S. B. (1995) 'Resistance and the Problem of Ethnographic Refusal', *Comparative Studies in Society and History* 37(1): 173–93.

Ostrom, E. (1990) *Governing the Commons: the Evolution of Institutions for Collective Action*. New York: Cambridge University Press.

Ostrom, E. (1998) 'A Behavioral Approach to the Rational Choice Theory of Collective Action', *American Political Science Review* 92(1): 1–22.

Otero, G. (1989) 'Agrarian Reform in Mexico: Capitalism and the State', in W. C. Thiesenhusen (ed.) *Searching for Agrarian Reform in Latin America*, pp. 276–304. Boston, MA: Unwin Hyman.

Padoch, C. and C. Peters (1993) 'Managed Forest Gardens in West Kalimantan, Indonesia', in S. Potter, J. I. Cohen and D. Janczewski (eds) *Perspectives on Biodiversity: Case Studies of Genetic Resource Conservation and Development*, pp. 167–76. Washington, DC: AAAS Press.

Peluso, N. L. (1992) *Rich Forests, Poor People: Resource Control and Resistance in Java*. Berkeley, CA: University of California Press.

Peluso, N. L. (1993) 'Coercing Conservation? The Politics of State Resource Control', *Global Environmental Change* 1993: 199–217.

Peluso, N. L. (1996) 'Fruit Trees and Family Trees in an Anthropogenic Forest: Ethics of Access, Property Zones, and Environmental Change in Indonesia', *Comparative Studies in Society and History* 38: 510–48.

Pendleton, M. R. (1998) 'Taking the Forest: The Shared Meaning of Tree Theft', *Society and Natural Resources* 11: 39–50.

Pincetl, S. (1993) 'Some Origins of French Environmentalism: An Exploration', *Forest and Conservation History* 37(1): 80–9.

Pinkerton, E. (ed.) (1989) *Co-operative Management of Local Fisheries*. Vancouver: University of British Columbia Press.

Poffenberger, M. (1994) 'The Resurgence of Community Forest Management in Eastern India', in D. Western, R. M. Wright and S. Strum (eds) *Natural Connections: Perspectives in Community-Based Conservation*, pp. 53–79. Washington, DC: Island Press.

Rangan, H. (1996) 'From Chipko to Uttaranchal: Development, Environment, and Social Protest in the Garwhal Himalayas, India', in R. Peet and M. Watts (eds) *Liberation Ecologies: Environment, Development, Social Movements*, pp. 205–66. London and New York: Routledge.

Ribot, J. C. (1996) 'Participation Without Representation', *Cultural Survival Quarterly* 20(3): 40–4.

Rocheleau, D. and D. Edmunds (1997) 'Women, Men and Trees: Gender, Power and Property in Forest and Agrarian Landscapes', *World Development* 25(8): 1351–71.

Romm, J. (1993) 'Sustainable Forestry, an Adaptive Social Process', in H. Aplet, N. Johnson, Jeffrey T. Olson and A. Sample (eds) *Defining Sustainable Forestry*, pp. 280–93. Washington, DC: The Wilderness Society and Island Press.

Sanchez Pego, M. A. (1995) 'The Forestry Enterprise of the Indigenous Community of Nuevo San Juan Parangaricutiro, Michoacan, Mexico', in N. Forster (ed.) *Case Studies of Community-Based Forestry Enterprises in the Americas: Presented at the Symposium 'Forestry in the Americas: Community-Based Management and Sustainability' University of Wisconsin-Madison, February 3–4, 1995*, pp. 137–60. Madison, WI: University of Wisconsin Madison Land Tenure Center and Institute for Environmental Studies.

Sanderson, S. R. W. (1984) *Land Reform in Mexico: 1910–1980*. Orlando, FL: Academic Press.

Sarin, M. (1996) 'From Conflict to Collaboration: Institutional Issues in Community Management', in M. Poffenberger and B. McGean (eds) *Village Voices, Forest Choices: Joint Forest Management in India*, pp. 167–209. Delhi: Oxford University Press.

Scott, J. C. (1976) *The Moral Economy of the Peasant*. New Haven, CT: Yale University Press.

Scott, J. C. (1985) *Weapons of the Weak: Everyday Forms of Peasant Resistance*. New Haven, CT: Yale University Press.

Scott, J. C. (1998) *Seeing Like a State: How Certain Schemes to Improve the Human Condition Have Failed*. New Haven, CT: Yale University Press.

Simonian, L. (1995) 'Defending the Land of the Jaguar: A History of Conservation in Mexico'. Austin, TX: University of Texas Press.

Styles, B. T. (1993) 'Genus *Pinus*: A Mexican Purview', in T. P. Ramamoorthy, R. Bye, A. Lot and J. Fa (eds) *Biological Diversity of Mexico: Origins and Distribution*, pp. 397–420. New York: Oxford University Press.

Thompson, E. P. (1975) *Whigs and Hunters: The Origins of the Black Act*. London: Allen Lane.

Thompson, E. P. (1991) *Customs in Common*. New York: The New Press.

Vandergeest, P. (1996) 'Mapping Nature: Territorialization of Forest Rights in Thailand', *Society and Natural Resources* 9: 159–75.

Western, D. and R. M. Wright (1994) 'The Background to Community-Based Conservation', in D. Western, R. M. Wright, and S. C. Strum (eds) *Natural Connections: Perspectives in Community-Based Conservation*, pp. 1–12. Washington, DC: Island Press.

World Bank (1995) 'Mexico Resource Conservation and Forest Sector Review'. Report No. 13114–ME, Natural Resources and Rural Poverty Operations Division Country Department II Latin America and the Caribbean Regional Office. Washington, DC: The World Bank.

World Bank (1996) *The World Bank Participation Sourcebook*. Environmentally Sustainable Development Publications. Washington, DC: The World Bank.

World Bank (1997) *World Development Report 1997: The State in a Changing World*. New York: Oxford University Press.

Zabin, C. (1989) 'Grassroots Development in Indigenous Communities: A Case Study from the Sierra Juarez in Oaxaca, Mexico'. Berkeley, CA: University of California at Berkeley.

Zimmerer, K. S. (1994) 'Human Geography and the "New Ecology": The Prospect and Promise of Integration', *Annals of the Association of American Geographers* 84 (1): 108–25.

Zimmerer, K. S. and K. R. Young (1998) *Nature's Geography: New Lessons for Conservation in Developing Countries*. Madison, WI: University of Wisconsin Press.

13 Remote Sensibilities: Discourses of Technology and the Making of Indonesia's Natural Disaster

Emily E. Harwell

'All I have to say about the fire crisis is — Alhamdulillah (Thanks be to God). Because now everyone understands that we need GIS.'[1]

Indonesian official from the Agency of Forest Protection and Nature Conservation (PHPA) at a 'Discussion on the Impacts of Forest Fires' in Jakarta, organized by an environmental NGO consortium.

'In my grandfather's time, there was a drought that lasted seven years ... the rice harvest failed then too, but we ate honey mixed with the boiled bark of the bangkris tree. There was game to hunt and tubers to dig ... But this time there is only smoke and wind and ash left to us. And we are just waiting to die.'

East Kalimantan Benuaq Dayak elder, commenting on the impacts of the widespread 1997–8 forest fires on his community.

The Indonesian forest fires of 1997–8 were an unprecedented disaster. The fires' novelty, however, lay not in the occurrence of drought or fire in the Indonesian rainforest, but in the way that the resulting disaster captured the imagination of people all over the world. Images of fires and choking smoke appeared in news stories world-wide. Web sites multiplied. An entire industry of fire analysis projects sponsored by international donor agencies sprang up around the disaster.

What was it that caught the world's eye? Was it evocative images of loss — precious rainforests, inhabited by exotic and endangered species, consumed by fire? Or was it the God's Eye view of satellite photos, white shrouds of smoke hanging over the red and black island of Borneo? Was it the futility of local people swatting at the flames with brooms? Or the imagined face of greed and corruption evoked by the crony capitalism of the world's second longest reigning dictator? Eyes from many different places turned to Indonesia to assess this disaster, but they saw very different things. This article demonstrates that what is seen is a product of interpretation,

1. Geographical Information Systems — computerized systems of display and analysis of geographically referenced information.

and that the interpretations chosen by different actors reveal (and support) different relations of power in society.

Certainly the two speakers in the epigraph reflect such differently positioned discourses. In addition to differing interests and experiences of disaster, their remarks reflect different scales of analysis; one remote, the other proximate in location to the fires. Yet, more is revealed than a simple difference in scale. The speakers' discourses not only describe, but emerge from and rhetorically reproduce, two different types of disaster — a difference with serious consequences for the futures of both forests and forest dwellers. These divergent interpretations of disaster arise from different forms of representation, of marshalling and interpreting evidence, of asking some questions and not others. This is not to suggest that there is no *real* fire outside of anyone's perception, but rather to draw attention to the diverse ways in which fires, as physical occurrences, are translated into 'disasters'. These translations unfold neither randomly nor in a predetermined fashion, but are the contingent products of particular political-economic and social moments.

In 1997–8, a convergence of these moments placed GIS at the centre of these translations of fires into discourses of disaster. Analysis using remotely sensed[2] images played a primary role in every government, donor agency and non-governmental organization's fire analysis project, however diverse their ideological underpinnings, with the notable exception of local farmers. Yet even identical images[3] produced wildly different interpretations (see Table 1). In October 1997, estimates of total area burned ranged from 96,000 ha (Ministry of Forestry) to 300,000 ha (Ministry of Environment) to 1.7 million ha (WALHI, an environmental NGO).[4] Clearly, the data did not speak for themselves, but were made to speak, and to say very different things. Nor did the diverse analyses emerge in isolation, but in response to other discourses in competition for the real facts and different versions of proof. Competing analyses, therefore, largely served to *increase* uncertainty about the extent of damages, rather than decrease it. To the extent that particular agendas are served by particular analyses, the uncertainty in facts will not necessarily be banished by more research, but can serve as useful windows onto the interactions of competing interests.

While disagreement in interpretation is quite ordinary, what was extraordinary about the 1997–8 fires was that technology had become the *lingua franca* in which competing discourses of disaster addressed each other. In fact, the wide access to these satellite images and increased capacity to

---

2. 'Remote sensing' is the general term to designate a variety of aerial and satellite images.
3. These data were derived from 'real time' NOAA web sites, updated with each satellite orbit.
4. These estimates were based only on damages from April to September 1997, seven months before fires were finally extinguished by the return of monsoon rains. See Table 1 for other estimates.

Table 1. Estimates of Area Burned

Organization	Affiliation	Estimated area burned (ha.)	Time frame	Area covered
Ministry of Forestry[1]	Government (National)	96,000	Jul 97–Oct 97	Indonesia
Directorate General of Forest Protection (PHPA)[2]	Government (National)	263,991	Jul 97–Dec 97	Indonesia
Ministry of Environment[3]	Government (National)	330,772	Jul 97–Oct 97	Indonesia
East Kalimantan Environmental Impact Agency (BAPEDAL-DA)[4]	Government (Provincial)	442,634	Jan 98–Apr 98	East Kalimantan
Indonesian Forum for the Environment (WALHI)[5]	NGO (National)	1,714,000	Jul 97–Oct 97	Indonesia
Forest Fire Prevention and Control Project[6]	Donor (EU)	2,300,000	Jul 97–Oct 97	South Sumatra
Integrated Forest Fire Management Programme[7]	Donor (Germany)	4,500,000	Jul 97–May 98	East Kalimantan
Center for Remote Imaging, Sensing and Processing (CRISP), National Univ. of Singapore[8]	Academic	8,170,000	Jul 97–May 98	Kalimantan and Sumatra

Sources:
1. *Jakarta Post* (7/10/97)
2. DitJen PHPA (1998)
3. MLH (1997)
4. BAPEDAL-DA Kaltim (1998)
5. WALHI (1997)
6. Ramon and Wall (1998)
7. Siegert and Hoffmann (1998)
8. Liew (pers. comm.)

analyse them was the principal thing that set these fires apart from others in Indonesia's history. This technology allowed NGOs to pinpoint the location of the vast majority of fire within designated state forest, putting an abrupt end to the government's standard strategy of blaming subsistence farmers for fires and smoke that have repeatedly plagued Indonesia and its neighbours. What are the implications of this radical, anti-hegemonic counter-appropriation of GIS technology — a technology frequently characterized in social critiques as the handmaiden of surveillant military states? Does it herald a new dawn of recolonization of science for liberation, of the sort that Harding (1998), Haraway (1991) and other advocates of feminist, multicultural science have envisioned?

This article describes ambiguous intertextual conversations between a variety of speakers — government officials, activists, development workers, and farmers — and the nature of the disaster (and the solution) asserted by each. It examines discursive shifts (and their consequences) produced by the new emphasis on GIS technologies, as well as investigating discursive elements that persist regardless of these reconfigurations. The complexity described here argues for a nuanced view of so-called interest groups, their motivations, and methods of inquiry. This fine-grained analysis not only produces a richer understanding of social and political-economic causes underlying the fires, but in exposing the flexibilities inherent in discourses of technology, suggests new avenues toward more democratic resource use.

This analysis is concerned with scale, but more than that, it interrogates the consequences of analysing scenes at particular scales. It begins in Indonesia, by briefly outlining the political ecological background of the 1997–8 fires, including the history of forest policy, and the oil palm sector in particular. It shows how GIS technology played a novel and central role in the reaction to the fires, both domestically and internationally, and suggests that what is missing from current scholarly accounts of the fires, and from remote assessments in general, is a textured understanding of social landscapes and the role they play in creating fire hazards.

In addressing this need for fine detail, a subsequent section of this article outlines an ethnography of disaster, tracing how government officials, donor agencies, environmental activists and farmers represent the fires and their causes, before returning to a theoretical focus on GIS as a tool of representation. Finally, much has changed in Indonesia since the last fire season, and the article closes with a brief epilogue considering the role of the fire disaster in the dramatic changes in the Indonesian political landscape in 1998, musing on the possibilities for forests and forest-dependent people in an age of reform.

THE POLITICAL ECOLOGY OF FOREST DISASTER IN KALIMANTAN

Beginning in July 1997 and coinciding with the extended El Niño drought, fires tore through Indonesia's tinder dry forests, particularly devastating the islands of Kalimantan (Indonesian Borneo) and Sumatra. Smoke from the fires reached Thailand, the Philippines, Singapore and Malaysia, whose vociferous complaints forced Indonesia's President Suharto to publicly apologize to his neighbours (although pointedly not to Indonesia's own citizens, who were far more affected by the smoke). This apology was a novel event in the non-interventionist ASEAN (Association of Southeast Asian Nations) diplomatic community, but it was followed up with little effective action to either put out the fires or prevent more from starting. In most provinces, the fires were finally extinguished by the return of the rains in late November 1997. In East Kalimantan, however, the drought dragged torturously on until May 1998.

The impacts of the fires are still being calculated and debated, and we can expect little agreement in the estimates. While the amount and type of destruction depends on the observer's position, however, all would agree that the fires were indeed catastrophic. Forests and wildlife were decimated. An estimated 12.5 million people were exposed to hazardous pollution levels (UNDP, 1998), the effects of which may prove chronic. Standardized API (Air Pollution Index) levels were recorded in Sarawak (Malaysian Borneo) at 849 in September 1997, and an astonishing 1000 in East Kalimantan in April 1998 (Schweithelm, 1998).[5] Acute medical impacts included a range of respiratory complications, and even deaths (527 reported; UNDP, 1998). The long-term impacts on human health may never be known.

Economic costs were also devastating. Businesses, schools and airports closed, and in Sarawak a state of emergency was declared, with residents advised to pour water over their balconies to dispel the smoke. Rural livelihoods were crippled as fires wiped out orchards and agricultural crops, although these losses are yet to be systematically calculated. Conservative estimates of macro-economic losses for Indonesia alone range from US$1.1 bn (UNDP, 1998) to US$4.4 bn (EEPSEA/WWF, 1998).[6] Even the actual area of damage is still a very contentious issue, but preliminary satellite analysis has produced estimates as high as 8.17 million ha burned in Sumatra and Kalimantan during 1997–8 (Soo Chin Liew/CRISP, pers. comm.).

Although damages were severe, the emphasis on damage in fire analysis portrays the disaster as an acute, discrete event, obscuring long-term processes that set the stage for fires to spread quickly across an already devastated social and physical landscape. Some might argue that an unrecognized forest disaster had already begun long before the arrival of El Niño.

Fire Hazards: Palm Oil and Social Landscapes

Indonesia is home to the second largest expanse of tropical rainforest in the world. These forests are environments not of monotonous constancy, but of flux. In some parts of Kalimantan, forests receive seasonal rainfall of 2000–5000 mm annually,[7] but typically also experience one or two

5. API of above 100 is considered unhealthy and above 300 hazardous.
6. The exchange rates used in the estimates varied dramatically because the economic collapse caused the free fall of the rupiah, Indonesia's currency. The WWF calculation was Rp2500/US$1, the UNDP report used a standard Rp5000/US$1. It is also worth noting that preliminary WWF estimates put the costs of the fires, including tourism losses, health effects, legal costs and lost productivity, at US$ 20 bn (WWF, 1997: 12).
7. In the extremely wet areas of the upper Kapuas River basin, annual rainfall is sometimes as high as 8000 mm in association with La Niña climate shifts. However, rainfall patterns vary throughout the island — East Kalimantan is the driest with annual rainfall around 2000 mm in the coastal regions (King, 1996).

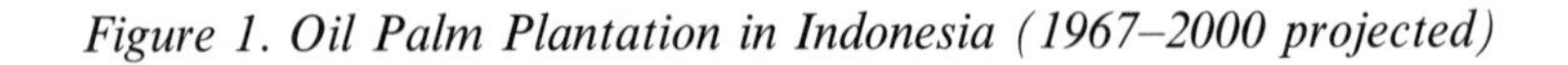

Figure 1. Oil Palm Plantation in Indonesia (1967–2000 projected)

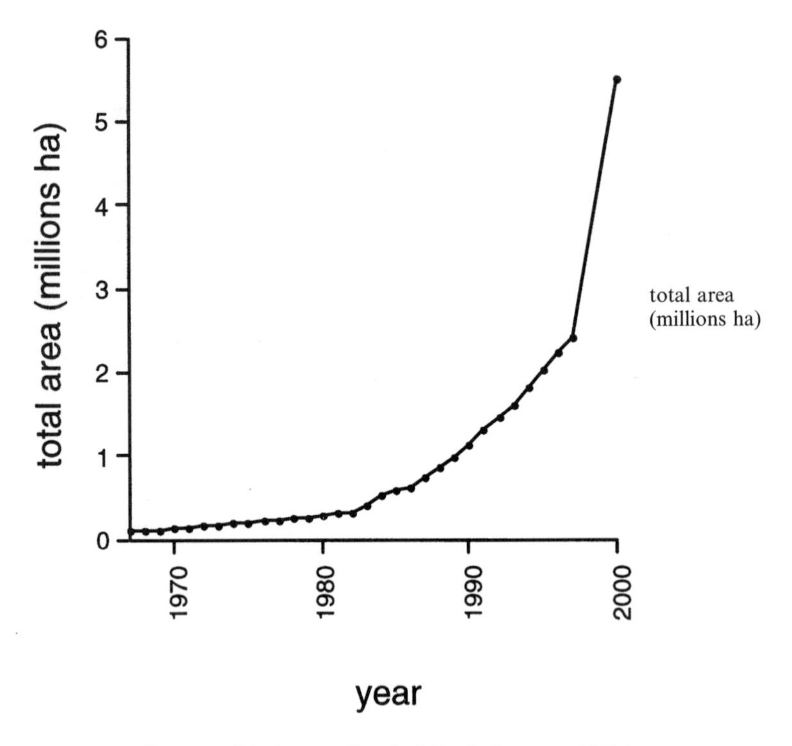

dry seasons per year, some up to 3 months long (MacKinnon et al., 1996). Further, climatic fluxes associated with El Niño Southern Oscillation (ENSO) periodically cause prolonged drought followed by extreme flooding. Local residents report that every two to three years, the expansive lakes on the Upper Kapuas and Lower Mahakam Rivers of Kalimantan become so dry that residents ride bicycles across the dry lake beds.

The role of ENSO climate fluctuations in establishing the conditions for widespread fires is undeniable. Without extended dry periods, moist tropical forest simply will not burn.[8] However, drought is nothing extraordinary, even in the so-called 'ever wet' tropical rainforest. In fact, local swidden agriculturalists have not only survived these extended droughts, their live-

---

8. See Brookfield and Byron (1993), King (1996) and Salafsky (1994) for more on the association between ENSO events and fire in Borneo.

lihoods have long taken advantage of these dry periods to burn off cleared vegetation from dry rice fields.[9] The diversified economy of subsistence and commercial crops, involving both agricultural and forest products, has historically sustained the indigenous Dayaks through periodic economic and climatic oscillation.

However, the perceived mobility, low technology and low returns per unit land of this form of agriculture have long been demonized by the Indonesian government as primitive and in need of modernization. Since the 1960s Suharto's New Order economic development initiatives aggressively promoted capitalization of large-scale 'modern' natural resource production. Paradoxically, this rush to forest exploitation (and the weak law enforcement engendered by its characteristic crony capitalism) allowed plantation owners to utilize on a colossal scale the same 'primitive' method of fire as a cheap and efficient means of clearing land in order to establish their oil palm plantations.[10] That this irony is recognized by neither the government nor the international financial institutions that serve as financial partners is testimony to the power of discursive practice in manipulating both landscapes and futures. These state policies have, therefore, produced not only the physical but also the political economic conditions for fires.

Other scholars have detailed the increasing role of Indonesian forest products as a nationalized commodity, and forest territory as a form of state control over national space, resources and populations.[11] Gellert (1998) shows how successive stages of Indonesian forest (mis)management

---

9. Space does not permit an in-depth examination of the dynamics of Dayak swidden agricultural practice. The reader is referred to the works of Colfer and Dudley (1993), Dove (1985), Geddes (1954), Padoch (1982, 1994) and Vayda et al. (1980) for further reading on this subject.
10. The oil from the pressed fruit of the African oil palm (*Elaeis guineensis*) is used as a cooking oil, in margarine, and as an additive in soap, cosmetics and other products.
11. Recent publications of note are Barber (1998), Barr (1998), Dauvergne (1994), McCarthy (this volume), Peluso (1994), Peluso and Vandergeest (forthcoming), Potter and Lee (1998), Sunderlin and Resosudarmo (1996), who all give excellent background to the recent political economy of forest management in Indonesia. Barber (1998), in particular, shows how the synergistic effects of increased value of forest products, forest degradation and inequitable distribution of forest resources have all led to a state of forest scarcity that has produced dangerous levels of social conflict. For further study, the reader is referred to Peluso et al.'s (1995) exhaustive review of the scholarly literature on Southeast Asian forestry since the 1950s. However, with the exception of Gellert (1998), the most recent fires have still not been widely examined in scholarly publications. There has been an explosion of attention in the popular press (notably, economic popular journals) and 'grey' literature. In addition, there have been several in-depth analyses of the 1982 'Great Fires of Borneo' (including hauntingly prophetic warnings), including Adriawan and Moniaga (1986), Brookfield and Byron (1993), King (1996), Leighton and Wirawan (1986), Mayer (1996) and Salafsky (1994).

— including the development of the plywood industry, pulp and paper industry, and most recently, the rapid expansion of the oil palm plantation sector — increasingly compromised the ecological integrity of the forest and consolidated both business interests and state control over forest territory. These processes of increasing commodification and territorialization of Indonesia's forests contributed to the outbreak of fires and the lack of political will to extinguish or to enforce the 'Zero Burn' legislation enacted in 1995 prohibiting the use of fire in commercial land clearing.[12]

Although not new to Indonesia, the rate of expansion in the oil palm plantation sector during the past decade is indeed remarkable (see Figure 1). Indonesian exports of palm oil in 1997 were over US$1 bn, compared to US$100 m in 1986 (DirJenBun, 1997). Indonesian government plans call for Indonesia to be the largest palm oil producer in the world by 2005 and by the year 2000 to have converted 5.5 million ha to oil palm alone (Potter and Lee, 1998).[13] This area would be twice the 1997 total, ten times the 1985 total (DirJenBun, 1997).[14]

The expansion of the oil palm sector has created ecological and social landscapes vulnerable to fire. The ecological impacts of large-scale forest disturbance by the logging and plantation sectors show a marked increase in flammability, including decreasing moisture content of the soil and vegetation, and increasing shrubby understory and fuel load (Leighton and Wirawan, 1986; Mackie, 1984). Further, these changes in forest cover and forest control have produced tremendous social changes. In particular, the state has generated epidemic social conflict through its appropriation for plantation of lands which local people have controlled for generations, and which they consider to be both part of their customary territory and under descent group ownership. This conflict made itself visible both in the use of arson by plantation owners to stake claims to locally-held lands, and in the destruction of oil palm by local people embittered by the injustice of the plantation operations. Many locals commented that if plantations or timber concessions caught fire, the owners should not count on local help to extinguish the flames, indicating that the roots of the disaster lie with struggles over ownership of land and forest resources. 'Why should we

---

12. This legislation was passed in response to Singapore's and Malaysia's vehement complaints about the smoke coming from Indonesia during the 1994 drought.
13. The global production of palm oil was 16 million tons in 1997. Indonesia's production alone is projected to be 12.5 million tons by 2005, or 41.4 per cent of the global total (CIC, 1997).
14. This figure does not include conversion to other plantation crops, timber concessions, or resettlement programmes. Indonesia is already the world's largest plywood exporter (Gellert, 1998), and Bob Hasan has publicly stated his ambition to make Indonesia the largest paper producing country in the world. Timber is Indonesia's second largest export (after oil), totalling US$5.5 bn. in 1994 (The Economist Intelligence Unit, 1995).

help?' snorted one resident, 'Let it burn — that land was lost to us a long time ago'. In fact, many locals implied that resentful farmers may have been the ones to set plantations alight. For local farmers, then, the forest disaster began when their forest resources were destroyed for timber and plantations from which they would receive no benefit.

Meanwhile, when their own rubber and fruit orchards were threatened, local people reported carrying water great distances on their backs, forgoing food and sleeping in the forest in an attempt to fight or prevent fires. Some even lost their lives, overcome by heat and smoke, trying to save their gardens. Clearly, forest flammability has as much to do with the social as with the ecological landscape.

GIS Technology and New Discourses of Disaster

Novel uses of satellite imagery, GIS and Internet technologies played a central role in the translation of the fires of 1997–8 into disaster. In August 1997, following the first two months of drought, then Environment Minister Sarwono Kusumaatmadja issued stern warnings of prosecution for plantation owners who used fire to clear their land. In an unprecedented move in October 1997, Sarwono attempted to make good his threats by bringing charges to the Minister of Forestry, Djamaludin Suryohadikusomo, against 179 companies. These companies had been identified using remote sensing images and GIS, overlain with plantation concession boundaries, to determine the sites of fires (or 'hotspots'— an English word now incorporated into the Indonesian lexicon). This was the first such use of GIS in investigating the sites and causes of fires, and changed irrevocably the familiar tactics of laying blame for the fires. Availability of the satellite photos on the Internet, along with the overlays of the forest classifications and concession boundaries, showed that the vast majority of hotspots originated on plantation land. This not only broadened attention to the crisis in the international media, but opened the door for the NGO community to apply pressure on the government and to widely publicize the mismanagement of Indonesia's forests, even bringing their own suits against the companies for destroying forests and endangering lives.

Among those accused by Sarwono were companies owned by Suharto's children as well as his close associates — including the tycoons Liem Sioe Liong and Mohammad 'Bob' Hasan. Shortly thereafter, Hasan, denying the charges, told the press that the fires were due solely to the ENSO drought, and only NGOs with 'links to communists' thought otherwise (*The Guardian*, 8/11/97). Given Indonesia's bloody history of eradicating communism and political opposition, this statement had a threatening ring. By November, charges against 163 of the 179 companies had been dropped, including those against companies held by Liem, Hasan and members of Suharto's family.[15]

Furthermore, the majority of companies with charges still outstanding are jointly owned by foreign investors, primarily Singaporean and Malaysian, supporting the government's contention that 'foreigners' were largely to blame.[16] Although these 'foreign' companies must by law have Indonesian partners, labelling them outsiders allowed some legal action while creating a nationalist smokescreen for Indonesia's own timber barons.[17]

Meanwhile, Djamaludin, although never completely abandoning the attribution of some blame to small farmers, joined Sarwono in his pledge to prosecute plantation owners (at that time still under the jurisdiction of the Ministry of Agriculture). Remarkably, Djamaludin officially took full responsibility for the disaster, even tearfully offering to resign from his post.[18] Despite the radical new uses of technology, the degree of real change in the political economic structure of Indonesian forest management remains to unfold in the post-New Order regime (see epilogue). However, the fall of the New Order in May 1998 was partially brought about by the occurrence of the fires in a constellation of events that rocked Indonesia in 1997, which brought it firmly to the world's attention.

Triple Crisis and Out-of-Sight Local Impacts

The experience of the fires as a disaster for local people was greatly exacerbated by Indonesia's economic collapse in late 1997. The lengthy drought

---

15. These charges were dropped by the Ministry of Forestry, but since the fall of Suharto, there have been some dramatic changes. In June 1999, Dayak residents of South Kalimantan (assisted by WALHI) won an independent suit against seven oil palm plantations owned by the Salim Group (controlled by Liem Sioe Liong and his son). The firms were ordered to pay Rp150 m (at that time equal to US$18,000) in compensation for damages to local property caused by the company's land clearing fires (*Jakarta Post* 1/6/99). Nevertheless, although there are reports of some plantation licences being revoked, each company holds a multitude of different licences for different activities (for appropriating land, for clearing land, for planting, etc.) and each can be reissued annually. Therefore, there is little reason to believe that even those companies who supposedly had their licences revoked will be shutting down operations.
16. There were even veiled threats by former Environment Minister Juwono Sudarsono (who held his post only for a few brief months before the fall of Suharto's government necessitated a cabinet reshuffle), of a class action suit against these countries (*Dow Jones Newswire* 24/4/98).
17. This feigned nationalism intensified with tensions over IMF loan conditions, which were painted as neo-colonialism by the privileged business community. Hasan in particular reacted with indignation to the IMF attempts to discipline the Indonesian economy (which included the dissolution of his own timber and rattan cartels). Following Suharto's submission to IMF reforms, Hasan defiantly declared 'This is the Republic of Indonesia, not the Republic of IMF!' (*Far Eastern Economic Review,* 14/5/98).
18. In Suharto's final 1998 Cabinet, Sarwono and Djamaludin were both replaced.

destroyed many local rice crops. Following the return of the rains, waves of pests wrought havoc on those that remained. Fallback cash crops of rubber, rattan, and fruits either were destroyed by fire (especially in East Kalimantan) or failed to produce due to the long drought and smoke.[19]

With no means to generate cash, the final blow of rising staple prices due to currency devaluation drove farmers to despair.[20] Rice prices rose from Rp1000 to Rp7000 per kilo. Prices of other staples including cooking oil, sugar, coffee, salt, and noodles also skyrocketed. Even though government stores of discounted rice (*beras dolog*) were periodically made available at Rp5000 (in December 1998) local people said they could not buy, since they had lost their cash crops. These government stores of rice have since been exhausted. None of the traditional famine foods (tubers and other forest starches) could be found in the forest after the large scale conversion to plantation. Even the cultivated cassava roots typically eaten during lean times were badly shrunken by drought or lost to fire.

These devastating impacts are unregistered by satellite cameras, and as a result have gone largely unnoticed in the feverish tallying of disaster damages, as it is undertaken in the offices of Jakarta, Singapore and other international capitals. However, this is more than a simple question of inappropriate scale. The silence on the subject of rural welfare in the calculation of damages is particularly worrisome because these analyses are largely intended to demonstrate the unsustainability of current forest policies. What does it mean when local peoples' losses are not considered relevant in such an argument? Given the angry crowds that have taken to the streets in popular protest in Indonesia in recent months (see epilogue), it is worth considering for whom Indonesia's forest management is most unsustainable.

The absence of government response was predictable given the remote technological epistemology largely employed by the government and its donor partners, and served to create a unique type of disaster. The government's disaster gives prominence to problems it is uniquely qualified to address ('under-development') and suppresses those for which it bears responsibility (misguided development and social injustice).

AN ETHNOGRAPHY OF DISASTER

Whatever mental pictures people may have had of Borneo, these are likely to have changed dramatically since vivid images of disaster were projected

19. The smoke reduced photosynthesis by blocking sunlight and coating the leaves with a thick layer of ash.
20. Fuel prices also rose overnight as part of the IMF's demand for removal of subsidies to the national petroleum company.

around the globe. This section outlines how diverse images of Borneo as a site of disaster articulate with wider processes and serve diverse interests. What follows is a ground level analysis of these discourses of disaster — how different speakers made sense of the fires, what discursive tools they used to narrate the scene of the disaster, and why such rhetorical scenes matter for creating Bornean futures.[21]

The Indonesian Government

During the ten months of burning, there was no unified government response. President Suharto, his Ministers of Forestry, Environment, Social Welfare, Agriculture,[22] and Transmigration and Resettlement of Forest Nomads[23] all contradicted each other on the causes and damages produced by the fires, while their staff had still different versions. Individuals varied widely in their willingness to publicly stray from the accepted official explanations.

Whatever the public comments of the government's high profile leaders, privately lower officials felt they were being made scapegoats who would suffer the consequences of their bosses' self-indulgent loose tongues. One provincial official, asked about the impacts of the fire in his province, replied testily, 'It's all very well for Ministers to talk so boldly, they will be replaced in the next cabinet anyway ... Who is the victim when the press yells "fire" and "arson"? *We are* — the local officials. We cannot manoeuvre within the national system, yet we take the blame'. Jakarta officials blamed corrupt local officials who protected wrongdoers in the plantation sector in exchange for kickbacks. In turn, provincial, regency and district officials privately blamed the centralized government for essentially the same thing on a larger scale — making policy benefiting business élites (including central government officials) and refusing to prosecute offenders. In an election

---

21. I refer to these peopled landscapes as 'scenes' (not simply localities, the term used by Appadurai, 1996, and Gupta and Ferguson, 1997) in order to indicate not only their physically-, but also their temporally-situated nature and the presence of visible players. Further, these scenes have a place in a larger narrative.
22. During the fires of 1997, the Ministry of Agriculture's jurisdiction included the plantation sector. In 1998, this responsibility was assumed by the Ministry of Forestry (revised to Ministry of Forestry and Plantations), further consolidating that Ministry's power through control over forest resources.
23. This Ministry was in charge of the massive Million Hectare Rice Project — an ill-conceived transmigration project involving the relocation of thousands of residents from the overcrowded islands of Java, Madura and Bali to Central Kalimantan, to areas of peat soil where they were to cultivate rice and boost national rice self-sufficiency. Peat soil is not only very infertile, but also, when accidentally set alight (during the burning away of cleared forest), creates toxic thick smoke and can cause subterranean fires which burn for years and are virtually impossible to extinguish.

year,[24] everyone recognized the political consequences of this 'disaster' and all were trying to distance themselves from it, as well as minimize its severity. In fact, one local official in Ketapang, one of the most badly burned regencies in West Kalimantan, simply denied outright that there had been any fires there at all.

Amidst the contradictions and finger pointing, however, a few main themes emerged in government discourse in relation to the fires. These centred on the notions of nature, development and poverty.

Nature: To be Conquered or Feared?

In the overwhelming majority of official government statements, El Niño consistently took the blame for the disaster. Indonesian officials repeatedly emphasized climate as the cause of the fires, invariably asking, 'Who can we blame for the weather?' To prevent any discomforting answers to what was intended to be a rhetorical question, the Minister of Information officially forbade the domestic media to mention any cause but El Niño for the fires (*World Watch*, 2/98).[25]

President Suharto and his Minister for Social Welfare, Azwar Anas, focused exclusively on natural causes for the fires, and downplayed the fires' significance altogether. This denial strategy became untenable when neighbouring countries began to complain of smoke, with the Malaysian opposition party even threatening to sue Indonesia for damages (*The Economist*, 4/10/97: 32).[26] The public outcry from ASEAN neighbours provoked a sense of national shame, as Indonesia was portrayed in the international media as the backward (read: undeveloped) ASEAN cousin (see, for example, 'The Neighbor From Hell', *New Scientist* 4/10/97).[27]

---

24. During the New Order, the government party (GOLKAR) 'won' the majority of votes in every 'election' by exactly the same margin. It was in election years that many government officials could be re-assigned to either more or less desirable positions, according to their favour with the party. So in election years it was not accountability to the voters that mattered, but to one's superiors.
25. Likewise, the Malaysian government prohibited its own academics and journalists from implicating government forest policy in the smoke disaster, as it was tarnishing the country's image abroad (*Nature*, 13/11/97).
26. This threat is rather disingenuous, given Malaysia's heavy investment in the Indonesian palm oil sector. Although ASEAN has a long-standing policy of non-intervention in its member countries' national affairs, investment by Malaysian and Singaporean business no doubt significantly contributed to the rather toothless response of ASEAN countries to the issue of trans-border pollution from Indonesia's fires.
27. From *Suara Karya* Letters to the Editor: 'Inconsiderate forest burning has produced shame for Indonesia due to the catastrophes it has caused for neighboring countries. It is as if the Indonesian people are still *ancient and uncivilized*' (reprinted in the *Jakarta Post* 11/10/97, emphasis added). See also an editorial in *Bangkok Post* (14/11/97) 'Indonesia's shame won't blow away'.

The government's use of Nature to explain Indonesia's 'extraordinary disaster' attempts to conceal human responsibility and to delineate the crisis as an isolated event. The 'scientific' explanation for the fires routinely given by Indonesian officials became a uniformly repeated chorus: 'To have a fire, you need three things: 1) dry fuel (forests parched by El Niño), 2) oxygen (strong winds produced by El Niño), and 3) a spark'. The source of this final element was consistently left to the audience's imagination, but the clear absence of agency also implied natural origins. In fact, I was specifically told by more than one local forestry official that *tree limbs* rub together to create the spark that starts the fires. This astonishing outright denial of human agency is indicative of the government's overall steadfast refusal to analyse the real causes of the fires and to deflect attention away from its own responsibility.

However, government actors are not unified speakers. In other iterations, Nature is the basis of unexpected alliances between some government officials, and both Indonesian and international NGOs. Articulating with both environmentalist and nationalist rhetoric, this vision asserts that El Niño is more frequent and more severe due to increased global warming caused by high consumption in the industrialized world. Nature, then, is not 'uncontrollable' in this formulation, but has in fact been sullied by the First World at the Third World's expense. As one official put it, 'This is a global problem now, not just an Indonesian one'.[28] Drawing on environmentalist ideas of global processes and flows, this view can also be deployed as part of a more nationalist, anti-neocolonial stance, bringing together rarely allied Indonesian officials and environmental activists to shift the blame for the fires to the industrialized world. Thus, some officials' nationalistic position that the West has affected global climate contradicts the insistence of many other officials that the fires are the consequence of an uncontrollable nature — exposing the multiple audiences to which discourses speak, as well as the multiple motivations running through the government's response to the crisis.

Importantly, the discourse of natural disaster does more than obscure responsibility for the crisis: it works to naturalize and support the conditions that compromised the resilience of forests and human populations, reproducing the conditions for disaster.

Technology Dreams and Discourses of Poverty

Scholars of natural hazard have long argued that discourses of natural disaster can be read as reflections of wider political and social agendas

---

28. Indonesia's burning peat soils have been estimated to have released one billion tons of carbon into the atmosphere — more than that emitted by all the power stations and automobiles in Western Europe over an entire year ('Indonesia's Inferno will Make us all Sweat', *New Scientist*, 4/10/97).

(Cronon, 1992; Davis, 1998, forthcoming; Dove and Khan, 1995; Hewitt, 1983, 1995; Watts, 1983; Watts and Bohle, 1993). Many of these scholars suggest that increased occurrence of disaster in developing countries reveals the increasing vulnerability of populations with compromised coping mechanisms and little choice but to occupy marginal, high-risk lands. But more than simply reveal, responses to disasters also reproduce relations of power. Hewitt (1995) suggests that we read the increasing vulnerability of populations to climatic extremes, not as an indication of incomplete development — a deficiency for development to fix — but rather as a *consequence* of development initiatives which have radically restructured patterns of resource access and social relations. These works suggest that the erasure of human agency in discourse of 'natural' disaster is the predictable consequence of a hegemonic development agenda, which not only obscures its own role in the outcome of catastrophic events, but also in fact advocates its own perpetuation.

While Ministers Djamaludin and Sarwono publicly exhibited support for the prosecution of companies using fire to clear land, they seemed to firmly believe that high technology surveillance was the answer to the fire problem, rather than more preventive measures, such as reforms of state forest management policy or law enforcement checks and balances. Near the end of his term, although Sarwono made increasingly caustic public comments referring to the ignorance of the plantation sector[29] (emphasizing the need for higher technology land clearing, including bulldozing, tilling and chemical composting), he never mentioned a problem with the extent or pace of forest conversion. There were 140 million ha of state-designated 'permanent forest' (*hutan tetap*) in all of Indonesia in 1996, although even official statistics admit that only 91.7 million ha of that forest is 'still forested' (*hutan yang masih berhutan*, GOI, 1996). Even with a reported 3.9 million more hectares proposed for conversion to oil palm by 2005 in East Kalimantan alone (*Kaltim Post* 27/11/98), government ministers nevertheless clung to the notion that it was a technology deficit that contributed most to the fires.

The lack of technology and the resultant need for more development is a recurring theme in government discourse. This is unsurprising given the thirty-two year reign of a dictator referred to by his staff as 'The Father of Development' (*Bapak Pembangunan*). Development has been the imperative of Suharto's New Order government since its birth in the mid-1960s. It is, therefore, to be expected that government officials frequently cite 'poverty' (although sometimes in contradictory ways) as one of the leading causes of the fires in Indonesia. For my purposes here, multiple constructions of poverty operate at both national and international scales, and figure prominently in state and donor interpretation of the causes of the disaster

29. '*Kurang Ajar Orang Kebun Itu*' (*Indonesia Daily News Online* 14/3/98).

and its solutions. This case illustrates how disaster becomes, as Hewitt (1995: 117) has theorized, the antithesis of development, its polar opposite, as 'weeds to be eradicated' by development. In this view, the solution to disaster can only be more development — as measured (partially) by more technology to conquer more nature — and therefore, reduced poverty.

At the national level, rural populations who practise swidden cultivation (using fire instead of tillage or composting to clear old vegetation) are represented as materially and culturally backward, in need of further economic and cultural development.[30] Time and time again, in discussing causal factors of the fires, both government officials and donor agency staff voice frustration at what they perceive as the intractable rural culture that refuses to accept 'modern' agriculture and 'modern' Indonesian culture. The alleged recalcitrance of 'primitive' rural swiddeners provides an easy scapegoat for state officials wishing to protect élite interests in lucrative forest conversion and extraction. For development proponents, these cultivators represent an unreached audience of development. This use of rural 'poverty' as the cause of resource degradation suggests more development would provide the cure to what the state and development agencies regard as inefficient and culturally backward agriculture

Yet, at the same time, this directive promotes as 'development' a plantation sector using the very same 'backward' technology of fire to clear land. Significantly, in 1997 the government and many of their donor partners adopted the terms 'farmer' and 'slash and burn agriculture' to refer to plantations — a less obtrusive means of shifting blame from small 'farmers' to big 'farmers'. This move essentially lumps all bad practice together under the unassailable sector of agriculture, which should not be curbed but merely modernized. This allows development to be seen not as the problem, but the solution.

Furthermore, at the international scale the Indonesian government draws attention to international inequality of development (although it does not use the word 'poverty' to describe this inequality) through its claims that it cannot detect or fight fires effectively because it cannot afford satellite imagery processing or water-bomber aircraft. In this construction, appropriate responses to the fire problem must include a high tech fix of infrastructural aid, accessible to only a few (ASEAN, 1997; Soemarsono, 1997).

In December 1997, ASEAN members met to discuss the fire problem and produced an Action Plan, which, ironically, was virtually identical to those produced but unimplemented following the Great Borneo Fires of 1982–3.[31] The Plan highlights the goals of strengthening the roles of Singapore in monitoring, Malaysia in prevention and Indonesia in fire-fighting (ASEAN,

---

30. Dove (1993) outlines the economic contributions of this 'backward agriculture' — for example, 75 per cent of Indonesia's rubber production (the second largest in the world) comes from smallholdings of 1 ha or less.

31. These Action Plans went unimplemented despite warnings that future fires would burn hotter and spread faster (Schindele et al., 1989).

1997). However, in the spirit of ASEAN diplomacy, these duties are non-binding and offer little more than vague references. Non-intervention is the accepted stance of ASEAN nations, and the report notes that 'while some countries have already developed their rational policies and strategies, others are in the process of advancing them based on their own development needs, priorities and concerns' (ibid.: 2). No specific directives or methods are given as to how prevention is to be approached. Large-scale forest conversion for plantations is not mentioned anywhere in the document, which instead suggests that 'guidelines be established to discourage activities which can lead to forest fires' (ibid.). Instead, the plan focuses on establishing taskforces, operational procedures, and educational programmes. Regional monitoring and fire fighting capacity building receive the bulk of attention in the document, with monthly schedules for establishing monitoring capacity and infrastructure. This document highlights the approach to the fires as an event, not as a symptom of a larger problem, and the emphasis on high tech remote surveillance rather than a proximal (field-based) approach to data collection and remediation.

In summary, rather than parroting a monolithic discourse, the state was racked with internal conflicts. Individuals displayed diverse loyalties, and waged private and public battles within a larger landscape of political power. Overall, however, a strategy emphasizing the remote high technology analysis of the 'natural disaster' was promoted over one that investigates causes and impacts at the field level. This approach comfortably encompassed the two ill-matched discourses of uncontrollable nature and insufficient development to suggest that if the government had more technology (that is, if Indonesia were more modern) it could predict nature more precisely and respond more quickly. This explanation obfuscates the role of human agency in the cause of the fires as well as political will in the response.

In government discourse, Kalimantan, as a scene of the fires, was constructed through the aid of remote technology as wild and untamed frontier, populated by primitive and lawless people. Tellingly, former Environment Minister Juwono commented to the Western press, 'Kalimantan at this time is part of the Wild West, *part of a nation without government,* like parts of America in the nineteenth century' (*Reuters*, 14/4/98, emphasis added). According to this vision, if Kalimantan could be properly governed and developed (using improved technology), it could yield valuable resources and help bring Indonesia into modernity (implicitly — as America has done with its own 'lawless' lands, imagined to be populated by primitives and criminals, but rich in natural resources).

Donors

By early 1998, international donors had sponsored a wide array of fire analysis projects (twenty-six in total by that time) and they amassed a variety of assessments. Of primary interest were macroeconomic losses, from tourists who did not visit, planes that did not fly, destroyed timber and plantation crops, productivity lost from closed businesses. Conflicting estimates of burned forest area competed for the public eye. There were atmospheric indices of haze (as the smoke was euphemistically called), climate records of regional rainfall and temperature, and historical comparisons of global El Niño events. There was also great interest in the impacts of fires on orang-utans — how many were dead, wounded, displaced, what were the impacts on their reproduction, diet and habitat. At that time, this was the only data derived from field based research. Indeed, it was striking that the vast majority of impact data were derived from remote sensing — satellite photos showing area, thickness and content of smoke, and location of hotspots. No one had collected any systematic field data of impacts on local communities, and of the twenty-six donor projects underway, only two even had any plans to do so.[32]

This silence was more than oversight. It can also be seen as a reflection of a specific agenda that framed the research in particular ways, and which spoke to particular audiences. While international donors are rather more at liberty to push for reform (and many do) than appointed governmental officials, they are nevertheless constrained by the political climate within which they work. Furthermore, organizations such as the World Bank, the International Monetary Fund, and the Asian Development Bank can hardly be expected to introduce policy reforms which would reduce the commodity value or export of the timber and plantation sectors — sectors whose increasing output they have unfailingly supported. In fact, even as the fires — already widely recognized to be a consequence of logging and the expanding oil palm sector — continued to devastate East Kalimantan, the IMF included as a condition of its 'rescue package' loan (following the 1997 financial collapse) the further expansion of the oil palm sector and the inclusion of foreign investment in the forestry sector.[33] It was clear who the IMF intended to rescue.

Several bilateral donor employees told me that the current donor emphasis is on high profile 'capacity-building', not the sort of projects that involve long trips by canoe and interviews with rice farmers. Remote sensing and GIS fit the bill perfectly as a high prestige, highly priced, high modernity project that could provide both training and hardware for the host country, while minimally involving the donor in sticky political questions of policy reform, social conflict, and government corruption. As one employee observed, 'The government is very interested in GIS — not

32. A CIFOR/ICRAF/UNESCO collaboration and a WWF/WALHI/WRI collaboration, in addition to my own brief visits to three villages while consulting for the UNDP Environmental Emergency Project.

33. Foreign investment has been illegal in the forestry sector since the 1980s when Hasan began to develop Indonesia's plywood industry.

only for the fires, but for routine management as well, particularly in the forestry sector. It's the perfect example of sustainable development, not just hit-and-run relief aid'. Whatever 'routine management' might be planned by the Indonesian government, this new manipulation of information and space dovetails neatly with a longer trajectory of increasing technocratic control over territory, resources and populations by the state.

The image of Kalimantan constructed by donors' analysis is not far from the scene imagined by the Indonesian government. Kalimantan is viewed as an international resource, a source of priceless biodiversity that is part of a global heritage, as well as a source of valuable natural resources that could be used to generate foreign exchange. In this discourse, Kalimantan is seen as the home of impoverished people who must degrade their environment in order to survive. Indonesia, by this assessment, was in need of more efficient means of modernizing and monitoring this process of development for international benefit and to slow local (read: poor) people's mining of valuable resources. These solutions locate blame for degradation of the environment with poor local people rather than with wealthy extra-local people.

As donor emphasis supports the increasing prestige of GIS-processed images, alternative forms of inquiry are filtered out in favour of remote analyses. Even more worrisome, this trajectory of inquiry suggests that equity and local welfare have slipped from view as appropriate points of engagement for donors, in favour of remote macro-level assessments. The deliberate exclusion of local livelihoods from the official discussion of impacts and sustainability deserves scrutiny within donors' mission of improving local welfare. Many donor staff routinely defend this deeply political decision as the consequence of a necessary 'apolitical' stance, denying the fundamentally political nature of development interventions. Such omission reflects a persistent conflation of 'local people' with their government — the assumption that developing the state leads to the development of its citizens. 'Ground truthing' would reveal this assumption to be severely flawed.[34]

Activists

In New Order (and recent post New Order) Indonesia, political activism bears very high risks. In a dangerous and repressive political climate, where questions of resource management easily merge with issues of democracy and human rights, NGO advocacy groups have bravely mobilized as defenders of nature. Advocacy for local indigenous control of resources and a reduction in large-scale forest conversion to plantation is as much an argument for human rights as it is for the environment. Environmental

34. 'Ground truthing' is the cross-checking of remote data with direct observations from the field.

NGOs have organized their advocacy around two primary principles: 1) indigenous people (*masyarakat adat*) are traditional farmers living in harmony with nature; and 2) access to sufficient resources and a clean, healthy environment are components of basic human rights.

Given these tenets, the fire disaster served as the ideal nexus by which to simultaneously address problems of ecologically destructive and socially inequitable government forest practices. While the activists recognized the role of climate in bringing about the fires, many argued the inevitability of the disaster given the free rein of the forest and plantation sectors. Within this discourse, goals of ecological sustainability, human rights and democratic advocacy would all be served by reformed forest management policy.

A landmark legal suit was brought by WALHI (Indonesian Forum on the Environment, the country's largest and most vocal environmental NGO) against plantations implicated by satellite images to be using fire to clear land — fires that spread to villagers' nearby gardens and protected forest. WALHI brought the suit, not as a representative of local communities, but as advocates of the actual 'forests that cannot represent themselves' (*Suara Pembaruan*, 5/2/98).[35] The complex realities of environmental problems and political context in Indonesia make nature the point of convergence of multiple material, social and political activist agendas.

In this struggle for nature and democracy, Indonesian environmental NGOs have made productive new use of a radical, appropriated technology. The availability of Internet communications, satellite images on websites, and the increased ability to analyse these images using GIS technology enabled NGOs to document and publicize for the first time what they had always suspected: that it was not small farmers setting fire to rice fields but large plantation owners clearing forest who were largely to blame for the smoke and out-of-control flames.

However, NGO inquiry on the impacts of the fires, like government and donor assessments, has given priority to the prestige and international advocacy supported by computerized 'counter mapping' technologies.[36] GIS

---

35. Indonesian law prohibits advocacy groups from representing entire communities, requiring instead that individual cases be brought to the court. Under the menacing military rule of Suharto's New Order, few individuals would be brave enough to attempt such a suit. However, in the new atmosphere of reform, at least two of the accused companies in this suit, PT Inti Remaja Concern and PT Musi Hutan Persada have so far been found guilty of the charges of deliberate violation of the zero burn legislation, as well as criminal negligence for allowing their fires to burn out of control to surrounding forests (WALHI press release, 17/10/98). Samihim Dayak communities also successfully brought suits (with WALHI's assistance) against PT Laguna Madiri I, II, III; PT Langgeng Muara Makmur II, III; PR Paripurna Swakarsa I; and PT Swadaya Andika II (*Jakarta Post* 1/6/99).

36. 'Counter mapping' is the practice undertaken by NGOs to appropriate powerful GIS technologies to map local territories. These alternative maps are used to counter state claims to forests (Peluso, 1995).

is a powerful tool because of its perceived objectivity and modernity — it is powerful precisely because it is *not* an alternative representation, but one that is legible to politicians and administrators. Co-opting the scientific authority of mapping technology has proven to be a fruitful move for activists, especially those engaging in negotiations at national and international levels.

The Kalimantan of activist discourse was thus the rhetorical battleground for democracy struggles against the corrupt and repressive New Order, to be waged in Jakarta and internationally, and to be marshalled in establishing the legitimacy of NGO participation in policy dialogues and the development of an Indonesian civil society. Yet, the strategic appropriation of powerful technology comes at a cost. Narrowing evidence to remotely sensed data reinforces the hegemonic position of these kinds of data, necessarily filtering out some local voices, individual experiences and alternative representations of disaster, which can only come from field-level data.

Farmers

While the rest of the world peered at the smoke from satellites, from the ground Kalimantan had a very different look. 'The era of smoke' (*jaman asap*) is how many in Kalimantan remember the ten months during 1997–8 when a cloud of yellow smog hung over the island. People fainted and wheezed, coughing black phlegm; eyes burned and watered; soot accumulated in nostrils, ears and the corners of eyes. Most health impacts went undocumented, as medical care is largely inaccessible for many residents. Street lights remained lit throughout the twilit day. Farmers without wristwatches returned from their swidden fields early: unable to judge the time by the sun, they feared being caught far from home after nightfall. In near zero visibility, planes and buses crashed, ships plowed into each other, killing hundreds. Although the island sits astride the equator, the sun rarely appeared, and when it did, it was a frightening blood-red apparition that disappeared behind the smog as quickly as it had come. Apocalyptic imaginings were hard to resist.

The East Kalimantan village of Tanyut (a pseudonym), perched on the edge of the province's largest lake, usually serves as a dock for traders' barges, but in April of 1998 it could only be reached by *ketintin,* shallow canoes powered by long outboards. The expansive lake had been transformed by drought into a tiny stream so shallow that even *ketintin* passengers frequently had to drag their craft. Fisherfolk whose fishing grounds had evaporated planted rice in the middle of the dried lake bed. Yet, all these things amount to a normal dry season, not an unheard of event. It was not the drought, but the impact of the fires on livelihoods that was an unprecedented disaster.

The Benuaq Dayaks who have lived in Tanyut for generations make their living from a combination of subsistence and commercial agroforestry including swidden agriculture, rubber, pineapple and rattan gardens. In April 1998, the entire rice crop of Tanyut had been lost to the drought, and many of the gardens to fire. What remained was unproductive: latex did not flow and trees didn't bear fruit. Drought or excessive rains or pests have ruined harvests before, the villagers said, but they have always had other resources to fall back on — they could sell rattan or rubber, and eat fruits, game and vegetables from the forest and their gardens. This year many people had lost all their gardens to fire and for them there were no fallback resources. The fire had taken everything. Essentially, they had lost their life savings, invested in the landscape — and *that*, they said, was unprecedented. The fires had come, they all agreed, from the land clearing activity of the adjacent oil palm plantation. Some thought it was an accidental spreading from the huge rows of piled and dried vegetation set alight. But many accused plantation owners of more than criminal negligence — villagers believed they were victims of arson.

When the plantation first came in 1996, residents of Tanyut had decided as a group to refuse to sell their gardens.[37] When called individually to the camp to be persuaded — frequently by police wearing side-arms — residents refused to go, instead inviting the camp managers to come to the village and speak to them as a group. Knowing that plantation owners were well aware of the locals' desperate financial situation, many farmers whose lands abutted the plantation concession alleged that their gardens had been deliberately burned by the concessionaires. Even worse, locals knew their burned land would be nearly worthless if they did sell, since the plantation compensated only for trees planted on the property.[38] Farmers would be forced to beg for the concessionaire's 'goodwill' in giving them something in exchange for land they had cultivated since their great grandparents' time. By April 1998, some residents had decided to trade some of their ancestral land for rice — 1 ha for 125 kg of rice, about a month's worth of food, at

---

37. A neighbouring village, in contrast, sold nearly all their gardens and became labourers on the plantation. With the economic collapse, they were all laid off, some still owed back wages. The gardens of the few families that refused to sell remained islands within the sea of palm oil trees. Many of these islands were mysteriously burned, although none of the surrounding palm trees were affected. The owners of these island gardens accused the concessionaires of arson, but when they demanded compensation from the company, they were dismissed by the police as having insufficient evidence.
38. Under the national constitution, the state owns all land within its borders. There is some nominal exception made for lands under customary law, but these rights are only allowed insofar as they do not conflict with 'national interest'. Therefore, once land has been declared 'state forest' in the national interest (thus negating any customary claims), the state has the right to lease out that land to concessionaires. Concessionaires, in turn, are under no *legal* obligation to 'buy' the land from locals who see the land as theirs by customary right, although most offer some compensation for lands appropriated in order to ease local relations. The terms, however, are left to the discretion of the concessionaire, and are frequently agreed to under coercion by local officials or law enforcement personnel.

that time worth about US$20. This survival strategy, aside from threatening future food security by reducing available farmland, had begun to cause disagreement among the residents and threatened the community unity that in the past had enabled them to resist the plantation's encroachment. Plantation expansion, then, affected not only forests and farmland but the social landscape as well.

Disaster and Responsibility

In that same month, West Kalimantan looked lush by comparison. The rains had been falling continuously since December, and new green growth concealed blackened soil. Yet in the Pesaguan Dayak communities of Sibau (a pseudonym), one of the districts most affected by fire in the previous year, people worried. Their rice harvests were poor (they attributed this both to the extended drought and the waves of pests that followed) and many had lost rubber gardens. But most had relatives who shared rice (stored from previous harvests) or had access to 'share-tapped' rubber — trees owned by someone else that could be tapped in exchange for a portion of the profit.[39]

What concerned many of the Dayaks of Sibau most were not the material losses but that many of the fires had come from a neighbouring community.[40] People in that village, it was said, had stopped using the traditional methods of controlling fire, and local sanctions for uncontrolled fire had become unenforceable. In communities that depend on social reciprocity for nearly all aspects of labour relations and social security, this trend is of concern for more reasons than the immediate fire hazard. Most thought this declining social order had begun when the *Acacia* and oil palm plantations first came to Sibau. People were divided on whether they should resist selling their land. Some wanted to stand united against the plantation, while others considered that tack futile and wanted to get whatever they could from the deal. Bitter community divisions developed, and old fractures deepened. Residents complained that instead of settling disputes and protecting them from (what they considered) land theft, local police were in fact engaged by

39. In fact, for those who still had access to rubber and other forest products, the effects of drought and economic crisis were not all negative. The global shortage caused by the drought, combined with Indonesian currency devaluation, had increased global demand for Indonesian rubber. In turn, local prices for products also increased. For the producer, pepper was four times its pre-drought price, rubber seven times its former price. Prices further downstream (through several middlemen) were even higher.

40. There had been problems with runaway fires from nearby concessions when the plantation first began clearing forest, but that had been in 1994. In 1998, there were no large-scale land clearing activities nearby. There were reports that some of the oil palm had been burned deliberately by disgruntled day labourers, who had been cheated out of their wages and then laid off. There were no reports of these fires destroying any locals' property.

the plantations for protection and 'land procurement'. Locals complained that this fostered a disrespect for local law, and many people tried to get what they could from the land before it was taken from them.

In a country that in the last year had experienced ethnic war, political rioting, economic collapse, and wild climatic fluctuations, some Dayaks, and indeed people all over Indonesia, including some businessmen and government officials in Jakarta, began to believe that the disaster was a reflection of some cosmic dissonance.[41] This explanation points to human misdeeds as having identifiable environmental effects.[42] This representation of 'disaster' (*malapetaka* or *musibah*) turns on a recognition of cosmic dissonance, but does not deny political economics — rather, it interprets political-economic processes as part of a larger cosmology. Elders frequently tell stories of social and ritual misconduct leading to extended droughts and pestilence. But in a modern twist on local idioms of human misbehaviour, local people suggest that it was the greedy and unjust behaviour of concessionaires, politicians, and law enforcement officers involved in the conversion of forest to timber and plantation that resulted in ecological and social dysfunction. Here it is not poverty that sets in motion degradation of nature, but social injustice and moral turpitude.

This interpretation deserves serious consideration, since part of the reason that the fires of 1997–8 were so widespread was epidemic social conflict between plantation owners and local communities. Arson on both sides was used as a way of settling tenure disputes, or as a weapon by those who believed themselves to be the victims of injustice. Further, if it is social misdeeds that have led to Indonesia's disaster, then it follows that it is in human reform and responsibility that solutions must lie — solutions that would create quite a different Kalimantan from that pictured from satellites.

CONCLUSION: REMOTE SENSIBILITIES, VIRTUAL REALITIES

Fire in Indonesia is not new. But the interpretations and dialogues surrounding fire in 1997–8 were novel indeed, and contributed to the creation of new disasters, with new implications and new solutions. As I have shown, the role of remote sensing technologies, which have been used in different strategic ways, has been central to most of these new visions of disaster.

---

41. Many Javanese believed that Indonesia's hardships were an indication that President Suharto had lost his *wahyu,* or his divine right to rule.
42. Dayak legal principles are in fact based on this understanding of the wider consequences of human actions: under traditional law, ritual fines are incurred not as compensation or revenge but as a ritual cleansing of disharmony. If these rituals are not observed, Dayaks say, the repercussions are visited on the entire community in the form of pestilence, epidemics, climatic extremes, accidents and deaths. For more discussion on indigenous interpretations of sustainability, see Dove (1998), Dove and Kammen (1997).

In government discourse, GIS is held up as the path to modernity, governance and development, as an effective means for planning and surveillance of forest management. For donor organizations, GIS is promoted as an apolitical method of capacity building, as well as a means to document the economic unsustainability of current forest practice. For Indonesian activists, GIS provides powerful tools to argue for democratic reforms. What contexts made these new articulations of disaster possible and what can we make of these diverse possibilities for the high technologies used in their creation?

While Indonesia has historically held a hazy place in Western consciousness, 1997 marked a moment of sudden visibility. Since the award of the Nobel Prize in 1996 to East Timorese activists Bishop Carlos Filipe Ximenes Belo and Jose Ramos-Horta, international attention turned to human rights abuses of the Suharto government. But it was Indonesia's currency collapse under the Asian economic 'contagion' in 1997 that crystallized transnational concern over Suharto's crony capitalism and put Indonesia on the front burner of Western attention. With the fall of Suharto in 1998 and the first free elections in forty years in 1999, eyes are trained on Indonesia's hopeful transition to democracy. Combined with increasing international concern for conservation issues and the gathering influence and international linkages of the NGO community in Indonesia, these conditions concentrated international attention toward the Indonesian disaster in a way which did not apply to widespread fires in Brazil or Mexico during that same year. GIS technology, used in this context of increasing NGO strength and international concern with Suharto's policies, broadcast images of fire to remote locations and focused international attention on the disaster.

What consequences follow this expansion of remote representations of disaster? There has been much scholarly attention paid to the embeddedness of representations (especially those associated with discourses of science and technology) within particular cultural and political economic contexts (Haraway, 1991; Kuhn, 1962; Latour, 1987). Aside from being passively situated within a context, however, representations also actively *reproduce* particular value systems, and principles of inclusion and exclusion (Fyfe and Law, 1988). Decoding the social work done by GIS has been the foundation of the critical literature on the topic, largely spearheaded by Curry (1995), Pickles (1991, 1995), Sheppard (1993, 1995), Taylor (1990) and others. These writers draw particular critical attention to the applications of this new organization of spatial and demographic information for the military, whose surveillance and information control interests are most directly served through GIS. As Hall observes (1993, cited in Pickles, 1995: 21), 'Every map presages some form of exploitation ... Geopolitics, after all, is impossible without a cartographer, and that exercise of control over a distant domain marks a watershed in political power, confirming the notion that maps are not merely pictures of the world, but depictions of the world that can be shaped, manipulated, acted upon'.

While evidence presented in this article suggests that such concerns are valid, it can also be argued (as Pickles, 1995, does) that a critique that focuses exclusively on the role of GIS in serving hegemonic agendas is too polarizing. Such a view obscures the possibilities of remote high technology in democratic projects — fax messages from Chiapas and Tiananmen Square, e-mail linking transborder NGOs to a variety of indigenous movements, satellite photos and the Internet used to publicize plantation owners' roles in Indonesia's 'natural' disaster. These examples illustrate that high technology may also be appropriated to undermine hegemonic states' control of information.[43]

Therefore, while discourse makes use of specific forms of representation to serve particular agendas, these representations are not *necessary* parts of any particular discourse, rather they also contain the potential for innovation and flexibility. Their presence as part of a discourse is contingent, and in particular political and social moments may produce strange convergences and collaborations (Hall, 1996; Li, forthcoming; Tsing, 1998). This article has examined the contexts of diverse representations of disaster, as well as their consequences, but takes the analysis a step further by examining the contingency and slippage in these 'hegemonic' representations, specifically those of remote sensing and GIS, creatively appropriated to serve new agendas. Further, that GIS has become the battleground and the very means by which different viewpoints are argued is a consequence, not of the inherent or ontological superiority of GIS as a source of 'truth', but of specific historical and social moments. 'Hence although whichever propositions are true may depend on the data, the fact that they are even candidates for being true is a consequence of an historical event' (Hacking, 1989: 56).

In the unique moment of 1997–8, the emergence of GIS remote technology in Indonesia as a diverse tool for governance, as well as to assess fire causes and damages and to undertake prevention, is a contingent outcome with wide possibilities. Certainly, regardless of the trumpeting of an Internet 'Global Village', the fact remains that access to this information is limited to those with access to computers and telecommunications. In this regard, the Global Village remains a gated community. The centralized access to the technology and its potential for Foucauldian surveillance has clear militaristic applications, which have been amply outlined by GIS critics. Nevertheless, the proven ability for the technology to be appropriated for anti-hegemonic uses also suggests that simplifications of nature and populations through remotely created representations have complex capabilities. Rather than abandon the fields of science and technology to

43. The turn to globalized networks of remote sensing information, however, is not a deterritorialized process removed from local processes of translation and interpretation. Rather, the fires of 1997–8 suggest new forms of place-making from remote positions, whose effects are felt locally.

the interests of the powerful, both Harding (1998) and Haraway (1991) urge the 'recolonization of science', as a strategic move toward liberation and social justice.

What would such a recolonization entail? Would it truly be a subaltern science? Representations, as Paul Rabinow (1986: 231) reminds us, are 'both productive and permeative of social relations and the production of truth in our current regime of power'. What power relations are supported, what other truths are hidden as diverse ideological projects converge on remote technology as *the* means by which questions are investigated, and what will be the material effects of these shifts for social relations? In this article, discourses served not only to illuminate, but also (albeit sometimes unintentionally) to reproduce particular relations of power by naturalizing and removing them from view. Government discourses of 'natural' disaster conceal the role of its own policies in creating the disaster and effectively reproduce state and business élite control over forests. Donor discourses of poverty and technology obscure the politics encoded in development programmes, and recreate the role such programmes have played in creating vulnerable social and physical landscapes. Even activists' radical and fruitful appropriation of strategic science and technology in many ways reinforces the position of these tools and the consequent subordination of alternative views and methods.

While there are obvious advantages, there are also costs to advocating a remote sensibility. Kai Erikson (1976, 1994), in his attention to the local experience of trauma, argues that the overlooked definition of a disaster lies in the experience of the victims. Such an approach demands that local voices be heard and not reduced solely to the economic estimates of damage. Dayak experience of the disaster is concerned not only with losses to livelihood, but with disrupted social relations and a world out of balance. Deep senses of dread, anger and distrust of the plantation owners, of neighbours, of the state and even the forces of nature signal trauma that is linked not only to the fires themselves, but also to long processes of upheaval of social and forest landscapes.[44]

In addition to the silencing of local voices (in some cases, the very voices of those disempowered subjects for whom the discourse purports to speak), remote approaches to disaster events also obscure these on-going linkages of humans with their environment. Questions of just human behaviour, periodic human error and ecological flux are removed from view, banished from the everyday to the extraordinary for which there has been no planning and no responsibility. In focusing on monitoring of the disaster itself, causes and impacts are swept away. Remote disasters are unexpected events, sometimes tragedies, but they are not symptoms of deeper failings.

---

44. See Erikson (1994) for a fascinating variety of studies in 'trauma' and 'disaster', from floods to toxic spills and irradiation to homelessness and war.

What I am arguing for is a balance — a recolonization of science and technology without an abandonment of ethnographic fieldwork. Remote sensibilities used in an uncritical way, unbalanced with other forms of inquiry, engender more than an incomplete picture of the world and scenes of fires in Kalimantan. They serve to naturalize hegemonic agendas of manipulation of space, resources and populations. Remote technology, as many have shown for other forms of mapping and simplifications of the natural world (Harley, 1989; Scott, 1998; Thongchai, 1994), is a modality of power that most effectively serves the needs of urbanites, demographers, planners, administrators and strategists, and obfuscates alternative realities. Even if used widely by diverse institutions with diverse ideological projects, if not coupled with some more proximate form of fieldwork, remote sensibilities can only promote certain perspectives on disaster and home-places. By narrowing the kinds of data used in determining whether an event was a disaster, many alternatives are lost. In such a scenario, disaster will certainly continue to be our constant companion ... or at least the companion of those least empowered to make their experience of disaster heard.

EPILOGUE: FORESTS AND PEOPLE IN THE AGE OF *REFORMASI*

President Suharto's thirty-three year rule of Indonesia came to an abrupt end on 21 May 1998, when he assented to the demands for his resignation by university students and activists who had occupied the national parliament building for four days. Now the spirit of reform has taken hold and the slogan 'down with corruption, collusion and nepotism' (a trio in the Indonesian language now referred to simply as 'KKN') appears everywhere, even on the lips of the worst offenders of Suharto's government. The role of the fire and smoke disaster in the eruption of public discontent surrounding Suharto's presidency is undeniable. The material effects of rhetorical practice stand out boldly in these circumstances.

In this new era of reform (*reformasi*), a space has opened for alternative viewpoints on questions of state forest policies, native land rights and participation in resource management. NGOs seem to be able to operate more freely, and have developed a strong public presence, even organizing to advise the government on environmental issues (although it is not clear whether this advice will be taken). There are reasons to be optimistic about the futures of forests and forest dependent people.

Although the excitement of change is in the air, there is also cause for caution because much remains unchanged. Most of the beneficiaries of Suharto's administration are still in place and hold important positions in the government. The political influence of crony businessmen may now be more covert than in the past, when timber barons golfed and fished with the President, but it is doubtful that decades of relations will be cast aside so easily. Furthermore, the military is still very much in control, and its most powerful figures hold the majority of forest concessions in Indonesia. These military men are unlikely to go quietly if NGOs call for land to be returned to native control. In the chilling wake of unearthed bodies of political activists, there is still much cause to be wary.

Furthermore, hyperinflation and astronomic unemployment continue to cripple the nation's economy. The IMF, as a condition of its rescue package, demanded that economically inefficient natural resource monopolies, including Bob Hasan's timber cartel, be dissolved. This may increase economic efficiency by restoring timber prices to reflect those on the international market, but it may not be ecologically sustainable since the higher price may well fuel more cutting in pursuit of more profit. Moreover, the combination of currency devaluation (the rupiah fell from US$1 = 2400 in July 1997, to a low of US$1 = 15,000) and the high price for oil palm on the international market due to global shortages caused by El Niño, have increased demand for Indonesian oil palm. The quickest way for Indonesia to generate foreign exchange to pay back its three-year IMF loan is through the forestry and plantation sectors. Finally, what is currently happening in Indonesia is primarily an uprising of the intellectual urban élite — a revolution from the centre, not from the margins. This suggests that the development imperatives pursued under the New Order will most likely go uncontested under the post New Order. The issue at stake is to whom the spoils will go.[45]

As local people continue to feel the pinch of skyrocketing prices for staples, rioting is already widespread. An increase in land appropriation by plantations is unlikely to be well received, yet the military has recently demonstrated its renewed commitment to violence in the maintenance of order. All these factors point to a heightened fire risk. In July 1999, as the election ballots were being counted, the Internet carried the message that over 350 hotspots were already visible from space.

Acknowledgements

In addition to my own fieldwork in Kalimantan during 1994–7, I undertook three months of field research on the fires and institutional responses from March to May 1998 as a social impact analyst for the Environmental Emergency Project of the United Nations Development Program, in collaboration with the Indonesian Ministry of Environment. My dissertation fieldwork was funded by the National Science Foundation (Grant No SBR-9510422), the Fulbright Foundation, and the Social Science Research Council and sponsored by the Indonesian Institute of Science and the Universitas Tanjungpura. The analysis here is my own, however, and does not reflect the opinions of any of my sponsoring institutions. I wish to thank Arthur Blundell, Amity Doolittle, Michael Dove, Celia Lowe, Nancy Peluso, Hugh Raffles, Charles Zerner and anonymous reviewers for comments on earlier drafts. I am also grateful to Dr Soo Chin Liew of the Center for Remote Imaging, Sensing and Processing (CRISP) for permission to use unpublished burn scar estimates.

45. I am indebted to Celia Lowe for this point.

REFERENCES

Adriawan, Erwin and Sandra Moniaga (1986) 'The Burning of a Tropical Forest: Fire in East Kalimantan. What Caused this Catastrophe?', *The Ecologist* 16(6): 269–70.

Appadurai, Arjun (1996) *Modernity at Large: Cultural Dimensions of Globalism*. Minneapolis, MN: University of Minnesota Press.

Association of South East Asian Nations (ASEAN), Regional Haze Action Plan (1997) Joint Press Statement, Singapore.

Badan Pengendalian Dampak Lingkungan Daerah Tingkat I Kalimantan Timur (1998) 'Kebakaran Hutan dan Lahan Di Wilayah Propinsi Kalimantan Timur Tahun 1998 (01Januari s/d 22April 1998)'. Samarinda: Official Report.

Barber, Charles (1998) 'Forest Resource Scarcity and Social Conflict in Indonesia', *Environment* 40(4): 4–9.

Barr, Christopher (1998) 'Bob Hasan, the Rise of APKINDO, and the Shifting Dynamics of Control in Indonesia's Timber Sector', *Indonesia* (65): 1–35.

Brookfield, H. and Y. Byron (eds) (1993) *South-East Asia's Environmental Future: The Search for Sustainability*. Tokyo: United Nations University Press/Oxford University Press.

CIC Consulting Group (1997) 'Study on Oil Palm Industry and Plantation in Indonesia'. Jakarta: PT Capricorn Indonesia Consult, Inc.

Colfer, Carol J. P. and Richard Dudley (1993) 'Shifting Cultivators of Indonesia: Marauders or Managers of the Forest?'. Community Forestry Case Study Series. Rome: FAO.

Cronon, W. (1992) 'A Place for Stories: Nature, History, and Narrative', *The Journal of American History* (March): 1347–76.

Curry, M. R. (1995) 'Rethinking Rights and Responsibilities in Geographic Information Systems: Beyond the Power of Imagery', *Cartographic and Geographic Information Systems* 22(1): 58–60.

Dauvergne, P. (1994) 'The Politics of Deforestation in Indonesia', *Pacific Affairs* 66(4): 497–518.

Davis, Mike (1998) *Ecology of Fear: Los Angeles and the Imagination of Disaster*. New York: Metropolitan Books.

Davis, Mike (forthcoming) 'Late Victorian Holocausts: The El Niño Famines of the 1870s and 1890s', in Nancy Peluso and Michael Watts (eds) *Violent Environments*. Berkeley, CA: University of California Press.

Direktorat Jendral Perkebunan (1997) *Statistik Perkebunan Indonesia*. Jakarta.

Direktorat Jendral Perlindungan Hutan dan Pelestarian Alam (1998) 'Rekapitulasi Data Kebakaran Hutan Tahun 1997 per Fungsi per Propinsi Seluruh Indonesia (Hasil Klarifikasi)'. Bogor: Official report.

Dove, M. R. (1985) *Swidden Agriculture in Indonesia: The Subsistence Strategies of the Kalimantan Kantu'*. Berlin: Mouton.

Dove, M. R. (1993) 'Smallholder Rubber and Swidden Agriculture in Borneo: A Sustainable Adaptation to the Ecology and Economy of the Tropical Forest', *Economic Botany* 47(2): 136–47.

Dove, M. R. (1998) 'Living Rubber, Dead Land, and Persisting Systems in Borneo: Indigenous Representations of Sustainability', *Bijdragen* 154(1): 1–35.

Dove, M. R. and Mahmudul Huq Khan (1995) 'Competing Constructions of Calamity: The April 1991 Bangladesh Cyclone', *Population and Environment* 16(5): 445–71.

Dove, M. R. and D. K. Kammen (1997) 'The Epistemology of Sustainable Resource Use: Managing Forest Products, Swiddens, and High-Yielding Variety Crops', *Human Organization* 56(1): 91–101.

Economic and Environment Program for Southeast Asia and The World Wide Fund for Nature (EEPSEA/WWF) (1998) 'The Indonesian Fires and Haze of 1997: The Economic Toll'. Unpublished report, 29 May 1998.

Economist Intelligence Unit (1995) 'Country Report: Indonesia 4th Quarter 1994–95'. London: The Economist Intelligence Unit.

Erikson, Kai (1976) *Everything in Its Path*. New York: Simon and Schuster.
Erikson, Kai (1994) *A New Species of Trouble: The Human Experience of Modern Disasters*. New York: Norton.
Fyfe, G. and J. Law (1988) *Picturing Power: Visual Depiction and Social Relations*. New York: Routledge.
Geddes, W. R. (1954) *The Land Dayaks of Sarawak*. London: Her Majesty's Stationery Office.
Gellert, Paul (1998) 'A Brief History and Analysis of Indonesia's Forest Fire Crisis', *Indonesia* (65): 63–85.
Government of Indonesia (1996) 'National Forest Inventory of Indonesia: Forest Resources Statistics Report'. Jakarta: Ministry of Forestry.
Gupta, Akhil and James Ferguson (1997) *Beyond Culture: Space Identity and the Politics of Difference*. Durham, NC: Duke University Press.
Hacking, Ian (1989) 'Language, Truth and Reason', in R. Hollis and S. Lukes (eds) *Rationality and Relativism*, pp. 185–203. Cambridge, MA: MIT Press.
Hall, Stuart (1993) *Mapping the Next Millennium: How Computer-Driven Cartography is Revolutionizing the Face of Science*. New York: Vintage Books.
Hall, Stuart (1996) 'On Postmodernism and Articulation: An Interview with Stuart Hall', edited by Lawrence Grossman, in David Morley and Chen Kuang-Hsing (eds) *Stuart Hall: Critical Dialogues in Cultural Studies*, pp. 131–50. London: Routledge.
Haraway, Donna (1991) *Simians, Cyborgs, and Women: The Reinvention of Nature*. London: Free Association Books.
Harding, Sandra (1998) *Is Science Multicultural? Post-colonialisms, Feminisms and Epistemologies*. Bloomington and Indianapolis, IN: Indiana University Press.
Harley, J. B. (1989) 'Deconstructing the Map', *Cartographica* 26(2): 1–20.
Hewitt, K. (ed.) (1983) *Interpretations of Calamity*. Boston, MA: Allen and Unwin.
Hewitt, K. (1995) 'Sustainable Disasters? Perspectives and Powers in the Discourse of Calamity', in J. Crush (ed.) *Power of Development*, pp. 115–28. New York: Routledge.
King, Victor (1996) 'Environmental Change in Borneo: Fire, Drought and Rain', in Michael Parnwell and Raymond Bryant (eds) *Environmental Change in Southeast Asia: People, Politics and Sustainable Development*, pp. 165–89. New York: Routledge.
Kuhn, Thomas (1962) *The Structure of Scientific Revolutions*. Chicago, IL: University of Chicago Press.
Latour, B. (1987) *Science in Action: How to Follow Scientists and Engineers Through Society*. Milton Keynes: Open University Press.
Leighton, Mark and Nengah Wirawan (1986) 'Catastrophic Drought and Fire in Borneo Associated with the 1982–83 El Niño Southern Oscillations Event', in Gillian Prance (ed.) *Tropical Rainforests and the World Atmosphere*, pp. 75–102. Washington, DC: American Association for the Advancement of Science.
Li, Tania M. (forthcoming) 'Constituting Tribal Space: Indigenous Identity and Resource Politics in Indonesia', *Comparative Studies in Society and History*.
Mackie, C. (1984) 'The Lessons Behind East Kalimantan's Forest Fires', *Borneo Research Bulletin* (16): 63–74.
MacKinnon, K., G. Hatta, H. Halim and A. Mangalik (1996) *The Ecology of Kalimantan: Indonesian Borneo*. Ecology of Indonesia Series Vol. III. Singapore: Periplus Editions.
Mayer, Judith (1996) 'Impacts of the East Kalimantan Forest Fires of 1982–92 on Village Life, Forest Use, and Land Use', in Christine Padoch and Nancy Peluso (eds) *Borneo in Transition: People, Forests, Conservation, and Development*, pp. 187–218. Kuala Lumpur: Oxford University Press.
Mentri Negara Lingkungan Hidup (1997) 'Sambutan Dialog Nasional Dampak Kebakaran Hutan Dan Lahan 1997'. Jakarta: October 22, 1997. Proceedings published by KONPHALINDO.
Padoch, Christine (1982) *Migration and Its Alternatives Among the Iban of Sarawak*. Leiden: Royal Institute of Linguistics and Anthropology.

Padoch, Christine (1994) 'The Woodlands of Tae: Traditional Forest Management in Kalimantan', in William Bentley and Marcia Gowen (eds) *Forest Resources and Wood-Based Biomass Energy as Rural Development Assets*, pp. 1–17. New Delhi: Oxford Publishing.

Peluso, Nancy (1994) *Rich Forests, Poor People: Resource Resistance and Control in Java.* Berkeley, CA: University of California Press.

Peluso, Nancy (1995) 'Whose Woods are These? Counter-mapping Forest Territories in Kalimantan, Indonesia', *Antipode* 27(4): 383–406.

Peluso, N., P. Vandergeest and L. Potter (1995) 'Social Aspects of Forestry in Southeast Asia: A Review of Postwar Trends in the Scholarly Literature', *Journal of Southeast Asian Studies.* 26(1): 196–218.

Peluso, N. and P. Vandergeest (forthcoming) 'Genealogies of Forest Law and Customary Rights in Indonesia, Malaysia and Thailand'. Unpublished manuscript.

Pickles, John (1991) 'Geography, GIS, and the Surveillant Society', *Papers and Proceedings of the Applied Geography Conferences* (14): 80–91.

Pickles, John (ed.) (1995) *Ground Truth: The Social Implications of Geographic Information Systems.* New York: Guilford Press.

Potter, Lesley and Justin Lee (1998) 'Tree Planting in Indonesia: Trends, Impacts and Directions'. Consultants' Report to Center for International Forestry (CIFOR) and USAID. University of Adelaide.

Rabinow, Paul (1986) 'Representations are Social Facts: Modernity and Post-modernity in Anthropology', in James Clifford and George Marcus (eds) *Writing Culture: The Poetics and Politics of Ethnography*, pp. 234–61. Berkeley, CA: University of California Press.

Ramon, J. and D. Wall (1998) 'Fire and Smoke Occurrence in Relation to Vegetation and Land Use in South Sumatra Province, Indonesia, with Particular Reference to 1997'. Jakarta: Unpublished paper of EU/Forest Fire Prevention and Control Project.

Salafsky, N. (1994) 'Drought in the Rain Forest: Effects of the 1991 El Niño Southern Oscillation Event on a Rural Economy in West Kalimantan, Indonesia', *Climatic Change* (7): 373–96.

Schindele, W., W. Thomas and K. Panzer (1989) 'The Forest Fire 1982/83 in East Kalimantan, Part I: The Fire, the Effects, the Damage and Technical Solutions', FR Report No. 5. Feldkirchen: Deutsche Forstservice GmbH.

Schweithelm, James (1998) 'The Fire this Time: An Overview of Indonesia's Forest Fires in 1997/98'. World Wide Fund for Nature Indonesia Programme, Discussion Paper. Jakarta.

Scott, James (1998) *Seeing Like a State: How Certain Schemes for Improving the Human Condition Have Failed.* New Haven, CT: Yale University Press.

Sheppard, E. (1993) 'Automated Geography: What Kind of Geography for What Kind of Society?', *Professional Geographer* 45(4): 457–60.

Sheppard, E. (1995) 'GIS and Society: Towards a Research Agenda', *Cartographic and Geographic Information Systems* 22(1): 5–16.

Siegert, F. and A. A. Hoffmann (1998) 'Evaluation of the Forest Fires of 1998 in East Kalimantan, Indonesia using multi-temporal ERS-2 SAR images and NOAA-AVHRR data'. Paper presented at the International Conference on Data Management and Modelling Using Remote Sensing and GIS for Tropical Forest Land Inventory. Jakarta, Indonesia (26–29 October).

Soemarsono (Director General of PHPA) (1997) 'Langkah-Langkah Penanggulangan Kebakaran Hutan Selama Musim Kemarau Tahun 1997 di Indonesia'. Paper presented at National Dialogue on the the Impacts of the 1997 Forest Fires, Jakarta (22 October).

Sunderlin, William and Ida Aju Pradnja Resosudarmo (1996) 'Rates and Causes of Deforestation in Indonesia: Towards a Resolution of Ambiguities'. Center for International Forestry (CIFOR) Occasional Paper No. 9. Bogor, Indonesia: CIFOR.

Taylor, P. (1990) 'Editorial Comment: GKS', *Political Geography Quarterly* 9(3): 211–12.

Thongchai Winichakul (1994) *Siam Mapped: A History of the Geo-body of Siam.* Honolulu, HI: University of Hawaii Press.

Tsing, Anna (1998) 'Notes on Culture and Natural Resource Management'. Paper Presented to the UC Berkeley Environmental Politics Seminar (11 December).

United Nations Development Program (1998) 'Forest and Land Fires in Indonesia: Impacts, Factors and Evaluation'. Jakarta: Project Report published by the Indonesian State Ministry for Environment.

Vayda, A. P., C. J. P. Colfer and M. Brotokusomo (1980) 'Interactions Between People and Forests in East Kalimantan', *Impact of Science on Society* 30(3): 179–90.

WALHI (1997) 'Data Kebakaran Hutan: Hasil Investigasi POSKOBAR WALHI'. Paper presented at National Dialogue on the Impacts of the 1997 Forest Fires, Jakarta (22 October).

Watts, M. J.(1983) *Silent Violence: Food, Famine and Peasantry in Northern Nigeria*. Berkeley, CA: University of California Press.

Watts, M. J. and H-G. Bohle (1993) 'The Space of Vulnerability: the Causal Structure of Hunger and Famine', *Progress in Human Geography* 17(1): 43–87.

WWF International (1997) 'The Year the World Caught Fire'. Discussion Paper. Gland, Switzerland: WWF International.

Press Citations

Bangkok Post (14/11/97) 'Indonesia's Shame Won't Blow Away'.

Dow Jones Newswire (24/4/98) 'Indonesia to Prosecute Five Companies for Starting Fires'.

Far Eastern Economic Review (14/5/98) 'Fund under Fire'.

Indonesia Daily News Online (14/3/98) 'Sarwono Kusumaatmadja: Orang Kebun Itu Kurang Ajar'.

Jakarta Post (7/10/97) 'Djamaludin reports damage of forest fires to Suharto'.

Jakarta Post (11/10/97) Letters to the Editor.

Jakarta Post (1/6/99) 'Dayak People Win Lawsuit against Salim Group'.

Kaltim Post (27/11/98) '3,5 Juta Hektare Lahan Kaltim Disiapkan Untuk Perkebunan Sawit'.

Nature (13/11/97) 'Malaysia Backs "Gag" on Haze Scientists'.

New Scientist (4/10/97) 'The Neighbour from Hell'.

New Scientist (4/10/97) 'Indonesia's Inferno Will Make Us All Sweat'.

Reuters (14/4/98) 'Thickening Smog Sets Off SE Asian Alarm Bells'.

Suara Pembaruan (5/2/98) 'Dituduh Bakar Hutan, 11 PT Diseret ke Pengadilan'

The Economist (4/10/97) 'When the Smoke Clears in Asia'.

The Guardian (8/11/97) 'The Thousand Mile Shroud'.

WALHI Press Release (17/10/98) 'Gugatan Asap WALHI Dimenangkan oleh Pengadilan'.

World Watch (Jan/Feb 98) 'Indonesia Ablaze'.

14 From Timber to Tourism: Recommoditizing the Japanese Forest

John Knight

INTRODUCTION

Japan is one of the most urbanized societies in the world. An almost continuous urban belt has developed stretching from Tokyo in the east to Kita Kyushu in the west — a 'megalopolis' comprising some 78 million people or 63 per cent of the Japanese population (Karan, 1997: 23–5). Almost two-thirds of the national population live on just 3 per cent of Japan's land (ibid.). This integration affects rural space, as well as urban space, as the megalopolis absorbs much of the arable part of rural Japan. These rural areas in turn take on a quasi-urban character whereby local livelihoods come to be based on urban-type employment in conjunction with part-time farming. According to Kelly's formulation, the age of rural Japan is over, and in its place there exists only 'regional Japan' (Kelly, 1990) — a derivative space wholly subordinate to the megalopolis.

This development has profound implications for upland areas of Japan. One of the features of post-war Japan has been a national preoccupation with the rural village as a spiritual home, or *furusato*, an unchanging locus of affective belonging in a rapidly urbanizing society. But the era of the megalopolis heralds the disappearance of rural Japan, and this ongoing urbanization has the effect of valorizing remoter, upland areas as the new site of the Japanese pastoral.

This is the background to the present-day transformation of upland areas of Japan into sites of mass recreation. One form which this trend takes is large-scale resort development in which leisure space is created by felling forest, levelling hillsides, and building large-scale tourist facilities. Another centres on the forest itself as the prime tourist attraction. This article examines the process by which a forest dominated by industrial timber forestry is being transformed into a recreational forest, drawing on ethnographic data from the Kii Peninsula in western Japan, and on written sources on forest tourism from other regions of the country.

RECREATIONAL FORESTS

Plantation forestry is practised throughout the world. In the early 1990s, plantations or 'man-made forest' accounted for around 150 million ha of the world's 4 billion ha of forest — less than 4 per cent of the global forest area (Mather, 1993a: 4). In the mid-1980s around three-quarters of the industrial plantation area was located in temperate regions, with a quarter in the tropics and sub-tropics (Kanowski and Savill, 1992: 122). Timber plantations are set to become much more ubiquitous in the twenty-first century, especially in the tropics, as states attempt to realize a range of important resource-related and environmental objectives. Afforestation policies can help to secure long term wood supplies (for fuel, timber, paper) on a renewable basis; restore tree cover to bare land; slow rainwater flow and prevent flooding; re-utilize degraded farmland and other derelict land; and meet national targets for carbon fixation.

However, monocultural timber plantations have also attracted much criticism. There have been many examples of failures in industrial plantations, especially those in tropical areas, due to their biological instability, to mismatches between site and species, and to poor management (Sargent, 1992a: 9–12). One consequence of this critique of plantation management is the emergence of a trend towards the diversification of plantations. Timber plantations in the future can no longer be simple monocultural 'tree farms', 'green deserts' or 'biological deserts', but must meet certain minimum standards of biodiversity (Spellerberg and Sawyer, 1996).

Many erstwhile monocultural plantations are set to become new kinds of forest — variously characterized as 'multi-purpose forests', 'multi-objective forests', 'post-industrial forests', and so forth. If 'industrial forests' are 'geared to the production of wood for industrial purposes', 'post-industrial forests' offer 'a wider range of social and environmental benefits' (Mather, 1993b: 217–18). The 1992 Convention on Biological Diversity gave a further fillip to the establishment of forests which have environmental, recreational and other non-industrial sources of value (Spellerberg and Sawyer, 1996).

Recreational forests have become common in recent decades. Forests in Europe, North America and Australia are being strategically re-defined and modified — and even new forests created — as sites of recreation which provide a range of outdoor leisure opportunities for national populations (Bell and Evans, 1997; Cloke et al., 1996; Romano, 1995; Selin and Lewis, 1994). This shift is sometimes characterized in terms of a change in management ethos away from 'commodity production' towards 'non-commodity resource management' (Farnham et al., 1995). Yet the move towards 'tourist forests' is often explicitly couched in commercial and marketing language — with forests defined as a touristic 'resource' to be exploited, and forest administrators and managers charged with the task of 'marketing the forest' to metropolitan populations (Sogar and Oostdyck, 1994).

One of the obstacles to this new trend of forest 'marketing' is the forest's previous history of commoditization, particularly as a site of timber growing. In many cases, this new, ubiquitous landscape legacy of conifer plantations is the object of public complaint on aesthetic grounds, in terms of both their monocultural uniformity and the visual impact of clearcuts (Bass, 1992: 55; Gillmor, 1993: 46–7; Lindhagen, 1996; Lucas, 1997). The rise of forest tourism often leads to restrictions placed on industrial forestry, such as reduced clearfelling (Bostedt and Mattsson, 1995). The emergence of forest tourism in Japan, in a landscape dominated by industrial forestry, illustrates the tensions that can exist between these two different forms of forest commoditization.

JAPANESE FORESTS

Two-thirds of the Japanese land area consists of mountain forest. In Japan forest is synonymous with mountain, as mountains are generally wooded and forest elevated. The word *yama*, written with the Chinese character for mountain, applies to both mountain and forest. Japan consists of two main ecological zones corresponding to two kinds of indigenous forest: beech forest (*bunarin*), a deciduous forest distributed towards the northeast of the country and at higher elevations (700–1,500 metres); and Lucidophyllous or shiny-leafed forest (*shoyojurin*), a warm temperate forest widely found to the southwest and at lower elevations (below 500 metres), which forms part of a wider East Asian ecological zone. Both types of forest are found in the upland interior of the southern Kii Peninsula, and some local writers proclaim the special importance of this area's natural environment as the place of the most southerly Japanese beech forest and even the site where Japan's two ecologies meet (Ue, 1994: 4–6).

However, since the late nineteenth century, much beech and Lucidophyllous forest has been replaced by industrial forest — first by fuelwood forest (for charcoal) and then by conifer forest (for timber). In the 1990s around 45 per cent of the Japanese forest area consisted of timber plantations of cypress (*hinoki*, *Chamaecyparis obtusa*) and cryptomeria (*sugi*, *Cryptomeria japonica*). The scale of plantation forestry greatly increased after the war. Japan's mountainous landscape had lost much of its tree cover, resulting in floods, landslides and river silting. In response, the government launched a nationwide reforestation campaign, involving public tree-planting ceremonies attended by the emperor who took the lead in planting conifer saplings himself, along with many other local tree-planting events. The campaign succeeded in restoring tree cover to hillsides and mountainsides throughout the archipelago. The total plantation area established since the war exceeds ten million ha.

One of the hopes for post-war reforestation was that it would secure the future of upland areas through economic development based on a modern forestry industry. Through large-scale state-supported investment, the mountain forests would become the place of modern, technologically advanced, mechanized forest work. Mountain forests would be made more accessible through the development of an enhanced network of roads, the introduction of mechanized saws would increase the efficiency of forest labourers, and advances in science would boost productivity and product quality in forestry. The combination of scientific, technical and logistical improvements suggested that forestry would bring mountain villagers a prosperous future.

Yet despite the achievements in reforestation — in restoring tree cover to the mountains — Japanese forestry is in decline. In the course of its post-war recovery, Japan experienced a shortage of usable wood products, and from the 1960s began importing large quantities of wood. Wood imports have continued to grow ever since, and by the 1990s domestic timber growers were complaining that their product had been displaced by wood imports from southeast Asia, Siberia and North America. In the mid-1990s, imports accounted for some 80 per cent of the wood consumed in Japan. In this same period, rural Japan underwent large scale outmigratory depopulation resulting in a forestry labour shortage and the neglect of much of the post-war plantation forest. At least one-third of the nation's timber plantations suffer from complete neglect, while a much larger proportion of the plantation forest suffers from partial neglect and, therefore, commercial devaluation. Another consequence of rural depopulation and forestry decline has been the depreciation of forest land. It is against this background that new forms of forest land use have emerged.

Forest Tourism in Japan

Many Japanese people visit forests. A 1993 survey found that 70 per cent of respondents in their twenties and thirties had visited 'the mountains, a forest, valley or other natural area for a non-work purpose during the past year' (Fujitake, 1993). In a 1994 survey of urban workers in Tokyo, three-quarters of those surveyed expressed a positive preference for visiting village areas; two-thirds of these indicated a preference for forest hiking and rambling and one-third for fishing and forest gathering activities (Inoue, 1996: 39). National park areas which include much primary forest, from Shiretoko in the north to Yakushima in the south, have become major tourist destinations (Mitsuda and Geisler, 1992: 33–4).

The recreational use of forests is something promoted by the Japanese government: '560,000 hectares of national forests at about 1,100 sites throughout the country are designated as forest recreation areas, which include forests for nature observation and outdoor sports' (NLAPO, 1991: 54). Forests are becoming the site of a variety of recreational activities, including hiking, gathering (nuts, herbs, mushrooms, flowers etc.), camping, bird-watching, fishing, canoeing, mountaineering and paragliding. There is also a growing number of wildlife parks (including monkey parks, bear parks and deer parks) and other nature parks in Japan. Some leisure parks have established 'ecology camps' in which tourists learn about forest flora and fauna.

Forests are being re-designed to enhance their touristic appeal. Aya-machi in Miyazaki Prefecture promotes itself to tourists as a site of Lucidophyllous forest, the natural vegetation of southwest Japan (Takeuchi, 1993: 116–18), while other areas, especially in northeastern Japan, appeal to would-be visitors as sites of native beech forest. But forest tourism extends to secondary forests. In Japan, the pine is one of the most popular trees, and secondary pine forests have considerable touristic appeal. Japanese people express a preference for an airy forest where they 'can walk, play and enjoy flowers', and the pine forests provide an ideal environment for this kind of recreation (Kamada et al., 1991: 61). There are also examples of explicitly non-native forest tourism — such as *doitsu no mori*, 'the German forest', that has been re-created in Okayama Prefecture.

On the southern Kii Peninsula, visitors are encouraged to try hiking through the forest and tracing the footsteps of ancient pilgrims. Skyline roads have been built along mountain ridges in order to facilitate scenic drives. 'Forest showering' or *shinrin'yoku*, a kind of group-hike through the forest, is promoted as a recreational activity in many areas of Japan (see Kamiyama, 1983; Mishima, 1994), including municipalities on the Kii Peninsula. Also known as the 'forest remedy' (*shinrin ryoho*), it is believed to be a health-enhancing pastime which benignly stimulates the workings of the main internal body organs (Kamiyama, 1983: 12).

More generally, the forest forms a pervasive motif in tourism in upland areas of Japan. Guest-houses and restaurants are named after the forest (for example, *Mori no sato*, 'The Village in the Forest'); forest imagery is ubiquitous (scenic forest posters and framed forest photographs are common in coffee houses and guest-houses); a range of souvenirs, made from forest wood, are on offer; small wooden or ceramic models of forest animals are sold as souvenirs; guest-houses display stuffed animals (such as deer or pheasant) and caged forest songbirds in their lobby areas; dishes of wild boar meat, venison and forest herbs are served to guests; and books about the forests (on hunting, charcoal burning etc.) are on sale.

There is a revival of traditional skills and crafts associated with forest areas such as charcoal-burning, wood-turning, bamboo crafts, paper making, and dyeing. Tourism provides a stimulus for farming parts of the forest. One example of this is wildlife farming, for meat and medicinal wildlife products.[1]

---

1. Deer products include venison, deer 'ham' on the bone, deer horn alcohol (*tsunoshu*), and deer velvet (*rokujo*). Some proponents believe that deer farming could become a mainstay industry for depopulated upland areas in Japan (Drew et al., 1989: 345). Demand for wild boar meat, marketed as 'the taste of the wilds' (*yasei no aji*) or 'the taste of the forest' (*yama no aji*), has soared in the 1980s and 1990s (Takahashi, 1995: 40–55; see also Nagata, 1989: 346; Ue, 1994: 151–4). Venison and boar meat are served in *ryokan* tourist inns in rural areas and sold in mail order farm produce businesses as a quintessential winter taste

Forest plants, including medicinal herbs, are also cultivated in connection with tourism.[2]

Problems with Tourist Forests

This process of converting the forest into a tourist resource faces a number of problems. First of all, despite the fact that many rural areas appeal to past traditions of forest visiting, such as pilgrimage, in Japan the forest has a rather frightening image. The forest is the place where the souls of the dead go, while in folklore it features as a site 'of the terrifying and mysterious, of violent abductions and ghastly crimes' (Ivy, 1995: 108). This negative image of the forest extends to the local human population: itinerant charcoal-burners, hunters, and loggers have often been viewed as primitive, backward and dangerous, while remote mountain villagers have been seen variously as demons, as half-man half-bird monsters (*tengu*), as monkeys and as bears.[3] There are even recent reports of hikers in the forest hiding in fright on encountering foresters (Ue, 1980: 165).

The forest also poses real dangers to tourists. Traffic accidents occur on the narrow, winding forest roads, and herb and mushroom collectors slip on the steep mountain slopes or get lost in the mountains. Bears, feral dogs and poisonous snakes pose a physical threat to visitors; there are regular newspaper reports of tourists attacked by wild animals, including herb gatherers attacked by bears (see Maita, 1996: 107–8; Miyao, 1989: 198–201). Hunting can also be dangerous for tourists. In general, hunters travel to the remote forests to hunt, whereas tourists visit the proximate forests; overlap between the two is rare, and tends to involve, at most, hikers hearing the sound of distant gunfire. However, there are hunting accidents involving hikers who unwittingly stray into the scene of a hunt and get fired on by mistake, attacked by a hound, or chased and injured by a wounded wild boar.[4]

Another problem with forest tourism in Japan is topographical. The elevated character of Japan's forests permits a range of vertical recreations, including mountaineering, skiing and para-gliding. Ski grounds, in particular, have become common on the snowy Japan Sea side of the country.

---

of the Japanese countryside (e.g. *Asahi Shinbun* 15/12/92). There is also a demand for the gall of wild boar. In some areas, snakes are farmed to make snake tonic and health products.

2. In some areas, in an effort to incorporate wild plants into the tourist economy, perilla juice and wild grape wine (*yamabudo no wain*) are produced, to serve as a counterpart to the consumption of local wildlife products such as venison or wild boar dishes (*Asahi Shinbun* 24/3/93). In some cases, the cultivation of *kanpoyaku* herbs is emphasized (e.g. Nakase, 1993: 17).
3. On the stigmatization of upland dwellers as 'bears', see Sakurai (1990: 599), and as monkeys, see Himeda (1984: 52) and Nebuka (1991: 91).
4. See *Asahi Shinbun* (18/11/91; 6/12/93; 23/5/94).

However, the temperate climate on Japan's Pacific side, on which the Kii Peninsula lies, does not permit skiing. This mild climate does allow for year-round hiking, but hiking through the mountain forests of the region can be a physically demanding activity. Although the tourist areas of the Kii Peninsula are popular among older people, the elevated character of the land effectively rules out hiking for many of them. The forests do not really permit the kind of gentle exertion appropriate to the elderly visitor. The mountainous landscape of the Kii Peninsula also inhibits golf-course development, despite the proliferation of golf courses across Japan. The vertical spaces of the peninsula are inappropriate for the basically horizontal game of golf.

Another problem arises from the degradation of pine forests. Many pine trees have withered due to pine rot. The tall pale skeletons of rotted pines (which stick out from the forest) are an unpleasant sight, and some tourists wonder if this is the effect of acid rain. Moreover, since the 1960s, the exploitation of the pine forest (for timber, resin, forest litter, charcoal etc.) has largely ceased. The consequent accumulation of organic litter on the forest floor and the increase in undergrowth and scrub make the pine forests difficult to pass through and diminish their recreational value. Some rural areas are attempting to arrest this trend and restore local pine forests on account of their aesthetic value and wider national appeal to tourists (Kanzaki, 1988: 165). There is also a trend towards the restoration of grasslands to produce flower-laden landscapes in the spring and early summer. In an effort to brighten the forest, picnic spots and parks within the forest area have spread, as have airy nature trails.[5]

The new trend of forest tourism also encounters problems of forest access. Tourist activities clash with local activities, for instance, where both tourists and locals pick the same mushrooms, herbs or flowers. The depletion of such a limited good by one side is the cause of complaint or anger by the other. Although no longer central to local livelihoods, the forest continues to be a site of local productive and income-earning activities — forest-labouring, hunting, gathering etc. — as well as a site of local recreation. Wild herb gathering, an important source of local income as well as a popular tourist activity, leads to local complaints about tourists picking valuable wild herbs.

Tourism is a cause of environmental destruction. Resort development — golf courses, ski resorts, theme parks, and so on — leads to large-scale forest clearance, attracting much public criticism (Honda, 1993: 178–9). Tourist

5. However, one of the problems that arises from this trend of opening up the forest to tourism by creating open spaces is that the new grassland becomes a prime feeding ground for forest herbivores such as deer and serow and contributes to the increase in their numbers, with adverse consequences for local foresters and farmers. Indeed, the spread of golf courses, one of the primary forms of 'tourist grasslands', is also said to cause an increase in deer numbers (*Asahi Shinbun* 9/12/94).

skyline roads have certain negative environmental consequences. Road widening itself can lead to considerable forest clearance, while the new roads are said to undermine the scenic beauty of many remote spots (in addition to the flat road itself, the roadside hill-face often has to be sandblasted and fenced). Tourist skyline roads, as well as forest roads, also adversely affect wildlife by cutting-up and fracturing wildlife habitat and sealing off animal sub-populations from one another. The building of toll highways along mountain ridges, as part of tourist development, has partitioned and fragmented monkey habitat (Suzuki, 1972),[6] while new skyline roads (for example, on the southern Kii Peninsula) often cut through bear habitat. This disruption of wildlife habitat can exacerbate crop-raiding on village farms.[7] For some local people on the southern Kii Peninsula, tourism, far from bringing local benefits, makes village life that much harder.

Forest tourism in Japan thus encounters a range of different problems — concerning both tourist perceptions of the forest and local perceptions of the tourists. The next section focuses on an even more basic problem for forest tourism, one which centres on the industrial character of much of the national forest landscape.

FORESTRY AND FOREST TOURISM

Forest tourism can co-exist with forestry to a certain extent, and forestry can benefit tourism in a number of ways. First, forest tourism relies on forestry roads which, although established by the forestry industry (to enhance access to timber plantations and expedite timber extraction and transportation etc.) also permit the exploitation of the forest for tourism. Many forest roads have been widened to permit access to the remote forest by tourist coaches and other traffic.

Secondly, timber plantations can, up to a point, appeal to tourists. In Japan there is not the degree of intense antipathy to conifers reported elsewhere. Mature timber forests can be an object of aesthetic appreciation, especially for the forester (as in the expression *birin* or 'beautiful forest', applied to a fine timber plantation). Reflecting the Japanese forester's perspective, Totman characterizes the well-maintained plantations of Akita

---

6. The clash between tourism and wildlife conservation is illustrated in the recent controversy on the World Heritage island of Yakushima, where the proposed expansion of a forestry road to permit tourist buses to circle the island has led to a clash between tourist interests and the conservationists who oppose the road expansion plan (JPN, 1995: 1–3).
7. Another cause of wildlife crop-raiding is tourism-related provisioning of wild animals. The provisioning of wild monkeys for tourism, critics argue, has caused or at least exacerbated farm crop-raiding because it has lured the monkeys down from the forests to the edge of the village and has made them accustomed to village food. Impromptu provisioning of monkeys by tourists also occurs and can have similar effects. To a somewhat lesser extent, problems arise with other wild animals, such as deer, wild boar and bears.

Prefecture as 'the jewels of Japan' (Totman, 1985: 3). For many urban Japanese who visit rural areas, the plantation forest is viewed as *yama* (forest) or *midori* (greenery) and does not readily stand out as artificial. The background to such positive views is that the two main commercial conifers, cypress and cryptomeria, are indigenous species which occupy an important place in Japanese culture and history.[8]

More specifically, older plantation forests, which have undergone successive thinnings, become airy, light places attractive to visit. Indeed, the more advanced-aged plantations, with a small number of large trees and green expanses in-between, come to resemble parkland. In areas where older plantations predominate, deer congregate to feed on the herbaceous undergrowth that emerges, and become a wildlife spectacle for tourists, especially as the deer gradually become tame. Thus, in addition to formally established 'deer parks', a kind of deer parkland can develop in forestry areas (cf. Dizard, 1994: 12–13).

On the other hand, tourism can benefit forestry in certain ways. Tourism provides new markets for plantation wood — through the construction of prestigious tourist buildings made of local wood; through the use of (inferior and otherwise discardable) plantation wood to make tourist souvenirs; and through the creation of new non-wood products.[9] Tourism also revives the demand for charcoal products. In the 1990s, demand for charcoal in Japan increased, in part due to the rise in outdoor camping.[10]

Woodcraft production for tourism boosts demand for low grade wood such as wood thinnings (*kanbatsuzai*). At present, much of the wood grown in rural Japan goes to waste; the commercial value of thinned logs is so low that they are often just left lying on the ground in the forest. If wood thinnings could be made commercially valuable, this would provide an extra incentive for forest landowners to thin their timber plantations. Due to large-scale outmigration and the decline of forestry, unthinned forests have become a major problem. A tourist woodcrafts market for wood thinnings encourages plantation thinning and, in turn, can help to improve the quality of maturing timber stands. Local woodcraft industries provide a new source

---

8. They are mentioned in Japanese mythology — for example, as two of the trees planted by the *kami* spirit from his body hair, and from the timbers of which were built the boats (cryptomeria) with which the Japanese travelled to the outside world and prospered, and the houses (cypress) for people to live in (see Wada, 1978: 196). Many celebrated Japanese shrines and temples around the country have been built from cryptomeria and cypress.
9. One of the main plantation conifers, cypress, with its distinctive smell, is used to make a variety of products such as cypress milk cream, cypress skin oil (for skin diseases), cypress shampoo, and cypress air-fresheners.
10. It has also been proposed that conifer wood be used to make charcoal for camping use (Tanaka, 1996: 92–7). In addition to the use of charcoal as fuel, a Wakayama company markets charcoal bath salts, charcoal pillows, charcoal mattresses, charcoal slippers, and charcoal for cisterns etc.

of income for retired foresters and help to occupy other foresters who have developed vibration white-finger (through overuse of chainsaws) and can no longer perform outdoor manual labour.

Tourism can provide employment for retired foresters as tourist guides — for example as herb-picking or mushroom-gathering guides for tourists (Kamata and Nebuka, 1992: 415). The knowledge of the forest accumulated over decades by former charcoal-burners, loggers and plantation foresters can be put to use in tourism. Older foresters have typically acquired an impressively wide knowledge of forest flora and fauna and can use this to point out and name the distinctive plants as well as the wild animals of the forest (their tracks, trails and other signs). They can also explain charcoal-burning at the sites of old kilns; talk about the timber plantations and their care; and recount the various bits of folklore associated with particular places in the forest (magical trees, monster-sightings, ghost stories etc.).

In some cases, forestry is directly incorporated into tourism. Upland municipalities across Japan have established 'tourist forestry' (*kanko ringyo*) schemes in which visitors plant, weed, prune or thin timber plantations (see CRSC, 1995: 119–88).[11] Another example of the use of forestry as a theme for tourism is the rafting experience offered by some upland municipalities.[12]

However, forest tourism also clashes with forestry in a variety of different ways. First, the ubiquity of timber plantations poses a serious obstacle to forest tourism in Japan. As the existing network of roads has largely been built for forestry, it usually leads to plantation forest. The roads that admit one into the forest are themselves signs of its transformation. On much of the interior of the Kii Peninsula, a road to a natural forest is almost a contradiction in terms: aside from the protected watershed forests/upstream locales, forest roads were built to fell such forest and replace it with fast-growing conifers.

A variety of environmental beautification measures have been undertaken in recent years in order to offset the uniformity of the conifer plantations. Roadside flower-growing and tree-planting (with cherry blossoms, maples, camellia etc.) have been carried out by youth groups and other civic groups. There have also been specific municipal initiatives aimed at re-naturalizing local forests. In the 1990s, for example, the municipality of Hongu-cho purchased an area of forest land along a local hiking trail popular with tourists. Known as the *Kumano sosei no mori* or 'The Forest of Kumano Revival', the long-term aim is to create a mixed forest by interplanting non-

---

11. One example of 'tourist forestry' from the Kii Peninsula is the Mountain Spirit Sweat Tour (*yama no kami ase kake tsua*) which has been held annually in Hongu-cho since 1994 in which women tourists recreationally undertake forestry labour tasks such as undergrowth-clearing and plantation thinning.
12. An example is Kitayama-mura on the southern Kii Peninsula. Tourists board converted rafts of the kind which used to transport lumber downstream, and experience the thrill of being carried by the river down the river valleys, just as raftsmen did in the past. Indeed, these tourist rafts are steered by former forestry raftsmen.

conifer species with existing older conifer stands — although such limited measures hardly conceal the prevalence of the conifer plantations.

Even though Japan does not suffer the same degree of conifer-phobia as reported for many other countries, conifers are increasingly criticized. The spread of conifer plantations at the expense of natural forest is said to create an 'unhealthy' or 'sick' forest (Kawasaki, 1993: 6; Matsuzawa, 1989: 114), a development which has been likened to cancer (Azumane, 1993: 115–17). Plantations are criticized as dull and ugly (see Tanaka, 1996: 103) and as creating a 'black' landscape (Azumane, 1993: 116, 126; Taguchi, 1993: 179–80). Another conifer complaint is the loss of seasonality in the Japanese forest that has resulted from the spread of cryptomeria and cypress plantations. Japanese culture is famed for its seasonal sensibility, and the changing forest has been the staple of much literary symbolism and many a *haiku* poem. But in contrast to the earlier mixed forest, conifer plantations are unchangingly 'black' throughout the year.

Even more than the standing plantations themselves, their clearfelling — still commonly practised in Japanese forestry — is the object of much criticism and complaint. Clearfelling creates the eyesore of bare mountain-sides. It also leads to soil erosion and the silting up of rivers. Where the river is used as a tourist resource, for example for water sports such as canoeing, the river-shallowing effects of forestry pose serious problems (Noda, 1997: 37–8). Even routine plantation care attracts criticism. Uemura reports that foresters carrying out the thinning of timber stands find themselves accused of 'nature destruction' (*shizen hakai*) (Uemura, 1992: 200–1), while Tadaki tells a similar story about vine removal (Tadaki, 1988: 194). Tadaki argues that urban environmentalists crystallize an antipathy to forestry that is widespread in urban and suburban Japan (ibid.).

In addition to the plantations themselves, measures to protect them from pests disturb or offend recreational forest visitors. In Japan plantations need to be protected from wildlife pests such as deer, serow and bears, and one means to this end is provided by propane guns, the regular explosions from which serve to drive away the animals. Some tourists are disturbed by this noise which they mistake for actual gunfire. Similarly, the culling of deer to protect timber forests from increased deer numbers is often criticized and opposed by urban dwellers (*Asahi Shinbun* 8/1/95).

As forest tourism increases, the nearby urban population comes to feel more involved in forest issues. One of the consequences of the increased sharing of the forest with the urban population is that the latter become more assertive in their views on forest management. The forest can no longer be managed simply in the interests of timber forestry. The strategic elimination of those wildlife parts of the forest deemed harmful to the timber stands ceases to go unchallenged. The 'cute' deer which so appeal to tourists cannot be simply removed at the foresters' whim.

Another problem is that the Japanese conifer forests generate allergic reactions in many people. The pollen of cryptomeria, the main plantation conifer, has been identified by scientists as the principal allergen causing *kafunsho* — a kind of hay fever which affects large numbers of Japanese people in the spring. Besides putting off visitors, the *kafunsho* problem, much covered in the Japanese mass media, somewhat undermines the image of forest areas as places of healthy recreation.

Environmental Value of Tourist Forests

A further aspect of the transformation of Japanese forests is their environmental value. In Japan, the dominant point of contrast is between the reforested present and the deforested past (and between reforested Japan and deforested southeast Asia). The ubiquitous conifer plantations are, in the first instance, a testament to the national achievement of reforestation, in making the 'bald mountains' (*hageyama*) of the immediate post-war years into green forest once again. This perception has tended to be reinforced by the emergence of the global discourse of 'carbon forests' according to which standing forest *per se* is accorded environmental value as a carbon sink.

However, Japanese plantations are also coming to be contrasted negatively against the primary and secondary forests they have replaced. Timber plantations are criticized on environmental grounds, as a cause of soil erosion, as acidifying streams and affecting fish stocks, and as contributing to wildlife pestilence. To a certain extent, this environmental critique of plantations articulates with the demands for a recreational forest. For one of the justifications of forest tourism is that it can help ameliorate the adverse environmental effects of industrial forestry. By promoting a re-diversification of forest vegetation, tourism promises to bring environmental benefits. The vegetative diversity of the new forest is justified both as an act of beautification and as a measure of environmental restoration or improvement. Forest tourism is often justified in an environmental idiom.

Tourism could contribute to the conservation of forest wildlife. As constituents of the timber forest, wild animals such as the deer and bear are defined negatively as pests to be eradicated. In the tourist forest they are defined and valued differently. Although, to a certain extent, they may be deemed a threat to or problem for forest tourism (deer eat flowers, bears maul tourists), they are also a potential tourist asset. In Japan, timber forestry is seen as a major threat to the bear, but bear conservation is in part justified in terms of the integrity of the Japanese forest. In the 1990s the bear is often referred to in the Japanese mass media as the 'King of the Forest' (*mori no oja*) (*Asahi Shinbun* 17/7/94) or 'King of the Mountains' (*yama no oja*) (*Asahi Shinbun* 5/4/95) — a play on the lion as the 'King of the Jungle' (*janguru no oja*). Some of those involved in the tourist sector feel that it is the presence of the bear which makes the forest a place worth visiting. Similarly, proposals for wolf reintroduction in Japan are justified, in part, in terms of the animal's potential contribution to forest tourism, although there are also fears that the measure could be detrimental to tourism (for this debate, see Knight, 1998: 58). The touristic re-evaluation of forest wildlife can support wildlife conservation.

The rise in forest tourism in Japan is leading to a number of changes in the Japanese forest. Japanese forestry is modifying its practices, away from clearfelling towards selective felling and away from single-age toward mixed-age plantations. More aesthetically appealing tree species (such as cherry blossoms) are being planted, and aesthetic inter-planting of trees in old timber plantations is being practised. Plantations are cosmetically altered by the planting of a belt of other kinds of trees along their edges. Rural municipalities are designating parts of their local forests as conservation areas where old-growth forests are planned as a tourist resource for the future (Kanzaki, 1997: 269–70).

Yet tourism also has adverse environmental effects. The growth of outdoor tourism has resulted in an increase in forest fires, caused by camp fires or by unextinguished cigarette butts (Nakazawa, 1992: 298). More generally, the growth of domestic Japanese tourism has been destructive of the natural beauty of many of the areas visited (Tokuhisa, 1980: 148), while the recent large-scale resort developments have been highly destructive of forests (including forest clearance to make way for ski slopes and golf courses).

The future of primary forest in Japan is of particular concern. Primary forest is now extremely scarce — only around 5 per cent of the forest area. One of the justifications of plantation forestry in Japan, as elsewhere (Sargent, 1992b: 31–2), has been that timber plantations alleviate the logging pressure on primary forests (Tanaka, 1996: 154). But the logic of tourism tends to work the other way: the diminution of primary forest elevates its touristic scarcity value, leading to increased touristic demand for what remains. There must be fears that the large-scale tourism which 'World Heritage' forests such as Yakushima and Shirakami attract will be detrimental to them, especially where new roads are built to accommodate increased tourist numbers (see JPN, 1995).

CONCLUSION

Post-war Japanese society has long been marked by the self-imagery of 'economic animal', 'worker bees', 'workaholics', etc. While proud of their 'economic miracle', post-war Japanese have also tended to view their own variant of industrial modernity as inferior, in many ways, to that achieved elsewhere. In the 1980s and 1990s, after the end of the period of fast economic growth, one of the main preoccupations of government, academia and mass media in Japan has been with the national attainment of an improved quality of life and a more balanced lifestyle — balancing work with leisure, production with consumption, material needs with emotional needs etc. (Linhart, 1988; Najita, 1989: 17–19). Increasingly, 'post-postwar

Japan' (Najita, 1989: 5) is said to have transcended the mass culture of the post-war period and to be undergoing diversification, individualization, and maturation. Along with technology, leisure is one of the main spheres in which this transition to a full modernity is measured, and much national attention in Japan is given to the increase in leisure hours and recreational activity and to the shift away from 'passive' to 'active leisure' (Linhart, 1988: 297–305). Late twentieth century Japan believes itself to be becoming a 'post-industrial leisure society' (Hamilton-Oehrl, 1998: 248).

One of the most visible expressions of the post-industrial leisure society is the new order of recreational land uses in rural Japan. This trend extends to the forests of upland Japan, many of which are being re-defined as recreational spaces. This shift from industrial timber forests to recreational forests can be seen as an instance of recommoditization whereby the forest ceases to be simply the site of timber products and becomes the site of a new kind of product — a touristic product. The enormous scale of capital investment in the leisure industry has changed the face of much of upland Japan.

However, this picture must be qualified. First, the emergence of tourist forests is not a simple replacement of the timber forest. Although Japanese domestic forestry is in serious decline, there are hopes that in the next two decades the demand for domestic timber will pick up again. In other words, rather than a shift from timber forests to tourist forests, upland Japan will be a site of both. The change in the Japanese forest is from a single-purpose industrial forest to a multi-purpose forest as much as a switch from timber to tourism. But this multi-purpose forest of the twenty-first century, to the extent that timber forestry and forest tourism spatially overlap within it, is likely to continue to be the site of some of the tensions described in this article.

Secondly, the trend towards forest tourism should be kept in perspective. Many of the resort developments of the 1990s have not succeeded. Much of this resort expansion had more to do with the vicissitudes of the so-called 'bubble economy' than with careful strategic assessments of particular market opportunities. Moreover, the emergence of tourist forests in Japan should not obscure the fact that the forest recreations of many Japanese people take place in overseas forests — in other parts of Asia and beyond. Indeed, because of their industrial character, the Japanese forests would appear to be significantly disadvantaged as recreational sites compared with the forests of tropical southeast Asia or temperate North America.

Viewed in this way, the emergence of tourist forests in Japan would appear questionable over the long term on the grounds that, in the larger (international) market context of outdoor tourism, they represent an inferior tourist resource. However, the new tourist forests described here should be understood as more than simply a new market trend. The change represents a re-symbolization, and not just a recommoditization, of the Japanese forest. The post-war productionist forest was very much a site of national symbolism, as was clearly expressed in the national reforestation movement. In particular, the ceremonial presence of the tree-planting emperor represented a powerful ritual statement about the forest as the site for growing national strength through timber forestry.

However, the post-war era of mass conifer planting, monocultural forestry, and landscape uniformity is coming to an end, and the Japanese forest of the next century is set to change in accordance with a new complex of national objectives. The national reforestation campaign continues, along with the annual ceremonies featuring the emperor, but it has come to express the new national concern in Japan with multi-purpose forests. Significantly, in the tree-planting ceremonies of the 1990s, non-conifer tree species are also planted — including (local varieties of) cherries, maples, horse chestnuts, camphors and zelkovas — and the importance of vegetative diversity proclaimed.

The change is also expressed in the shift in emphasis in reforestation catchphrases and slogans over the years, from the national resourcist and hydrological *functions* of the post-war forest up until the 1980s, to the affective *relation* between the Japanese people and their forest in recent years. This new normative association of the Japanese forest and its diversity of growth with the Japanese people and their post-industrial well-being is well captured in a recent reforestation slogan: 'Towards Affluence of the Heart through Forest Greenery'.[13]

REFERENCES

Azumane, Chimao (1993) *SOS tsukinowaguma (Bear SOS)*. Morioka: Iwate Nipposha.

Bass, Stephen (1992) 'Building from the Past: Forest Plantations in History', in Caroline Sargent and Stephen Bass (eds) *Plantation Politics: Forest Plantations in Development*, pp. 41–75. London: Earthscan.

Bell, Morag and David M. Evans (1997) 'Greening "The Heart of England" — Redemptive Science, Citizenship, and "Symbol of Hope for the Nation"', *Environment and Planning D: Society and Space* 15: 257–79.

Bostedt, Goran and Leif Mattsson (1995) 'The Value of Forests for Tourism in Sweden', *Annals of Tourism Research* 22(3): 671–80.

Cloke, P., P. Milbourne and C. Thomas (1996) 'From Wasteland to Wonderland: Opencast Mining, Regeneration and the English National Forest', *Geoforum* 27(2): 159–74.

CRSC (Chiiki Ringyo Shinko Chosakai) (1995) *Yamazukuri, murazukuri, hitozukuri saizensen (Forest-making, Village-making and People-making — The Front Line)*. Tokyo: Nihon Ringyo Chosakai.

Dizard, Jan E. (1994) *Going Wild: Hunting, Animal Rights, and the Contested Meaning of Nature*. Amherst, MA: University of Massachusetts Press.

Drew, K. R., Q. Bai and E. V. Fadeev (1989) 'Deer Farming in Asia', in Robert J. Hudson, K. R. Drew and L. M. Baskin (eds) *Wildlife Production Systems: Economic Utilization of Wild Ungulates*, pp. 334–45. Cambridge: Cambridge University Press.

---

13. This (1994) catchphrase in Japanese is: *mori no midori de kokoro no yutakasa o.*

Farnham, T. J., C. P. Taylor and W. Callaway (1995) 'A Shift in Values: Non-Commodity Resource Management and the Forest Service', *Policy Studies Journal* 23(2): 281–95.

Fujitake, Akira (1993) 'The Green of Health', *Japan Update* No. 22.

Gillmor, Desmond (1993) 'Afforestation in the Republic of Ireland', in Alexander Mather (ed.) *Afforestation: Policies, Planning and Progress*, pp. 34–48. London and Florida: Belhaven Press.

Hamilton-Oehrl, Angelika (1998) 'Leisure Parks in Japan', in Sepp Linhart and Sabine Fruhstuck (eds) *The Culture of Japan as Seen through its Leisure*, pp. 237–50. New York: SUNY Press.

Himeda, Tadayoshi (1984) *Yama to sato no fokuroa (The Folklore of Mountain and Village)*. Tokyo: Haru Shobo.

Honda, Katsuichi (1993) 'Why Resorts?', in John Lie (ed.) *The Impoverished Spirit in Contemporary Japan: Selected Essays of Honda Katsuichi*, pp. 178–9. New York: Monthly Review Press.

Inoue, Kazue (1996) 'Toshi seikatsusha to guriin tsurizumu' (Urban Dwellers and Green Tourism), in Inoue Kazue, Nakamura Osamu and Yamazaki Mitsuhiro (eds) *Nihongata guriin tsurizumu (Japanese-style Green Tourism)*, pp. 25–40. Tokyo: Toshi Bunkasha.

Ivy, Marilyn (1995) *Discourses of the Vanishing: Modernity, Phantasm, Japan*. Chicago, IL: Chicago University Press.

Kamada, Mahito, Nobukazu Nakagoshi and Kunito Nehira (1991) 'Pine Forest Ecology and Landscape Management: A Comparative Study in Japan and Korea', in N. Nakagoshi and F. B. Golley (eds) *Coniferous Forest Ecology from an International Perspective*, pp. 43–62. The Hague: SPB Academic Publishing.

Kamata, Ko'ichi and Nebuka, Makoto (1992) 'Shirakami sanchi ni nokosareta mono to sono mirai' (Things Left in the Shirakami Mountains and their Future), in Nebuka Makoto (ed.) *Mori o kangaeru (Thinking About the Forest)*, pp. 411–20. Tokyo: Rippu Shobo.

Kamiyama, Keizo (1983) *Mori no fushigi (The Mystery of the Forest)*. Tokyo: Iwanami Shinsho.

Kanowski, Peter J. and Peter S. Savill (1992) 'Forest Plantations: Towards Sustainable Practice', in Caroline Sargent and Stephen Bass (eds) *Plantation Politics: Forest Plantations in Development*, pp. 121–55. London: Earthscan.

Kanzaki, Noritake (1988) *'Chi'iki okoshi' no fokuroa (The Folklore of 'Regional Revival')*. Tokyo: Gyosei.

Kanzaki, Noritake (1997) '"Furusato sosei" no kessansho' (The Balance Sheet of 'furusato revival'), in Komatsu Kazuhiko (ed.) *Matsuri to ibento (Festivals and Events)*, pp. 257–85. Tokyo: Shogakkan.

Karan, P. P. (1997) 'The City in Japan', in P. P. Karan and Kristin Stapleton (eds) *The Japanese City*, pp. 12–39. Lexington: The University Press of Kentucky.

Kawasaki, Tatsuo (1993) 'Sei, chimei ni nokoru "kuma"', in Gifuken Honyurui Dobutsu Chosa Kenkyukai (ed.) *Horobiyuku mori no oja: tsukinowaguma (The Imminent Extinction of the King of the Forest)*, pp. 5–6. Gifu: Gifu Shinbunsha.

Kelly, William W. (1990) 'Regional Japan: The Price of Prosperity and the Benefits of Dependency', *Daedalus* 119(3): 209–27.

Knight, John (1998) 'Wolves in Japan? An Examination of the Reintroduction Proposal', *Japan Forum* 10(1): 47–65.

Lindhagen, A. (1996) 'An Approach to Clarifying Public Preferences about Silvicultural Systems: A Case Study concerning Group Selection and Clear-cutting', *Scandinavian Journal of Forest Research* 11(4): 375–87.

Linhart, Sepp (1988) 'From Industrial to Post-industrial Society: Changes in Japanese Leisure-related Values and Behaviour', *Journal of Japanese Studies* 14(2): 271–307.

Lucas, O. W. R. (1997) 'Aesthetic Considerations in British forestry', *Forestry* 70(4): 343–9.

Maita, Kazuhiko (1996) *Yama de kuma ni au hoho (Ways to Encounter Bears in the Mountains)*. Tokyo: Yama to Keikokusha.

Mather, Alexander (1993a) 'Introduction', in Alexander Mather (ed.) *Afforestation: Policies, Planning and Progress*, pp. 1–12. London: Belhaven.

Mather, Alexander (1993b) 'Review', in Alexander Mather (ed.) *Afforestation: Policies, Planning and Progress*, pp. 207–19. London: Belhaven.
Matsuzawa, Yuzuru (1989) 'Toshika shakai no tenkai to sanson' (The Development of Urbanized Society and Mountain Villages)', in Uchiyama Setsu (ed.) *'Shinrin shakaigaku' sengen (Declaration on 'Forest Sociology')*, pp. 114–29. Tokyo: Yuikaku Sensho.
Mishima, Akio (1994) *Jomon sugi no keisho (Alarm Bells for Jomon Cryptomeria)*. Tokyo: Meiso Shuppan.
Mitsuda, Hisayoshi and Charles Geisler (1992) 'Imperilled Parks and Imperilled People: Lessons from Japan's Shiretoko National Park', *Environmental History Review* 16(2): 23–39.
Miyao, Takeo (1989) *Tsukinowaguma: owareru mori no sumibito junin (The Black Bear: A Fugitive Forest Dweller)*. Nagano: Shinano Mainichi Shinbunsha.
Nagata, Keijuro (1989) 'Sokatsu' (Summary), in Nagata Keijuro and Iwatani Sanshiro (eds) *Kaso sanson no saisei (The Rebirth of Depopulated Mountain Villages)*, pp. 321–59. Tokyo: Ochanomizu Shobo.
Najita, Tetsuo (1989) 'On Culture and Technology in Postmodern Japan', in Masao Miyoshi and H. D. Harootunian (eds) *Postmodernism and Japan*, pp. 3–20. Durham, NC: Duke University Press.
Nakase, Isao (1993) 'Tanba no mori arekore' (This and That about the Tanba Forest)', in Tanba no Mori Kyokai and Nakase Isao (eds.) *Mori-hito-machizukuri (Developing the Forest, People and the Town)*, pp. 11–26. Kyoto: Gakugei Shuppansha.
Nakazawa, Kazuhiko (1992) *Nihon no mori o sasaeru hitotachi (The People Who Maintain Japan's Forests)*. Tokyo: Shobunsha.
Nebuka, Makoto (1991) *Yama no jinsei: Matagi no mura kara (Mountain Life: From the Village of Matagi)*. Tokyo: NHK Books.
NLAPO (National Land Afforestation Promotion Organization) (1991) *Green Forever: Forests and People in Japan*. Tokyo: NLAPO.
Noda, Tomosuke (1997) 'Paddling the Shimanto', *Japan Quarterly* 44(3): 30–9.
Romano, Bernadino (1995) 'National Parks Policy and Mountain Depopulation: A Case Study in the Abruzzo Region of the Central Appenines, Italy', *Mountain Research and Development* 15(2): 121–32.
Sakurai, Tokutaro (1990) *Minkan shinko no kenkyu: Sakurai Tokutaro chosakushu 4 (Research on Folk Religion: The Collected Works of Sakurai Tokutaro Volume 4)*. Tokyo: Yoshikawa Kobunkan.
Sargent, Caroline (1992a) 'Introduction', in Caroline Sargent and Stephen Bass (eds) *Plantation Politics: Forest Plantations in Development*, pp. 9–15. London: Earthscan.
Sargent, Caroline (1992b) 'Natural Forest or Plantation?', in Caroline Sargent and Stephen Bass (eds) *Plantation Politics: Forest Plantations in Development*, pp. 16–40. London: Earthscan.
Selin, Steve and Franklin Lewis (1994) 'The USDA Forest Service and Rural Tourism Development', *Trends* 31(1): 14–17.
Sogar, David and Willem Oostdyck (1994) 'Marketing the Forest', *Annals of Tourism Research* 21(2): 399–401.
Spellerberg, I. F. and J. W. D. Sawyer (1996) 'Standards for Biodiversity: A Proposal based on Biodiversity Standards for Forest Plantations', *Biodiversity and Conservation* 5(4): 447–59.
Suzuki, Akira (1972) 'On the Problems of the Conservation of the Japanese Monkey on the Boso Peninsula, Japan', *Primates* 13(3): 333–6.
Tadaki, Yoshiya (1988) *Mori to ningen no bunkashi (A Cultural History of Forests and Man)*. Tokyo: NHK Books.
Taguchi, Kazuhiro (1993) 'Shokurin de dobutsu e no hairyo' (Consideration for Animals with Regard to Plantations), in Gifuken Honyurui Dobutsu Chosa Kenkyukai (ed.) *Horobiyuku mori no oja: tsukinowaguma (The Imminent Extinction of the King of the Forest)*, pp. 179–80. Gifu: Gifu Shinbunsha.
Takahashi, Shunjo (1995) *Yasei dobutsu to yaseika kachiku (Wild Animals and Feral Livestock)*. Tokyo: Daimeido.

Takeuchi, Kazuhiko (1993) 'Waga machi, mura o utsukushiku' (Making our Town/Village Beautiful), in Tanba no Mori Kyokai and Nakase Isao (eds) *Mori-hito-machizukuri (Developing the Forest, People and the Town)*, pp. 85–120. Kyoto: Gakugei Shuppansha.

Tanaka, Atsuo (1996) *'Mori o mamore' ga mori o korosu (To 'Save the Forest' is to Kill the Forest)*. Tokyo: Yosensha.

Tokuhisa, Tamao (1980) 'Tourism within, from and to Japan', *International Social Science Journal Quarterly* 32(1): 128–50.

Totman, Conrad (1985) *The Origins of Japan's Modern Forests: The Case of Akita*. Honolulu, HI: University of Hawaii Press.

Ue, Toshikatsu (1980) *Yamabito no ki (Diary of a Mountain Man)*. Tokyo: Chuko Shinsho.

Ue, Toshikatsu (1994) *Mori no megumi (The Blessing of the Forest)*. Tokyo: Iwanami Shoten.

Uemura, Takeshi (1992) *Kizukuri no joshiki hijoshiki (Commonsense and Foolishness in Using Wood)*. Tokyo: Gakugei Shuppansha.

Wada, Hiroshi (1978) 'Wakayama no minwa, densetu', in Sei'ichi Ando (ed.) *Wakayama ken no kenkyu 5: hogen, minzokuhen (Research in Wakayama Prefecture Vol.5: Dialect and Folklore)*, pp. 187–301. Osaka: Seibundo.

Newspapers

Asahi Shinbun (18/11/91) 'Tozan no josei o inoshishi osou' (Wild boar attacks woman mountaineer).

Asahi Shinbun (15/12/92) 'Tennen inoshishiniku o zenkoku ni okurimasu' (Sending natural boar meat to the whole nation).

Asahi Shinbun (24/3/93) 'Wain, shikaniku de machi okoshi da' (Revitalization through wine and venison). Hyogo.

Asahi Shinbun (6/12/93) 'Shika to machigai happo, Hyogo de sasatori no shufushibo' (Mistaken for a deer and shot, the death of a Hyogo housewife out collecting bamboo grass).

Asahi Shinbun (23/5/94) 'Inoshishi ni osoware, dansei haika kega' (Male hiker injured in attack by wild boar). Hyogo.

Asahi Shinbun (17/7/94) 'Tsukinowaguma no kodo ou' (Monitoring the behaviour of bears). Okayama.

Asahi Shinbun (9/12/94) 'Fueru shika o heraseruka' (Should the increasing numbers of deer be reduced?).

Asahi Shinbun (8/1/95) 'Shika no kujo ni shomei undo' (Petition against deer-culling). Tochigi edition.

Asahi Shinbun (5/4/95) 'Yama no oja' (King of the Mountains).

JPN (Japan Primate Newsletter) (1995) 'Road expansion in a World Heritage Area'. Primate Society of Japan (November): 1–3.

Notes on Contributors

Katrina Brown is Senior Lecturer in the School of Development Studies and Senior Research Fellow at CSERGE at the University of East Anglia. Her research interests include political ecology, environment and development, forest access, use and management.

Paul T. Cohen is a Senior Lecturer in Anthropology, Department of Anthropology, Division of Society, Culture, Media & Philosophy at Macquarie University, Sydney, NSW 2109, Australia. His research has focused on Thailand and Laos and on studies of agrarian change, the political economy of opiates, health and development, and ethnicity. His recent publications include 'Communal Irrigation, State, and Capital in the Chiang Mai Valley (Northern Thailand)' (with R. Pearson), *Journal of Southeast Asian Studies* 29(1), 1998; and 'Lue Ethnicity in National Context: A Comparative Study of Tai Lue Communities in Thailand and Laos', *Journal of the Siam Society* 86(1&2), 1998.

After studying in England and France, **François E. Ekoko** became a Fellow of the United Nations University/Institute of Advanced Studies (Tokyo, Japan), before joining the Centre for International Forestry Research (CIFOR) in 1997. He has worked as a consultant on issues of forest policy and environmental governance to a number of institutions including UNDP, UNEP, WRI and ATO. In October 1999 he joined UNDP/BDP/SEED, 304 East 45th Street, FF1042, New York, NY 10017, USA. His recent and forthcoming publications include 'Poverty and Deforestation in the Congobasin Rainforest' (1998), 'Forest Encounters: Synergy among Agents of Forest Change in Southern Cameroon' (1999), and 'Environmental Adjustment: The Case of Cameroon' (2000).

James Fairhead is Reader in the Department of Sociology and Anthropology, School of Oriental and African Studies, University of London (Thornhaugh Street, Russell Square, London WC1E 7HP, jf18@soas.a-c.uk). Besides individual interests, they have worked and published jointly on West African forestry issues since 1991. They are currently carrying out a major comparative research project on science-policy relations in West Africa and the Caribbean.

Richard Gauld is a PhD candidate in the Department of Geography, King's College, London, WC2R 2LS, UK, where he is co-editor of the Department's Occasional Paper series. He is currently researching the political dynamics of local resistance to protected areas in Western Europe using a political ecology approach.

Emily E. Harwell is a PhD candidate at the Yale School of Forestry and Environmental Studies, New Haven, CT 06511, USA. Her dissertation investigates the changing relationship of ethnic identity, customary law and the environment in West Kalimantan, Indonesia.

Daniel Klooster is Adjunct Assistant Professor of International Affairs in the School of International and Public Affairs, Columbia University, 420 West 118th Street, New York, NY 10027, USA. His research addresses community forestry, common property theory, social strategies for biological conservation, land-use change, and human dimensions of global environmental change.

John Knight is Lecturer in Social Anthropology at the Queen's University of Belfast, Belfast BT7 1NN, Northern Ireland. Since 1987 he has conducted regular field research in Japanese mountain villages. He has published widely on various subjects related to rural Japan, including rural development, forestry and wildlife. Among his recent publications are: 'Monkey Wars in Japan: The Natural Symbolism of Macaque Pestilence', *Journal of Asian Studies* 58(3), 1999; 'Selling Mother's Love: Mail Order Village Food in Japan', *Journal of Material Culture* 3(2), 1998; and 'A Tale of Two Forests: Reforestation as Discourse in Japan and Beyond', *JRAI (Man)* n.s. 3(4), 1997.

Melissa Leach is a social anthropologist and Fellow of the Institute of Development Studies, University of Sussex (Brighton BN1 9RE, UK, m.leach@ids.ac.uk) where she directs the Environment Group.

John F. McCarthy is a researcher at the Asian Research Centre, Murdoch University, Western Australia 6150. Since 1996, he has carried out research in the Leuser Ecosystem, Northern Sumatra, under a research agreement with the Centre for International Forestry Research (CIFOR). He is now completing his doctoral thesis with the title 'In the Shadow of the Forest: Institutional Arrangements on Sumatra's Rainforest Frontier'.

Anja Nygren is a research scientist at the Department of Anthropology, Box 59, 00014 University of Helsinki (anja.nygren@helsinki.fi). Her current research interests are nature-based conflicts, cultural representations of tropical forest-dwellers, and hybrid knowledge systems. Her recent publications include 'Local Knowledge in the Environment–Development Discourse', *Critique of Anthropology* 17(3): 267–88 (1999).

Sérgio Rosendo is a PhD student at the University of East Anglia. He is interested in community-based natural resource management and his current research is concerned with the politics of local institutions in extractive reserve management in Brazilian Amazonia. Both authors can be contacted at the School of Development Studies, University of East Anglia, Norwich NR4 7TJ, UK (k.brown@uea.ac.uk and s.rosendo@uea.ac.uk).

Jin Sato is a research associate at the Institute of Environmental Studies of the Graduate School of Frontier Sciences, the University of Tokyo, 7-3-1 Honho, Bunkyo-ku, Tokyo 113-8656, Japan. His field of interest is the political economy of natural resources, particularly the forces that work towards decentralized management of natural resources in developing countries.

K. Sivaramakrishnan is Assistant Professor, Department of Anthropology, University of Washington, 241 Denny Hall, Box 353100, Seattle, WA 98195, USA. His current research examines the relationships between conservation, democracy, and development, as well as the influence of environmental change and identity politics on livelihood security in dry land areas of India. He is the author of *Modern Forests: Statemaking and Environmental Change in Colonial Eastern India* (Stanford University Press, 1999).

Nandini Sundar is an Associate Professor of Sociology at the Institute of Economic Growth, University Enclave, Delhi 110 007, India. Her publications include *Subalterns and Sovereigns: An Anthropological History of Bastar, 1854–1996* (Delhi, Oxford University Press, 1997) and an edited volume (with Roger Jeffery) entitled *A New Moral Economy for India's Forests? Discourses of Community and Participation* (Delhi, Sage Publications, 1999).

Index